IB
STUDY
GUIDES

CHEMISTRY
for the IB Diploma
STANDARD AND HIGHER LEVEL

Geoff Neuss

OXFORD
UNIVERSITY PRESS

OXFORD

UNIVERSITY PRESS

Great Clarendon Street, Oxford OX2 6DP

Oxford University Press is a department of the
University of Oxford. It furthers the University's objective
of excellence in research, scholarship, and education by
publishing worldwide in

Oxford New York

Athens Auckland Bangkok Bogotá Bombay Buenos Aires
Cape Town Chennai Dar es Salaam Delhi Florence
Hong Kong Istanbul Karachi Kolkata Kuala Lumpur Madrid
Melbourne Mexico City Mumbai Nairobi Paris Saõ Paulo
Shanghai Singapore Taipei Tokyo Toronto Warsaw

with associated companies in Berlin Ibadan

Oxford is a registered trade mark of Oxford University Press
in the UK and in certain other countries

British Library Cataloguing in Publication Data

Data available

Typesetting, artwork and design by Hardlines, Charlbury, Oxon

Printed in Great Britain

Introduction and acknowledgements

This book is written specifically for students studying Chemistry for the International Baccalaureate Diploma although I hope that those following their own national system will also find it helpful. It comprehensively covers the new programme that will be examined from 2003 onwards. All the information required for each topic is set out in separate boxes with clear titles that follow faithfully the layout of the syllabus.

The first eleven topics cover both the core content needed by all students and the extra Higher Level material under the main topic headings. The difference between the two levels is clearly distinguished. Both Higher Level and Standard Level students must study at least two of the eight options and each option stands in its own right even if this has meant repeating some of the material.

Worked examples are included where they are appropriate and at the end of each main topic and option there are practice questions. The majority of these questions are taken from past IB examination papers and I would like to thank the International Baccalaureate Organisation for giving me permission to use them. The remaining questions are written to the same IB standard specifically for this book. Answers to all the questions are provided. These answers are not necessarily full 'model' answers but they do contain all the information needed to score each possible mark.

To help you, the student, gain the highest grade possible the final chapter is devoted to giving you advice on how to study and prepare for the final examination. It also advises you on how to excel at the internally assessed practical component of the course. For those who opt for Chemistry as the subject for their Extended Essay it gives advice and guidance on how to choose the topic and write your Essay. A comprehensive Extended Essay checklist is included to help you gain bonus points towards your IB Diploma.

I have been fortunate at Atlantic College to teach many highly motivated and gifted students who have often challenged me with searching questions. During my association with the International Baccalaureate, the European Baccalaureate, and the United World Colleges I have been privileged to meet, work alongside, and exchange ideas with many excellent Chemistry teachers who exude a real enthusiasm for their subject. Many of these have influenced me greatly – in particular, John Devonshire, a fellow teacher at Atlantic College and Jacques Furnemont, an Inspector of Chemistry in Belgium. I value greatly their advice, opinions, and knowledge. I would also like to pay tribute to two former Chief Examiners for the IB, Ron Ragsdale and Arden Zipp. The high regard in which IB Chemistry is held today owes much to both of them.

Many IB students and teachers have encouraged me to write this book. Ideally, I would have liked to include more background information and depth to illustrate each topic further. However, I have stuck rigorously to the syllabus to produce a book which contains all the necessary subject content in one easily accessible format. Paul Fairbrother, the Chemistry Subject Officer for the IB, has been particularly encouraging and helpful. I owe much to Nick Lee at St Clare's, Oxford and Chris Talbot from the Overseas Family School, Singapore for making constructive suggestions and corrections after spending many hours reading through the draft. Finally I should like to thank my wife Chris and my friend and colleague John for their patience and unstinting support throughout.

Dr Geoffrey Neuss

CONTENTS

(Italics denote topics which are exclusively Higher Level.)

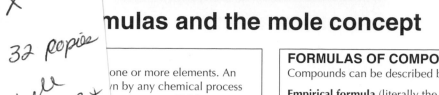

...mulas and the mole concept

(handwritten notes in top-left: X 32 copies staple pages 1, 2, 3 + 5)

...one or more elements. An ...wn by any chemical process ...are just over 100 known ...n element is called an atom.

...ts

Number		Symbol	Relative atomic mass
1	hydrogen	H	1.01
2	helium	He	4.00
3	lithium	Li	6.94
4	beryllium	Be	9.01
5	boron	B	10.81
6	carbon	C	12.01
7	nitrogen	N	14.01
8	oxygen	O	16.00
9	fluorine	F	19.00
10	neon	Ne	20.18
11	sodium	Na	22.99
12	magnesium	Mg	24.31
13	aluminium	Al	26.98
14	silicon	Si	28.09
15	phosphorus	P	30.97
16	sulfur	S	32.06
17	chlorine	Cl	35.45
18	argon	Ar	39.95
19	potassium	K	39.10
20	calcium	Ca	40.08

COMPOUNDS

Some substances are made up of a single element although there may be more than one atom of the element in a particle of the substance. Oxygen is diatomic, that is, a molecule of oxygen contains two oxygen atoms. A compound contains more than one element. For example, a molecule of water contains two hydrogen atoms and one oxygen atom. Water is a compound not an element because it can be broken down chemically into its constituent elements: hydrogen and oxygen.

FORMULAS OF COMPOUNDS

Compounds can be described by different chemical formulas.

Empirical formula (literally the formula obtained by experiment)
This shows the simplest whole number ratio of atoms of each element in a particle of the substance. It can be obtained by either knowing the mass of each element in the compound or from the percentage composition by mass of the compound. The percentage composition can be converted directly into mass by assuming 100 g of the compound are taken.

Example: A compound contains 40.00% carbon, 6.73% hydrogen and 53.27% oxygen by mass, determine the empirical formula.

	Amount /mol	Ratio	
C	$40.00/12.01 = 3.33$	1	Empirical formula
H	$6.73/1.01 = 6.66$	2	$= CH_2O$
O	$53.27/16.00 = 3.33$	1	

Molecular formula
For molecules this is much more useful as it shows the actual number of atoms of each element in a molecule of the substance. It can be obtained from the empirical formula if the molar mass of the compound is also known.
Methanal CH_2O ($M_r = 30$), ethanoic acid $C_2H_4O_2$ ($M_r = 60$) and glucose $C_6H_{12}O_6$ ($M_r = 180$) are different substances with different molecular formulas but all with the same empirical formula CH_2O. Note that subscripts are used to show the number of atoms of each element in the compound.

Structural formula
This shows the arrangement of atoms and bonds within a molecule and is particularly useful in organic chemistry.

The three different formulas can be illustrated using ethene:

CH_2	C_2H_4	$\underset{H}{\overset{H}{\diagdown}} C = C \underset{H}{\overset{H}{\diagup}}$
		(can also be written $H_2C=CH_2$)
empirical formula	molecular formula	structural formula

MOLE CONCEPT AND AVOGADRO'S CONSTANT

A single atom of an element has an extremely small mass. For example an atom of carbon-12 has a mass of 1.993×10^{-23} g. This is far too small to weigh. A more convenient amount to weigh is 12.00 g. 12.00 g of carbon-12 contains 6.02×10^{23} atoms of carbon-12. This number is known as Avogadro's constant (N_A or L).
Chemists measure amounts of substances in moles. A mole is the amount of substance that contains L particles of that substance. The mass of one mole of **any** substance is known as the **molar mass** and has the symbol M. For example, hydrogen atoms have $\frac{1}{12}$ of the mass of carbon-12 atoms so a mole of hydrogen atoms contains 6.02×10^{23} hydrogen atoms and has a mass of 1.01 g. In reality elements are made up of a mixture of isotopes.
The **relative atomic mass** of an element A_r is the weighted mean of all the naturally occurring isotopes of the element relative to carbon-12. This explains why the relative atomic masses given for the elements above are not whole numbers. The units of molar mass are g mol^{-1} but relative molar masses M_r have no units. For molecules **relative molecular mass** is used. For example, the M_r of glucose, $C_6H_{12}O_6 = (6 \times 12.01) + (12 \times 1.01) + (6 \times 16.00) = 180.18$. For ionic compounds the term **relative formula mass** is used.
Be careful to distinguish between the words **mole** and **molecule**. A molecule of hydrogen gas contains two atoms of hydrogen and has the formula H_2. A mole of hydrogen gas contains 6.02×10^{23} hydrogen molecules made up of two moles (1.20×10^{24}) of hydrogen atoms.

Chemical reactions and equations

PROPERTIES OF CHEMICAL REACTIONS

In all chemical reactions:
- new substances are formed.
- bonds in the reactants are broken and bonds in the products are formed resulting in an energy change between the reacting system and its surroundings.
- there is a fixed relationship between the number of particles of reactants and products resulting in no overall change in mass – this is known as the stoichiometry of the reaction.

CHEMICAL EQUATIONS

Chemical reactions can be represented by chemical equations. Reactants are written on the left hand side and products on the right hand side. The number of moles of each element must be the same on both sides in a balanced chemical equation,
e.g. the reaction of nitric acid (one of the acids present in acid rain) with calcium carbonate (the main constituent of marble statues).

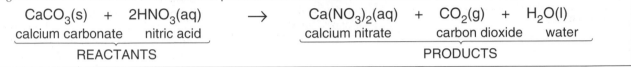

$$CaCO_3(s) \ + \ 2HNO_3(aq) \ \longrightarrow \ Ca(NO_3)_2(aq) \ + \ CO_2(g) \ + \ H_2O(l)$$

calcium carbonate nitric acid	calcium nitrate carbon dioxide water
REACTANTS	PRODUCTS

COEFFICIENTS

The coefficients in front of each species give information on the molar ratio. In the above example two moles of nitric acid react with one mole of calcium carbonate to produce one mole of calcium nitrate, one mole of carbon dioxide, and one mole of water.

STATE SYMBOLS

Because the physical state that the reactants and products are in can affect both the rate of the reaction and the overall energy change it is good practice to include the state symbols in the equation.

(s) – solid   (l) – liquid   (g) – gas   (aq) – in aqueous solution

→ OR ⇌

A single arrow → is used if the reaction goes to completion. Sometimes the reaction conditions are written on the arrow:

e.g. $\quad C_2H_4(g) \ + \ H_2(g) \ \xrightarrow{\text{Ni catalyst, } 180\,°C} \ C_2H_6(g)$

Reversible arrows are used for reactions where both the reactants and products are present in the equilibrium mixture:

e.g. $\quad 3H_2(g) \ + \ N_2(g) \ \underset{250 \text{ atm}}{\overset{\text{Fe(s), } 550\,°C}{\rightleftharpoons}} \ 2NH_3(g)$

IONIC EQUATIONS

Because ionic compounds are completely dissociated in solution it is sometimes better to use ionic equations. For example, when silver nitrate solution is added to sodium chloride solution a precipitate of silver chloride is formed.

$Ag^+(aq) + NO_3^-(aq) + Na^+(aq) + Cl^-(aq) \longrightarrow AgCl(s) + Na^+(aq) + NO_3^-(aq)$

$Na^+(aq)$ and $NO_3^-(aq)$ are spectator ions and do not take part in the reaction. So the ionic equation becomes:

$Ag^+(aq) + Cl^-(aq) \longrightarrow AgCl(s)$

From this we can deduce that any soluble silver salt will react with any soluble chloride to form a precipitate of silver chloride.

X

...tions

...red by weighing to obtain

...000 g

When weighing a substance the mass should be recorded to show the accuracy of the balance. For example, exactly 16 g of a substance would be recorded as 16.00 g on a balance weighing to + or – 0.01 g but as 16.000 g on a balance weighing to + or – 0.001 g.

MEASUREMENT OF MOLAR QUANTITIES
In the laboratory moles can conveniently be measured using either mass or volume depending on the substances involved.

LIQUIDS
Pure liquids may be weighed or the volume recorded.
The density of the liquid $= \frac{mass}{volume}$ and is usually expressed in g cm^{-3}.

SOLUTIONS
Volume is usually used for solutions.

1.000 litre = 1.000 dm^3 = 1000 cm^3

Concentration is the amount of solute (dissolved substance) in a known volume of solution (solute plus solvent). It is expressed either in g dm^{-3} or more usually in mol dm^{-3}. A 1.00 mol dm^{-3} solution of sodium hydroxide contains 40.00 g of sodium hydroxide in one litre of solution. A 25.0 cm^3 sample of this solution contains

$1.00 \times \frac{25.0}{1000} = 2.50 \times 10^{-2}$ mol of NaOH

GASES
Mass or volume may be used for gases. Avogadro's Law states that equal volumes of different gases at the same temperature and pressure contain the same number of moles. From this it follows that if the temperature and pressure are specified one mole of any gas will occupy the same volume. This is known as the **molar volume of a gas** and is equal to 22.4 dm^3 at 273 K and 1 atmosphere pressure. This is sometime quoted as 24.0 dm^3 under standard conditions of temperature and pressure (298 K, 1atm).

CALCULATIONS FROM EQUATIONS
Work methodically.
Step 1. Write down the correct formulas for all the reactants and products.
Step 2. Balance the equation to obtain the correct stoichiometry of the reaction.
Step 3. If the amounts of all reactants are known work out which are in **excess** and which one is the limiting reagent. By knowing the **limiting reagent** the maximum **yield** of any of the products can be determined.
Step 4. Work out the number of moles of the substance required.
Step 5. Convert the number of moles into the mass or volume.
Step 6. Express the answer to the correct number of significant figures and include the appropriate units.

WORKED EXAMPLE
Calculate the volume of hydrogen gas evolved at 273 K and 1 atm pressure when 0.623 g of magnesium reacts with 27.3 cm^3 of 1.25 mol dm^{-3} hydrochloric acid.

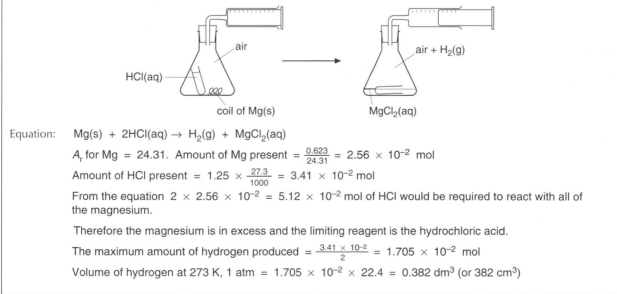

Equation: $Mg(s) + 2HCl(aq) \rightarrow H_2(g) + MgCl_2(aq)$

A_r for Mg = 24.31. Amount of Mg present $= \frac{0.623}{24.31} = 2.56 \times 10^{-2}$ mol

Amount of HCl present $= 1.25 \times \frac{27.3}{1000} = 3.41 \times 10^{-2}$ mol

From the equation $2 \times 2.56 \times 10^{-2} = 5.12 \times 10^{-2}$ mol of HCl would be required to react with all of the magnesium.

Therefore the magnesium is in excess and the limiting reagent is the hydrochloric acid.

The maximum amount of hydrogen produced $= \frac{3.41 \times 10^{-2}}{2} = 1.705 \times 10^{-2}$ mol

Volume of hydrogen at 273 K, 1 atm $= 1.705 \times 10^{-2} \times 22.4 = 0.382$ dm^3 (or 382 cm^3)

IB QUESTIONS – STOICHIOMETRY

1. What is the mass in grams of one molecule of ethanoic acid CH_3COOH?

   A. 0.1　**B.** 3.6×10^{25}　**C.** 1×10^{-22}　**D.** 60

2. Which is not a true statement?

   A. One mole of methane contains four moles of hydrogen atoms

   B. One mole of ^{12}C has a mass of 12.00 g

   C. One mole of hydrogen gas contains 6.02×10^{23} atoms of hydrogen

   D. One mole of methane contains 75% of carbon by mass

3. A pure compound contains 24 g of carbon, 4 g of hydrogen and 32 g of oxygen.

   No other elements are present. What is the empirical formula of the compound?

   A. $C_2H_4O_2$　**B.** CH_2O　**C.** CH_4O　**D.** CHO

4. Which one of the following statements about SO_2 is/are correct?

   I. One mole of SO_2 contains 1.8×10^{24} atoms

   II. One mole of SO_2 has a mass of 64 g

   A. Both I and II　**B.** Neither I nor II　**C.** I only　**D.** II only

5. What is the empirical formula for the compound $C_6H_5(OH)_2$?

   A. C_6H_6O　**B.** $C_6H_5O_2H_2$　**C.** C_6H_7O　**D.** $C_6H_7O_2$

6. A carbohydrate has the empirical formula CH_2O. Its relative molar mass is between 160 and 200. What is its molecular formula?

   A. CH_2O　**B.** $C_5H_{10}O_5$　**C.** $C_6H_{12}O_6$　**D.** $C_8H_{16}O_8$

7. Phosphorus burns in oxygen to produce phosphorus pentoxide P_4O_{10}.

   What is the sum of the coefficients in the balanced equation?

   $\_P_4(s) + \_O_2(g) \rightarrow \_P_4O_{10}(s)$

   A. 3　**B.** 5　**C.** 6　**D.** 7

8. Magnesium reacts with hydrochloric acid according to the following equation:

   $Mg(s) + 2HCl(aq) \rightarrow MgCl_2(aq) + H_2(g)$

   What mass of hydrogen will be obtained if 100 cm^3 of 2.00 mol dm^{-3} HCl are added to 4.86 g of magnesium?

   A. 0.2g　**B.** 0.4g　**C.** 0.8g　**D.** 2.0g

9. Butane burns in oxygen according to the equation below.

   $2C_4H_{10}(g) + 13O_2(g) \rightarrow 8CO_2(g) + 10H_2O(l)$

   If 11.6 g of butane is burned in 11.6 g of oxygen which is the limiting reagent?

   A. Butane　　**C.** Neither

   B. Oxygen　　**D.** Oxygen and butane

10. When 250 cm^3 of 3.00 mol dm^{-3} HCl(aq) is added to 350 cm^3 of 2.00 mol dm^{-3} HCl(aq) the concentration of the solution of hydrochloric acid obtained in mol dm^{-3} is:

    A. 2.42　**B.** 1.45　**C.** 2.90　**D.** 2.50

11. A 55.0 cm^3 sample of salt solution has a sodium chloride concentration of 0.135 mol dm^{-3}.

    How many moles of sodium chloride are present?

    A. 55.0×0.135　**C.** $\dfrac{1}{0.135} \times \dfrac{55.0}{1000}$

    B. $55.0 \times \dfrac{1000}{0.135}$　**D.** $\dfrac{55.0}{1000} \times 0.135$

12. Sulfuric acid and sodium hydroxide react together according to the equation:

    $H_2SO_4(aq) + 2NaOH(aq) \rightarrow Na_2SO_4(aq) + 2H_2O(l)$

    What volume of 0.250 mol dm^{-3} NaOH is required to neutralise exactly 25.0 cm^3 of 0.125 mol dm^{-3} H_2SO_4?

    A. 25.0 cm^3　**B.** 12.5 cm^3　**C.** 50 cm^3　**D.** 6.25 cm^3

13. Aspirin, $C_9H_8O_4$, is made by reacting ethanoic anhydride, $C_4H_6O_3$ ($M_r = 102.1$), with 2-hydroxybenzoic acid ($M_r = 138.1$), according to the equation:

    $2C_7H_6O_3 + C_4H_6O_3 \rightarrow 2C_9H_8O_4 + H_2O$

    (a) If 15.0 g 2-hydroxybenzoic acid is reacted with 15.0 g ethanoic acid, determine the limiting reagent in this reaction.

    (b) Calculate the maximum mass of aspirin that could be obtained in this reaction.

    (c) If the mass obtained in this experiment was 13.7 g, calculate the percentage yield of aspirin.

14. 14.48 g of a metal sulfate with the formula M_2SO_4 were dissolved in water. Excess barium nitrate solution was added in order to precipitate all the sulfate ions in the form of barium sulfate. 9.336 g of precipitate was obtained.

    (a) Calculate the amount of barium sulfate $BaSO_4$ precipitated.

    (b) Calculate the amount of sulfate ions present in the 14.48 g of M_2SO_4.

    (c) What is the relative molar mass of M_2SO_4?

    (d) Calculate the relative atomic mass of M and hence identify the metal.

2 ATOMIC THEORY

The atom

COMPOSITION OF ATOMS

The smallest part of an element is an atom. It used to be thought that atoms are indivisible but they can be broken down into many different sub-atomic particles. All atoms, with the exception of hydrogen, are made up of three fundamental sub-atomic particles – protons, neutrons, and electrons.

The hydrogen atom, the simplest atom of all, contains just one proton and one electron. The actual mass of a proton is 1.672×10^{-24} g but it is assigned a relative value of 1. The mass of a neutron is virtually identical and also has a relative mass of 1. Compared to a proton and a neutron an electron has negligible mass with a relative mass of only $\frac{1}{1840}$. Neutrons are neutral particles. An electron has a charge of 1.602×10^{-19} coulombs which is assigned a relative value of –1. A proton carries the same charge as an electron but of an opposite sign so has a relative value of +1. All atoms are neutral so must contain equal numbers of protons and electrons.

SUMMARY OF RELATIVE MASS AND CHARGE

Particle	Relative mass	Relative charge
proton	1	+1
neutron	1	0
electron	$\frac{1}{1840}$	–1

SIZE AND STRUCTURE OF ATOMS

Atoms have a radius in the order of 10^{-10} m. Almost all of the mass of an atom is concentrated in the nucleus which has a very small radius in the order of 10^{-14} m. All the protons and neutrons (collectively called nucleons) are located in the nucleus. The electrons are to be found in energy levels or shells surrounding the nucleus. Much of the atom is empty space.

MASS NUMBER A
Equal to the number of protons and neutrons in the nucleus.

ATOMIC NUMBER Z
Equal to the number of protons in the nucleus and to the number of electrons in the atom. The atomic number defines which element the atom belongs to and consequently its position in the Periodic Table.

SHORTHAND NOTATION FOR AN ATOM OR ION

$$^{A}_{Z}X^{n+/n-}$$

CHARGE
Atoms have no charge so n = 0 and this is left blank. However by losing one or more electrons atoms become positive ions, or by gaining one or more electrons atoms form negative ions.

EXAMPLES

Symbol	Atomic number	Mass number	Number of protons	Number of neutrons	Number of electrons
$^{9}_{4}Be$	4	9	4	5	4
$^{40}_{20}Ca^{2+}$	20	40	20	20	18
$^{37}_{17}Cl^{-}$	17	37	17	20	18

ISOTOPES

All atoms of the same element must contain the same number of protons, however they may contain a different number of neutrons. Such atoms are known as isotopes. Chemical properties are related to the number of electrons so isotopes of the same element have identical chemical properties. Since their mass is different their physical properties such as density and boiling point are different.

Examples of isotopes: $^{1}_{1}H$ $^{2}_{1}H$ $^{3}_{1}H$ $^{12}_{6}C$ $^{14}_{6}C$. $^{35}_{17}Cl$ $^{37}_{17}Cl$.

RELATIVE ATOMIC MASS

The two isotopes of chlorine occur in the ratio of 3:1. That is, naturally occurring chlorine contains 75% $^{35}_{17}Cl$ and 25% $^{37}_{17}Cl$. The weighted mean molar mass is thus:

$$\frac{(75 \times 35) + (25 \times 37)}{100} = 35.5 \text{ g mol}^{-1}$$

and the relative atomic mass is 35.5. Accurate values to 2 d.p. for all the relative atomic masses of the elements are given in Table 5 of the IB Data Booklet. These are the values which must be used when performing calculations in the examinations.

MASS SPECTROMETER

Relative atomic masses can be determined using a mass spectrometer. A *vaporized* sample is injected into the instrument. Atoms of the element are *ionized* by being bombarded with a stream of high energy electrons in the ionization chamber. In practice the instrument is set so that only ions with a single positive charge are formed. The resulting unipositive ions pass through holes in parallel plates under the influence of an electric field where they are *accelerated*. The ions are then *deflected* by an external magnetic field.

The amount of deflection depends both on the mass of the ion and its charge. The smaller the mass and the higher the charge the greater the deflection. Ions with a particular mass/charge ratio are then recorded on a *detector* which measures both the mass and the relative amounts of all the ions present.

DIAGRAM OF A MASS SPECTROMETER

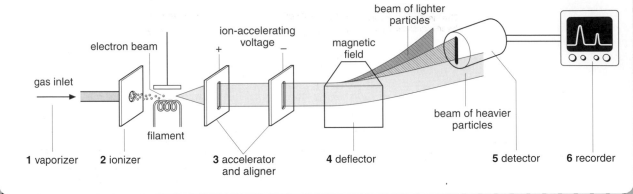

THE MASS SPECTRUM OF NATURALLY OCCURRING LEAD

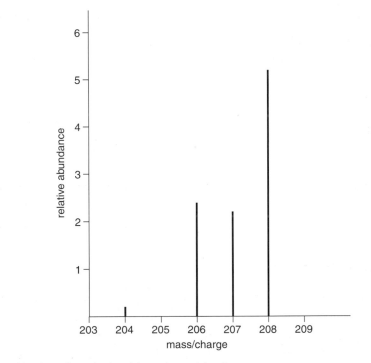

The relative atomic mass of lead can be calculated from the weighted average:

Isotopic mass	Relative abundance	% relative abundance
204	0.2	2
206	2.4	24
207	2.2	22
208	5.2	52

$$\text{relative atomic mass} = \frac{(2 \times 204) + (24 \times 206) + (22 \times 207) + (52 \times 208)}{100} = 207.2$$

Emission spectra

THE ELECTROMAGNETIC SPECTRUM

Electromagnetic waves can travel through space and, depending on the wavelength, also through matter. The velocity of travel c is related to its wavelength λ and its frequency f. Velocity is measured in m s^{-1}, wavelength in m and frequency in s^{-1} so it is easy to remember the relationship between them:

$$c = \lambda \times f$$

$$(\text{m s}^{-1}) \quad (\text{m}) \quad (\text{s}^{-1})$$

Electromagnetic radiation is a form of energy. The smaller the wavelength and thus the higher the frequency the more energy the wave possesses. Electromagnetic waves have a wide range of wavelengths ranging from low energy radio waves to high energy γ-radiation. Visible light occupies a very narrow part of the spectrum.

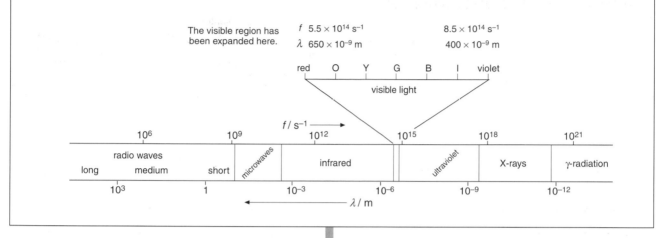

ATOMIC EMISSION SPECTRA

White light is made up of all the colours of the spectrum. When it is passed through a prism a **continuous spectrum** of all the colours can be obtained.

When energy is supplied to individual elements they emit a spectrum which only contains emissions at particular wavelengths. Each element has its own characteristic spectrum known as a **line spectrum** as it is not continuous.

The visible hydrogen spectrum

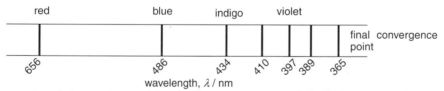

Note that the spectrum consists of discrete lines and that the lines converge towards the high energy (violet) end of the spectrum. A similar series of lines at even higher energy also occurs in the ultraviolet region of the spectrum and several other series of lines at lower energy can be found in the infrared region of the spectrum.

EXPLANATION OF EMISSION SPECTRA

When energy is supplied to an atom electrons are excited (gain energy) from their lowest (ground) state to an excited state. Electrons can only exist in certain fixed energy levels. When electrons drop from a higher level to a lower level they emit energy. This energy corresponds to a particular wavelength and shows up as a line in the spectrum. When electrons return to the first level (n = 1) the series of lines occurs in the ultraviolet region as this involves the largest energy change. The visible region spectrum is formed by electrons dropping back to the n = 2 level and the first series in the infrared is due to electrons falling to the n = 3 level. The lines in the spectrum converge because the energy levels themselves converge.

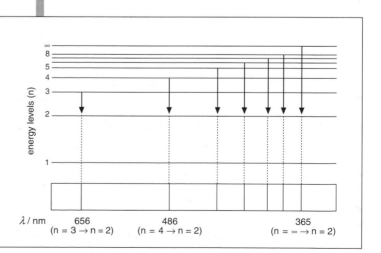

Electron arrangement

EVIDENCE FROM IONIZATION ENERGIES

The first ionization energy of an element is defined as the energy required to remove one electron from an atom in its gaseous state. It is measured in kJ mol^{-1}.

$$X(g) \rightarrow X^+(g) + e$$

A graph of first ionization energies plotted against atomic number shows a repeating pattern.

It can be seen that the highest value is for helium, an atom that contains two protons and two electrons. The two electrons are in the lowest level and are held tightly by the two protons. For lithium it is relatively easy to remove an electron, which suggests that the third electron in lithium is in a higher energy level than the first two. The value then generally increases until element 10, neon, is reached before it drops sharply for sodium. This graph provides evidence that the levels can contain different numbers of electrons before they become full.

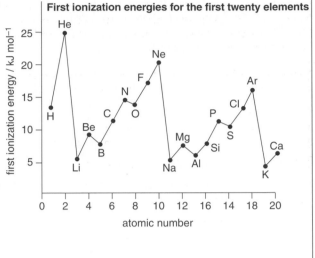

First ionization energies for the first twenty elements

Level	Maximum number of electrons
1 (K shell)	2
2 (L shell)	8
3 (M shell)	8 (or 18)

ELECTRON ARRANGEMENT

The arrangement of electrons in an atom is known as its electronic configuration. Each energy level or shell is separated by a dot (or a comma). The electrons in the highest main energy level (outermost level) are known as the **valence electrons**.

Element	Electron configuration	Element	Electron configuration
H	1	Na	2.8.1
He	2 (first level full)	Mg	2.8.2
Li	2.1	Al	2.8.3
Be	2.2	Si	2.8.4
B	2.3	P	2.8.5
C	2.4	S	2.8.6
N	2.5	Cl	2.8.7
O	2.6	Ar	2.8.8 (third level full)
F	2.7	K	2.8.8.1
Ne	2.8 (second level full)	Ca	2.8.8.2

EVIDENCE FOR SUB-LEVELS

The graph already shown above was for the first ionization energy for the first twenty elements. Successive ionization energies for the same element can also be measured, e.g. the second ionization energy is given by:

$$X^+(g) \rightarrow X^{2+}(g) + e$$

As more electrons are removed the pull of the protons holds the remaining electrons more tightly so increasingly more energy is required to remove them, hence a logarithmic scale is usually used. A graph of the successive ionization energies for potassium also provides evidence of the number of electrons in each main level.

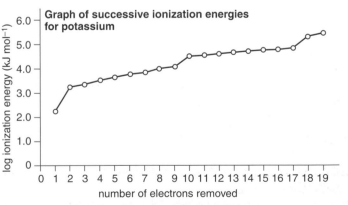

Graph of successive ionization energies for potassium

By looking to see where the first 'large jump' occurs in successive ionization energies one can determine the number of valence electrons (and hence the group in the Periodic Table to which the element belongs).

If the graph for first ionization energies is examined more closely then it can be seen that the graph does not increase regularly. This provides evidence that the main levels are split into sub-levels.

Sub-levels and orbitals

TYPES OF ORBITAL

Electrons are found in orbitals. Each orbital can contain a maximum of two electrons each with opposite spins. The first level contains just one orbital, called an s orbital. The second level contains one s orbital and three p orbitals. The 2p orbitals are all of equal energy but the sub-level made up of these three 2p orbitals is slightly higher in energy than the 2s orbital. This explains why the first ionization energy of B is lower than Be as a higher energy 2p electron is being removed from the B compared with a lower energy 2s electron from Be.

Principal level (shell)	Number of each type of orbital				Maximum number of electrons in level
	s	p	d	f	
1	1	–	–	–	2
2	1	3	–	–	8
3	1	3	5	–	18
4	1	3	5	7	32

The relative position of all the sub-levels for the first four main energy levels is shown.

Note that the 4s sub-level is below the 3d sub-level. This explains why the third level is sometimes stated to hold 8 or 18 electrons.

Electrons with opposite spins tend to repel each other. When orbitals of the same energy (degenerate) are filled the electrons will go singly into each orbital first before they pair up to minimize repulsion. This explains why there is a regular increase in the first ionization energies going from B to N as the three 2p orbitals each gain one electron. Then there is a slight decrease between N and O as one of the 2p orbitals gains a second electron before a regular increase again.

Relative energies of sub-levels within an atom

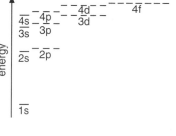

SHAPES OF ORBITALS

An electron has the properties of both a particle and a wave. Heisenberg's uncertainty principle states that it is impossible to know the exact position of an electron at a precise moment in time. An orbital describes the three-dimensional shape where there is a high probability that the electron will be located.

s orbitals are spherical and the three p orbitals are orthogonal (at right angles) to each other.

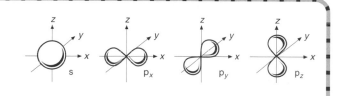

AUFBAU PRINCIPLE

The electronic configuration can be determined by following the aufbau (building up) principle. The orbitals with the lowest energy are filled first. Each orbital can contain a maximum of two electrons. Orbitals within the same sub-shell are filled singly first – this is known as Hund's rule,

e.g.　F $1s^2 2s^2 2p^5$

　　　V $1s^2 2s^2 2p^6 3s^2 3p^6 4s^2 3d^3$.

To save writing out all the lower levels the configuration may be shortened by building on the last noble gas configuration, e.g. V is more usually written:

　　[Ar] $4s^2 3d^3$.

(When writing electronic configurations check that for a neutral atom the sum of the superscripts adds up to the atomic number of the element.)

Sometimes boxes are used to represent orbitals so the number of unpaired electrons can easily be seen,

e.g.

C $1s^2 2s^2 2p^2$

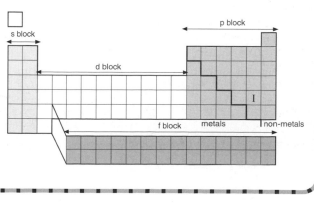

ELECTRONIC CONFIGURATION AND THE PERIODIC TABLE

An element's position in the Periodic Table is related to its valence electrons so the electronic configuration of any element can be deduced from the Table, e.g. iodine ($Z = 53$) is a p block element. It is in group 7 so its configuration will contain $ns^2 np^5$. If one takes H and He as being the first period then iodine is in the fifth period so n = 5. The full configuration for iodine will therefore be:

$1s^2 2s^2 2p^6 3s^2 3p^6 4s^2 3d^{10} 4p^6 5s^2 4d^{10} 5p^5$ or [Kr] $5s^2 4d^{10} 5p^5$

IB QUESTIONS – ATOMIC THEORY

1. Which of the following particles contain more electrons than **neutrons**?

   I. $_1^1H$ II. $_{17}^{35}Cl^-$ III. $_{19}^{39}K^+$

   A. I only
   B. II only
   C. I and II only
   D. II and III only

2. The atom with the same number of neutrons as ^{54}Cr is

   A. ^{50}Ti **B.** ^{51}V **C.** ^{53}Fe **D.** ^{55}Mn

3. All isotopes of tin (Sn) have the same

   I. number of protons
   II. number of neutrons
   III. mass number

   A. I only
   B. II only
   C. III only
   D. I and III only

4. Which one of the following sets represents a pair of isotopes?

   A. $_6^{14}C$ and $_7^{14}N$
   B. O_2 and O_3
   C. $_{16}^{32}S$ and $_{16}^{32}S^{2-}$
   D. $_{82}^{206}Pb$ and $_{82}^{208}Pb$

5. The atomic and mass numbers for four different nuclei are given in the table below. Which two are isotopes?

	atomic number	mass number
I.	101	258
II.	102	258
III.	102	260
IV.	103	259

   A. I and II
   B. II and III
   C. II and IV
   D. III and IV

6. Which species contains 16 protons, 17 neutrons and 18 electrons?

   A. $^{32}S^-$ **B.** $^{33}S^{2-}$ **C.** $^{34}S^-$ **D.** $^{35}S^{2-}$

7. Spectra have been used to study the arrangements of electrons in atoms. An emission spectrum consists of a series of bright lines that converge at high frequencies. Such emission spectra provide evidence that electrons are moving from

   A. lower to higher energy levels with the higher energy levels being closer together.
   B. lower to higher energy levels with the lower energy levels being closer together.
   C. higher to lower energy levels with the lower energy levels being closer together.
   D. higher to lower energy levels with the higher energy levels being closer together.

8. Which electron transition in a hydrogen atom releases the most energy?

   A. $n = 2 \rightarrow n = 1$ **C.** $n = 6 \rightarrow n = 3$
   B. $n = 4 \rightarrow n = 2$ **D.** $n = 7 \rightarrow n = 6$

9. An element has the electronic configuration 2.7. What would be the electronic configuration of an element with similar chemical properties?

   A. 2.6   **B.** 2.8   **C.** 2.7.1   **D.** 2.8.7

10. An element with the symbol Z has the electron configuration 2.8.6. Which species is this element most likely to form?

    A. The ion Z^{2+} **C.** The compound H_2Z
    B. The ion Z^{6+} **D.** The compound Z_6F

HL

The following diagram should be used to answer question 11.

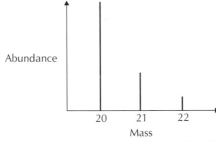

11. According to the mass spectrum above, the relative atomic mass of the element shown is best expressed as

    A. 20.0.
    B. between 20.0 and 21.0.
    C. 21.0.
    D. between 21.0 and 22.0.

12. The first four ionisation energies (kJ mol^{-1}) for a particular element are 550, 1064, 4210 and 5500 respectively. This element should be placed in the same Group as

    A. Li   **B.** Be   **C.** B   **D.** C

13. Which ionisation requires the most energy?

    A. $Na(g) \rightarrow Na^+(g) + e^-$
    B. $Na^+(g) \rightarrow Na^{2+}(g) + e^-$
    C. $Mg(g) \rightarrow Mg^+(g) + e^-$
    D. $Mg^+(g) \rightarrow Mg^{2+}(g) + e^-$

14. Which one of the following atoms in its ground state has the greatest number of unpaired electrons?

    A. Al   **B.** Si   **C.** P   **D.** S

15. All of the following factors affect the value of the ionisation energy of an atom **except** the

    A. mass of the atom.
    B. charge on the nucleus.
    C. size of the atom.
    D. main energy level from which the electron is removed.

The Periodic Table and physical properties (1)

THE PERIODIC TABLE

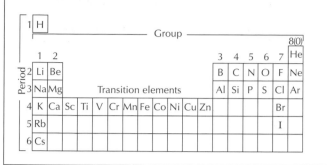

In the Periodic Table elements are placed in order of increasing atomic number. Elements with the same number of valence electrons are placed vertically in the same **group**. The groups are numbered from 1 to 8 (or 0). Some groups have their own name:
- Group 1 – alkali metals
- Group 7 – halogens
- Group 8 or 0 – noble gases (sometimes also called rare gases or inert gases).

Elements with the same outer shell of valence electrons are placed horizontally in the same **period**. The transition elements are located between groups 2 and 3.

ATOMIC RADIUS

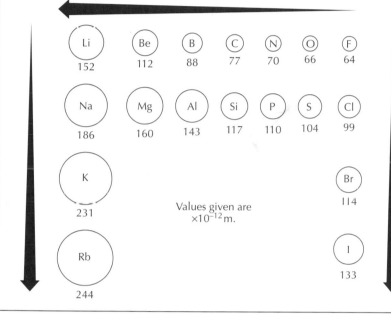

Values given are ×10⁻¹² m.

The atomic radius is the distance from the nucleus to the outermost electron. Since the position of the outermost electron can never be known precisely, the atomic radius is usually defined as half the distance between the nuclei of two bonded atoms of the same element.

As a group is descended the outermost electron is in a higher energy level, which is further from the nucleus, so the radius increases.

Across a period electrons are being added to the same energy level, but the number of protons in the nucleus increases. This attracts the energy level closer to the nucleus and the atomic radius decreases across a period.

IONIC RADIUS

It is important to distinguish between positive ions (**cations**) and negative ions (**anions**). Both cations and anions increase in size down a group as the outer level gets further from the nucleus.

Cations contain fewer electrons than protons so the electrostatic attraction between the nucleus and the outermost electron is greater and the ion is smaller than the parent atom. It is also smaller because the number of electron shells has decreased by one. Across the period the ions contain the same number of electrons (**isoelectronic**), but an increasing number of protons, so the ionic radius decreases.

Anions contain more electrons than protons so are larger than the parent atom. Across a period the size decreases because the number of electrons remains the same but the number of protons increases.

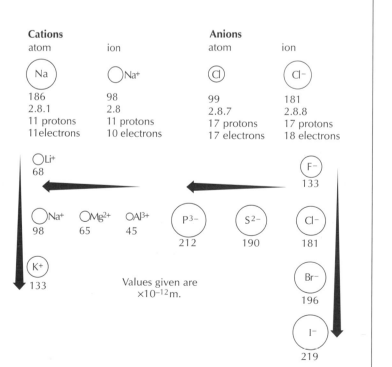

Values given are ×10⁻¹² m.

The Periodic Table and physical properties (2)

PERIODICITY

Elements in the same group tend to have similar chemical and physical properties. There is a change in chemical and physical properties across a period. The repeating pattern of physical and chemical properties shown by the different periods is known as **periodicity**.

These periodic trends can clearly be seen in atomic radii, ionic radii, ionization energies, electronegativities and melting points.

MELTING POINTS

Melting points depend both on the structure of the element and on the type of attractive forces holding the atoms together. Using period 3 as an example:

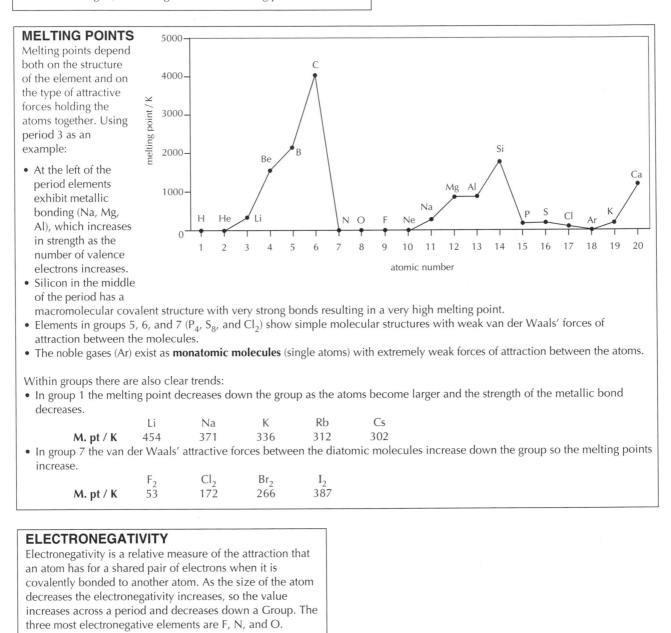

- At the left of the period elements exhibit metallic bonding (Na, Mg, Al), which increases in strength as the number of valence electrons increases.
- Silicon in the middle of the period has a macromolecular covalent structure with very strong bonds resulting in a very high melting point.
- Elements in groups 5, 6, and 7 (P_4, S_8, and Cl_2) show simple molecular structures with weak van der Waals' forces of attraction between the molecules.
- The noble gases (Ar) exist as **monatomic molecules** (single atoms) with extremely weak forces of attraction between the atoms.

Within groups there are also clear trends:
- In group 1 the melting point decreases down the group as the atoms become larger and the strength of the metallic bond decreases.

	Li	Na	K	Rb	Cs
M. pt / K	454	371	336	312	302

- In group 7 the van der Waals' attractive forces between the diatomic molecules increase down the group so the melting points increase.

	F_2	Cl_2	Br_2	I_2
M. pt / K	53	172	266	387

ELECTRONEGATIVITY

Electronegativity is a relative measure of the attraction that an atom has for a shared pair of electrons when it is covalently bonded to another atom. As the size of the atom decreases the electronegativity increases, so the value increases across a period and decreases down a Group. The three most electronegative elements are F, N, and O.

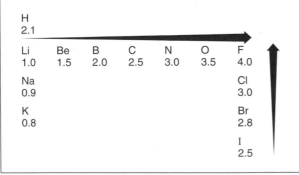

The Periodic Table and chemical properties

CHEMICAL PROPERTIES OF ELEMENTS IN THE SAME GROUP

Group 1 – the alkali metals

Lithium, sodium, and potassium all contain one electron in their outer shell. They are all reactive metals and are stored under liquid paraffin to prevent them reacting with air. They react by losing their outer electron to form the metal ion. Because they can readily lose an electron they are good reducing agents. The reactivity increases down the group as the outer electron is in successively higher energy levels and less energy is required to remove it.

They are called alkali metals because they all react with water to form an alkali solution of the metal hydroxide and hydrogen gas. Lithium floats and reacts quietly, sodium melts into a ball which darts around on the surface, and the heat generated from the reaction with potassium ignites the hydrogen.

$$2Li(s) + 2H_2O(l) \rightarrow 2Li^+(aq) + 2OH^-(aq) + H_2(g)$$

$$2Na(s) + 2H_2O(l) \rightarrow 2Na^+(aq) + 2OH^-(aq) + H_2(g)$$

$$2K(s) + 2H_2O(l) \rightarrow 2K^+(aq) + 2OH^-(aq) + H_2(g)$$

They all also react readily with chlorine and bromine to form ionic salts, e.g.

$$2Na(s) + Cl_2(g) \rightarrow 2Na^+Cl^-(s)$$

$$2K(s) + Br_2(l) \rightarrow 2K^+Br^-(s)$$

Group 7 – the halogens

The halogens react by gaining one more electron to form halide ions. They are good oxidizing agents. The reactivity decreases down the group as the outer shell is increasingly at higher energy levels and further from the nucleus. This, together with the fact that there are more electrons between the nucleus and the outer shell, decreases the attraction for an extra electron.

Chlorine is a stronger oxidizing agent than bromine, so can remove the electron from bromide ions in solution to form chloride ions and bromine. Similarly both chlorine and bromine can oxidize iodide ions to form iodine.

$$Cl_2(aq) + 2Br^-(aq) \rightarrow 2Cl^-(aq) + Br_2(aq)$$

$$Cl_2(aq) + 2I^-(aq) \rightarrow 2Cl^-(aq) + I_2(aq)$$

$$Br_2(aq) + 2I^-(aq) \rightarrow 2Br^-(aq) + I_2(aq)$$

Test for halide ions

The presence of halide ions in solution can be detected by adding silver nitrate solution. The silver ions react with the halide ions to form a precipitate of the silver halide. The silver halides can be distinguished by their colour. These silver halides react with light to form silver metal. This is the basis of photography.

$$Ag^+(aq) + X^-(aq) \rightarrow AgX(s) \qquad \text{where } X = Cl, Br, \text{ or } I$$

AgCl white

AgBr cream

AgI yellow

$$\text{light} \downarrow$$

$$Ag(s) + \tfrac{1}{2}X_2$$

CHANGE FROM METALLIC TO NON-METALLIC NATURE OF THE ELEMENTS ACROSS PERIOD 3

Metals tend to be shiny and are good conductors of heat and electricity. Sodium, magnesium, and aluminium all conduct electricity well. Silicon is a semi-conductor and is called a **metalloid** as it possesses some of the properties of a metal and some of a non-metal. Phosphorus, sulfur, chlorine, and argon are non-metals and do not conduct electricity. Metals can also be distinguished from non-metals by their chemical properties. Metal oxides tend to be basic, whereas non-metal oxides tend to be acidic.

Sodium oxide and magnesium oxide are both basic and react with water to form hydroxides,

e.g. $\quad Na_2O(s) + H_2O(l) \rightarrow 2NaOH(aq)$

Aluminium is a metal but its oxide is amphoteric, that is, it can be either basic or acidic depending on whether it is reacting with an acid or a base.

The remaining elements in period 3 have acidic oxides. For example, sulfur trioxide reacts with water to form sulfuric acid.

$$SO_3(g) + H_2O(l) \rightarrow H_2SO_4(aq)$$

OXIDES OF PERIOD 3 ELEMENTS

The oxides of sodium, magnesium, and aluminium are all ionic. This accounts for their high melting points and electrical conductivity when molten. Silicon dioxide has a diamond-like macromolecular structure with a high boiling point. At the other end of the period the difference in electronegativities between the element and oxygen is small, resulting in simple covalent molecular structures with low melting and boiling points.

The acid–base properties of the oxides are also linked to their structure. The oxides of the electropositive elements are very basic and form solutions that are alkaline.

$$Na_2O(s) + H_2O(l) \rightarrow 2Na^+(aq) + 2OH^-(aq)$$

$$MgO(s) + H_2O(l) \rightarrow Mg(OH)_2$$

The amphoteric nature of aluminium oxide can be seen from its reactions with hydrochloric acid and sodium hydroxide.

Acting as a base: $Al_2O_3(s) + 6HCl(aq) \rightarrow 2AlCl_3(aq) + 3H_2O(l)$

Acting as an acid: $Al_2O_3(s) + 2NaOH(aq) + 3H_2O(l) \rightarrow 2NaAl(OH)_4(aq)$
sodium aluminate

Silicon dioxide behaves as a weak acid. It does not react with water but will form sodium silicate with sodium hydroxide.

$$SiO_2(s) + 2NaOH(aq) \rightarrow Na_2SiO_3(aq) + H_2O(aq)$$

The oxides of phosphorus, sulfur, and chlorine are all strongly acidic.

$$SO_2(g) + H_2O(l) \rightarrow H_2SO_3 \text{ sulfurous acid}$$

$$P_4O_{10}(s) + 6H_2O(l) \rightarrow 4H_3PO_4(aq) \text{ phosphoric acid}$$

$$Cl_2O_7(l) + H_2O(l) \rightarrow 2HClO_4(aq) \text{ perchloric acid}$$

Oxides of period 3 elements

Formula	Na_2O	MgO	Al_2O_3	SiO_2	P_4O_{10} (P_4O_6)	SO_3 (SO_2)	Cl_2O_7 (Cl_2O)
State at 25 °C	Solid	Solid	Solid	Solid	Solid (Solid)	Liquid (Gas)	Liquid (Gas)
Melting point / °C	1275	2852	2027	1610	24	17	−92
Boiling point / °C	−	3600	2980	2230	175	45	80
Electrical conductivity in molten state	Good	Good	Good	Very poor	None	None	None
Structure	Ionic			Covalent macromolecular	Simple covalent molecular		
Reaction with water	Forms NaOH(aq), an alkaline solution	Forms Mg(OH)$_2$, weakly alkaline	Does not react	Does not react	P_4O_{10} forms H_3PO_4, an acidic solution	SO_3 forms H_2SO_4, a strong acid	Cl_2O_7 forms $HClO_4$, an acidic solution
Nature of oxide	Basic		Amphoteric		Acidic		

Chlorides of the third period (sodium → argon)

CHLORIDES OF PERIOD 3 ELEMENTS

The physical properties of the chlorides are related to the structure in the same way as the oxides. Sodium chloride and magnesium chloride are ionic – they conduct electricity when molten and have high melting points. Aluminium chloride is covalent and is a poor conductor. Unlike silicon dioxide, silicon tetrachloride has a simple molecular structure as do the remaining chlorides in the period. These molecules are held together by weak van der Waals' forces, which results in low melting and boiling points.

Sodium chloride dissolves in water to give a neutral solution, magnesium chloride gives a slightly acidic solution with water. All the other chlorides including aluminium chloride react vigorously with water to produce acidic solutions of hydrochloric acid together with fumes of hydrogen chloride.

$$2AlCl_3(s) + 3H_2O(l) \rightarrow Al_2O_3(s) + 6HCl(aq)$$

$$SiCl_4(l) + 4H_2O(l) \rightarrow Si(OH)_4(aq) + 4HCl(aq)$$

$$PCl_3(l) + 3H_2O(l) \rightarrow H_3PO_3(aq) + 3HCl(aq)$$

Chlorine itself reacts with water to some extent to form an acidic solution.

$$Cl_2(aq) + H_2O(l) \rightleftharpoons HCl(aq) + HClO(aq)$$

Chlorides of period 3 elements

Formula	NaCl	MgCl$_2$	Al$_2$Cl$_6$	SiCl$_4$	PCl$_3$ (PCl$_5$)	(S$_2$Cl$_2$)	Cl$_2$
State at 25 °C	Solid	Solid	Solid	Liquid	Liquid (Solid)	Liquid	Gas
Melting point / °C	801	714	178 (sublimes)	–70	–112	–80	–101
Boiling point / °C	1413	1412	–	58	76	136	–35
Electrical conductivity in molten state	Good	Good	Poor	None	None	None	None
Structure	Ionic		Simple covalent molecular				
Reaction with water	Dissolve easily		Fumes of HCl produced				Some reaction with water
Nature of solution	Neutral	Weakly acidic	Acidic				

HL d-block elements (first row)

THE FIRST ROW TRANSITION ELEMENTS

Element	(Sc)	Ti	V	Cr	Mn	Fe	Co	Ni	Cu	(Zn)
Electron configuration [Ar]	$4s^2 3d^1$	$4s^2 3d^2$	$4s^2 3d^3$	$4s^1 3d^5$	$4s^2 3d^5$	$4s^2 3d^6$	$4s^2 3d^7$	$4s^2 3d^8$	$4s^1 3d^{10}$	$4s^2 3d^{10}$

A transition element is defined as an element that possesses an incomplete d sub-level in one or more of its oxidation states. Scandium is not a typical transition metal as its common ion Sc^{3+} has no d electrons. Zinc is not a transition metal as it contains a full d sub-level in all its oxidation states. (Note: for Cr and Cu it is more energetically favourable to half-fill and completely fill the d sub-level respectively so they contain only one 4s electron).

Variable oxidation states

The 3d and 4s sub-levels are very similar in energy. When transition metals lose electrons they lose the 4s electrons first. All transition metals can show an oxidation state of +2. Some of the transition metals can form the +3 or +4 ion (e.g. Fe^{3+}, Mn^{4+}) as the ionization energies are such that up to two d electrons can also be lost. The M^{4+} ion is rare and in the higher oxidation states the element is usually found not as the free metal ion but either covalently bonded or as the oxyanion, such as MnO_4^-. Some common examples of variable oxidation states in addition to +2 are:

Cr(+3)	$CrCl_3$	chromium(III) chloride
Cr(+6)	$Cr_2O_7^{2-}$	dichromate(VI) ion
Mn(+4)	MnO_2	manganese(IV) oxide
Mn(+7)	MnO_4^-	manganate(VII) ion
Fe(+3)	Fe_2O_3	iron(III) oxide
Cu(+1)	Cu_2O	copper(I) oxide

CHARACTERISTIC PROPERTIES OF TRANSITION ELEMENTS

Formation of complex ions

Because of their small size d-block ions attract species that are rich in electrons. Such species are known as **ligands**. Ligands are neutral molecules or anions which contain a non-bonding pair of electrons. These electron pairs can form co-ordinate covalent bonds with the metal ion to form **complex ions**.

A common ligand is water and most (but not all) transition metal ions exist as hexahydrated complex ions in aqueous solution, e.g. $[Fe(H_2O)_6]^{3+}$. Ligands can be replaced by other ligands. A typical example is the addition of ammonia to an aqueous solution of copper(II) sulfate to give the deep blue colour of the tetraaminecopper(II) ion. Similarly if concentrated hydrochloric acid is added to a solution of $Cu^{2+}(aq)$ the yellow tetrachlorocopper(II) anion is formed. Note: in this ion the overall charge on the ion is −2 as the four ligands each have a charge of −1.

$$[CuCl_4]^{2-} \underset{H_2O}{\overset{Cl^-}{\rightleftharpoons}} [Cu(H_2O)_4]^{2+} \underset{H_2O}{\overset{NH_3}{\rightleftharpoons}} [Cu(NH_3)_4]^{2+}$$

The number of lone pairs bonded to the metal ion is known as the **co-ordination number**. Compounds with a co-ordination number of six are octahedral in shape, those with a co-ordination number of four are tetrahedral or square planar, whereas those with a co-ordination number of two are usually linear.

Co-ordination number	6	4	2
Examples	$[Fe(CN)_6]^{3-}$	$[CuCl_4]^{2-}$	$[Ag(NH_3)_2]^+$
	$[Fe(OH)_3(H_2O)_3]$	$[Cu(NH_3)_4]^{2+}$	

Coloured complexes

In the free ion the five d orbitals are degenerate (of equal energy). However, in complexes the d orbitals are split into two distinct levels. The energy difference between the levels corresponds to a particular wavelength or frequency in the visible region of the spectrum. When light falls on the complex, energy of a particular wavelength is absorbed and electrons are excited from the lower level to the higher level.

$Cu^{2+}(aq)$ appears blue because it is the complementary colour to the wavelengths that have been absorbed. The amount the orbitals are split depends on the nature of the transition metal, the oxidation state, the shape of the complex, and the nature of the ligand, which explains why different complexes have different colours. If the d orbital is completely empty, as in Sc^{3+}, or completely full, as in Cu^+ or Zn^{2+}, no transitions within the d level can take place and the complexes are colourless.

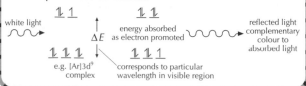

white light → energy absorbed as electron promoted → reflected light complementary colour to absorbed light

ΔE corresponds to particular wavelength in visible region

e.g. $[Ar]3d^9$ complex

Catalytic behaviour

Many transition elements and their compounds are very efficient catalysts. A catalyst increases the rate of a reaction by providing an alternative reaction pathway with a lower activation energy. Transition metal complexes are able to do this because they can exist in a variety of oxidation states. Reactants can be brought into close contact with each other by becoming reversibly attached to the catalyst. Some common examples include:

Iron in the Haber process
$$3H_2(g) + N_2 \underset{}{\overset{Fe(s)}{\rightleftharpoons}} 2NH_3(g)$$

Vanadium(V) oxide in the Contact process
$$2SO_2(g) + O_2(g) \underset{}{\overset{V_2O_5(s)}{\rightleftharpoons}} 2SO_3(g)$$

Nickel in hydrogenation reactions
$$C_2H_4(g) + H_2(g) \xrightarrow{Ni(s)} C_2H_6(g)$$

Manganese(IV) oxide with hydrogen peroxide
$$2H_2O_2(aq) \xrightarrow{MnO_2(s)} 2H_2O(l) + O_2(g)$$

IB QUESTIONS – PERIODICITY

1. In the periodic table, elements are arranged in order of increasing

   A. atomic number.   C. number of valence electrons.

   B. atomic mass.   D. electronegativity.

2. Which one of the following series is arranged in order of increasing value?

   A. The first ionisation energies of: oxygen, fluorine, neon.

   B. The radii of: H^- ion, H atom, H^+ ion.

   C. The electronegativities of: chlorine, bromine, iodine.

   D. The boiling points of: iodine, bromine, chlorine.

3. Which property increases with increasing atomic number for both the alkali metals and the halogens?

   A. Ionisation energies   C. Electronegativities

   B. Melting points   D. Atomic radii

4. Which set of reactants below is expected to produce the most vigorous reaction?

   A. $Na(s) + Cl_2(g)$ C. $K(s) + Cl_2(g)$

   B. $Na(s) + Br_2(g)$ D. $K(s) + Br_2(g)$

5. Which one of the following statements about the halogen group is correct?

   A. First ionisation energies increase from F to I.

   B. Fluorine has the smallest tendency to be reduced.

   C. Cl_2 will oxidise $I^-(aq)$.

   D. I_2 is a stronger oxidising agent than F_2.

6. Strontium is an element in Group 2 of the Periodic Table with atomic number 38. Which of the following statements about strontium is NOT correct?

   A. Its first ionisation energy is lower than that of calcium.

   B. It has two electrons in its outermost energy level.

   C. Its atomic radius is smaller than magnesium.

   D. It forms a chloride with the formula $SrCl_2$.

7. Which one of the following elements has the lowest first ionisation energy?

   A. Li   C. Mg

   B. Na   D. Al

8. Which element is most similar chemically to the element with 14 electrons?

   A. Al   C. Ge

   B. As   D. P

9. 0.01 mole samples of the following oxides were added to separate 1 dm^3 portions of water. Which will produce the most acidic solution?

   A. $Al_2O_3(s)$ C. $Na_2O(s)$

   B. $SiO_2(s)$ D. $SO_3(g)$

10. Which reaction occurs readily?

    I. $Br_2(aq) + 2I^-(aq) \rightarrow I_2(aq) + 2Br^-(aq)$

    II. $Br_2(aq) + 2Cl^-(aq) \rightarrow Cl_2(aq) + 2Br^-(aq)$

    A. I only   C. Both I and II

    B. II only   D. Neither I nor II

HL

---

11. In which region of the Periodic Table would the element with the electronic structure below be located?

    $1s^2 2s^2 2p^6 3s^2 3p^6 3d^{10} 4s^2 4p^6 4d^6 5s^2$

    A. group 6   C. s block

    B. noble gases   D. d block

12. A certain element has the electronic configuration $1s^2 2s^2 2p^6 3s^2 3p^6 4s^2 3d^3$. Which oxidation state(s) would this element most likely show?

    A. +2 only   C. +2 and +5 only

    B. +3 only   D. +2, +3, +4, +5

13. Which ion is colourless?

    A. $[Cr(H_2O)_6]^{3+}$ C. $[Cu(NH_3)_4]^{2+}$

    B. $[Fe(CN)_6]^{4-}$ D. $[Zn(H_2O)_4]^{2+}$

14. Which of the following chlorides give neutral solutions when added to water?

    I.   NaCl

    II. Al_2Cl_6

    III. PCl_3

    A. I only   C. II and III only

    B. I and II only   D. I, II and III

15. Based on melting points, the dividing line between ionic and covalent chlorides of the elements Mg to S lies between

    A. Mg and Al.   C. Si and P.

    B. Al and Si.   D. P and S.

16. The colours of the compounds of d-block elements are due to electron transitions

    A. between different d orbitals.

    B. between d orbitals and s orbitals.

    C. among the attached ligands.

    D. from the metal to the attached ligands.

Ionic bonding

IONIC BOND

When atoms combine they do so by trying to achieve an inert gas configuration. Ionic compounds are formed when electrons are transferred from one atom to another to form ions with complete outer shells of electrons. In an ionic compound the positive and negative ions are attracted to each other by strong electrostatic forces, and build up into a strong lattice. Ionic compounds have high melting points as considerable energy is required to overcome these forces of attraction.

The classic example of an ionic compound is sodium chloride Na^+Cl^-, formed when sodium metal burns in chlorine. Chlorine is a covalent molecule, so each atom already has an inert gas configuration. However, the energy given out when the ionic lattice is formed is sufficient to break the bond in the chlorine molecule to give atoms of chlorine. Each sodium atom then transfers one electron to a chlorine atom to form the ions.

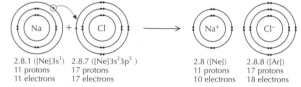

2.8.1 ([Ne]3s^1) 2.8.7 ([Ne]3s^{2}3p^5)   2.8 ([Ne])   2.8.8 ([Ar])
11 protons   17 protons   11 protons   17 protons
11 electrons 17 electrons   10 electrons 18 electrons

The charge carried by an ion depends on the number of electrons the atom needed to lose or gain to achieve a full outer shell.

Cations				Anions		
Group 1	**Group 2**	**Group 3**		**Group 5**	**Group 6**	**Group 7**
+1	+2	+3		−3	−2	−1
Li^+ Na^+ K^+	Mg^{2+} Ca^{2+}	Al^{3+}		N^{3-} P^{3-}	O^{2-} S^{2-}	F^- Cl^- Br^-

Thus in magnesium chloride two chlorine atoms each gain one electron from a magnesium atom to form $Mg^{2+}Cl^-_2$. In magnesium oxide two electrons are transferred from magnesium to oxygen to give $Mg^{2+}O^{2-}$. Transition metals can form more than one ion. For example, iron can form Fe^{2+} and Fe^{3+} and copper can form Cu^+ and Cu^{2+}.

FORMULAS OF IONIC COMPOUNDS

It is easy to obtain the correct formula as the overall charge of the compound must be zero.

lithium fluoride Li^+F^- magnesium chloride $Mg^{2+}Cl^-_2$ aluminium bromide $Al^{3+}Br^-_3$
sodium oxide $Na^+_2O^{2-}$ calcium sulfide $Ca^{2+}S^{2-}$ iron(III) oxide $Fe^{3+}_2O^{2-}_3$
potassium nitride $K^+_3N^{3-}$ calcium phosphide $Ca^{2+}_3P^{3-}_2$ iron(II) oxide $Fe^{2+}O^{2-}$

Note: the formulas above have been written to show the charges carried by the ions. Unless asked specifically to do this it is common practice to omit the charges and simply write LiF, $MgCl_2$, etc.

IONS CONTAINING MORE THAN ONE ELEMENT

In ions formed from more than one element the charge is often spread (delocalized) over the whole ion. An example of a positive ion is the ammonium ion NH_4^+, in which all four N–H bonds are identical. Negative ions are sometimes known as acid radicals as they are formed when an acid loses one or more H^+ ions.

hydroxide OH^-
nitrate NO_3^- (from nitric acid, HNO_3)
sulfate SO_4^{2-}
hydrogen sulfate HSO_4^- } (from sulfuric acid, H_2SO_4)

carbonate CO_3^{2-}
hydrogen carbonate HCO_3^- } from carbonic acid, H_2CO_3
ethanoate CH_3COO^-
(from ethanoic acid, CH_3COOH)

The formulas of the ionic compounds are obtained in exactly the same way. Note: brackets are used to show that the subscript covers all the elements in the ion.

sodium nitrate $Na^+NO_3^-$ calcium carbonate $Ca^{2+}CO_3^{2-}$ aluminium hydroxide $Al^{3+}(OH^-)_3$
ammonium sulphate $(NH_4^+)_2SO_4^{2-}$ magnesium ethanoate $Mg^{2+}(CH_3COO^-)_2$

IONIC OR COVALENT?

Ionic compounds are formed between metals on the left of the Periodic Table and non-metals on the right of the Periodic Table; that is, between elements in groups 1, 2, and 3 with a low electronegativity (electropositive elements) and elements with a high electronegativity in groups 5, 6, and 7. Generally the difference between the electronegativity values needs to be greater than about 1.8 for ionic bonding to occur.

	Al F	Al O	Al Cl	Al Br
Electronegativity	1.5 4.0	1.5 3.5	1.5 3.0	1.5 2.8
Difference in electronegativity	2.5	2.0	1.5	1.3
Formula	AlF_3	Al_2O_3	Al_2Cl_6	Al_2Br_6
Type of bonding	ionic	ionic	intermediate between ionic and covalent	covalent
M. pt / °C	1265	2050	Sublimes at 180	97

Covalent bonding

SINGLE COVALENT BONDS

Covalent bonding involves the sharing of one or more pairs of electrons so that each atom in the molecule achieves an inert gas configuration. The simplest covalent molecule is hydrogen. Each hydrogen atom has one electron in its outer shell. The two electrons are shared and attracted by both nuclei resulting in a directional bond between the two atoms to form a molecule. When one pair of electrons are shared the resulting bond is known as a single covalent bond. Another example of a diatomic molecule with a single covalent bond is chlorine, Cl_2.

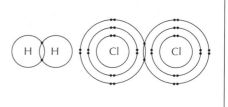

LEWIS STRUCTURES

In the Lewis structure (also known as electron dot structure) all the valence electrons are shown. There are various different methods of depicting the electrons. The simplest method involves using a line to represent one pair of electrons. It is also acceptable to represent single electrons by dots, crosses or a combination of the two. The four methods below are all correct ways of showing the Lewis structure of fluorine.

Sometimes just the shared pairs of electrons are shown, e.g. F–F. This gives information about the bonding in the molecule, but it is not the Lewis structure as it does not show all the valence electrons.

SINGLE COVALENT BONDS

methane, tetrafluoromethane, ammonia, water, hydrogen fluoride

The carbon atom (electronic configuration 2.4) has four electrons in its outer shell and requires a share in four more electrons. It forms four single bonds with elements that only require a share in one more electron, such as hydrogen or chlorine. Nitrogen (2.5) forms three single bonds with hydrogen in ammonia leaving one non-bonded pair of electrons (also known as a lone pair). In water there are two non-bonded pairs and in hydrogen fluoride three non-bonded pairs.

MULTIPLE COVALENT BONDS

In some compounds atoms can share more than one pair of electrons to achieve an inert gas configuration.

oxygen, nitrogen, carbon dioxide, ethene, ethyne

Note: electrons in the shared pair may originate from the same atom. This is known as a co-ordinate covalent bond, but in all other ways it is identical to a normal covalent bond. In some countries stress is laid upon trying to stick to the 'octet' rule so both sulfur dioxide and sulfur trioxide are shown as having a co-ordinate bond between sulfur and oxygen. In other countries sulfur 'expands its octet' to ten or twelve valence electrons and double bonds are shown between the sulfur and the oxygen. Both are acceptable in the International Baccalaureate.

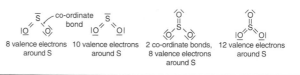

8 valence electrons around S   10 valence electrons around S   2 co-ordinate bonds, 8 valence electrons around S   12 valence electrons around S

BOND LENGTH AND BOND STRENGTH

The strength of attraction that the two nuclei have for the shared electrons affects both the length and strength of the bond. Although there is considerable variation in the bond lengths and strengths of single bonds in different compounds, double bonds are generally much stronger and shorter than single bonds. The strongest covalent bonds are shown by triple bonds.

		Length / nm	Strength / kJ mol^{-1}
Single bonds	Cl–Cl	0.199	242
	C–C	0.154	348
Double bonds	C=C	0.134	612
	O=O	0.121	496
Triple bonds	C≡C	0.120	837
	N≡N	0.110	944

e.g. ethanoic acid:

double bond between C and O shorter and stronger than single bond

BOND POLARITY

In diatomic molecules containing the same element (e.g. H_2 or Cl_2) the electron pair will be shared equally, as both atoms exert an identical attraction. However, when the atoms are different the more electronegative atom exerts a greater attraction for the electron pair. One end of the molecule will thus be more electron rich than the other end, resulting in a polar bond. This relatively small difference in charge is represented by δ+ and δ–. The bigger the difference in electronegativities the more polar the bond.

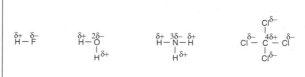

Shapes of simple molecules and ions

VSEPR THEORY

The shapes of simple molecules and ions can be determined by using the **valence shell electron pair repulsion (VSEPR)** theory. This states that pairs of electrons arrange themselves around the central atom so that they are as far apart from each other as possible. There will be greater repulsion between non-bonded pairs of electrons than between bonded pairs. Since all the electrons in a multiple bond must lie in the same direction, double and triple bonds count as one pair of electrons. Strictly speaking the theory refers to negative charge centres, but for most molecules this equates to pairs of electrons.

This results in five basic shapes depending on the number of pairs.

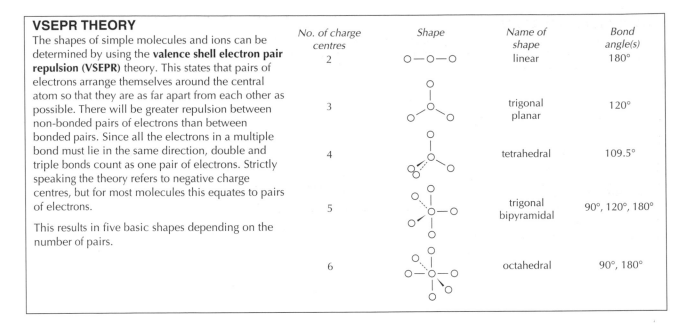

No. of charge centres	Shape	Name of shape	Bond angle(s)
2		linear	180°
3		trigonal planar	120°
4		tetrahedral	109.5°
5		trigonal bipyramidal	90°, 120°, 180°
6		octahedral	90°, 180°

WORKING OUT THE ACTUAL SHAPE

To work out the actual shape of a molecule calculate the number of pairs of electrons around the *central* atom, then work out how many are bonding pairs and how many are non-bonding pairs. (For ions the number of electrons which equate to the charge on the ion must also be included when calculating the total number of electrons.)

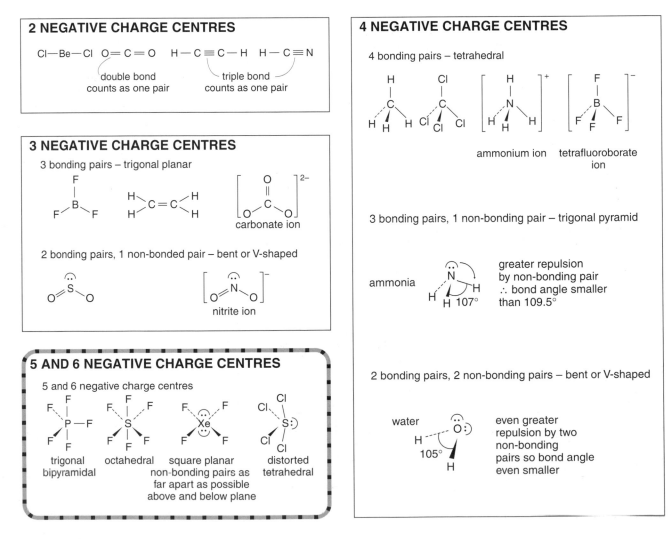

2 NEGATIVE CHARGE CENTRES

Cl—Be—Cl   O=C=O   H—C≡C—H   H—C≡N

double bond counts as one pair   triple bond counts as one pair

3 NEGATIVE CHARGE CENTRES

3 bonding pairs – trigonal planar

carbonate ion

2 bonding pairs, 1 non-bonded pair – bent or V-shaped

nitrite ion

5 AND 6 NEGATIVE CHARGE CENTRES

5 and 6 negative charge centres

trigonal bipyramidal   octahedral   square planar non-bonding pairs as far apart as possible above and below plane   distorted tetrahedral

4 NEGATIVE CHARGE CENTRES

4 bonding pairs – tetrahedral

ammonium ion   tetrafluoroborate ion

3 bonding pairs, 1 non-bonding pair – trigonal pyramid

ammonia   greater repulsion by non-bonding pair ∴ bond angle smaller than 109.5°   107°

2 bonding pairs, 2 non-bonding pairs – bent or V-shaped

water   105°   even greater repulsion by two non-bonding pairs so bond angle even smaller

Intermolecular forces

MOLECULAR POLARITY

Whether a molecule is polar, or not, depends both on the relative electronegativities of the atoms in the molecule and on its shape. If the individual bonds are polar then it does not necessarily follow that the molecule will be polar as the resultant dipole may cancel out all the individual dipoles.

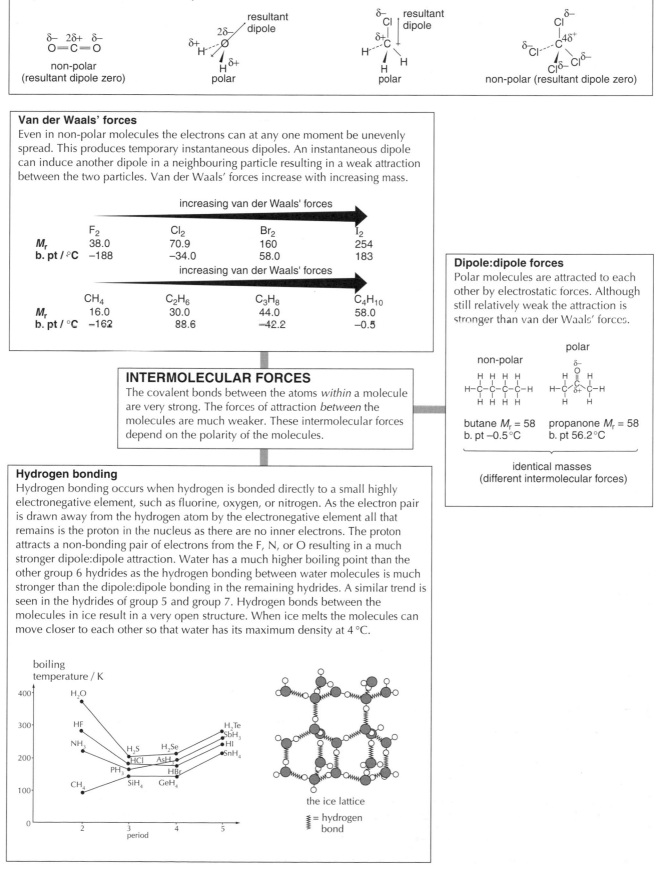

$$\overset{\delta-}{O}=\overset{2\delta+}{C}=\overset{\delta-}{O}$$
non-polar
(resultant dipole zero)

polar

polar

non-polar (resultant dipole zero)

Van der Waals' forces

Even in non-polar molecules the electrons can at any one moment be unevenly spread. This produces temporary instantaneous dipoles. An instantaneous dipole can induce another dipole in a neighbouring particle resulting in a weak attraction between the two particles. Van der Waals' forces increase with increasing mass.

increasing van der Waals' forces →

	F_2	Cl_2	Br_2	I_2
M_r	38.0	70.9	160	254
b. pt / °C	−188	−34.0	58.0	183

increasing van der Waals' forces →

	CH_4	C_2H_6	C_3H_8	C_4H_{10}
M_r	16.0	30.0	44.0	58.0
b. pt / °C	−162	88.6	−42.2	−0.5

INTERMOLECULAR FORCES

The covalent bonds between the atoms *within* a molecule are very strong. The forces of attraction *between* the molecules are much weaker. These intermolecular forces depend on the polarity of the molecules.

Dipole:dipole forces

Polar molecules are attracted to each other by electrostatic forces. Although still relatively weak the attraction is stronger than van der Waals' forces.

non-polar

butane $M_r = 58$
b. pt −0.5°C

polar

propanone $M_r = 58$
b. pt 56.2°C

identical masses
(different intermolecular forces)

Hydrogen bonding

Hydrogen bonding occurs when hydrogen is bonded directly to a small highly electronegative element, such as fluorine, oxygen, or nitrogen. As the electron pair is drawn away from the hydrogen atom by the electronegative element all that remains is the proton in the nucleus as there are no inner electrons. The proton attracts a non-bonding pair of electrons from the F, N, or O resulting in a much stronger dipole:dipole attraction. Water has a much higher boiling point than the other group 6 hydrides as the hydrogen bonding between water molecules is much stronger than the dipole:dipole bonding in the remaining hydrides. A similar trend is seen in the hydrides of group 5 and group 7. Hydrogen bonds between the molecules in ice result in a very open structure. When ice melts the molecules can move closer to each other so that water has its maximum density at 4 °C.

boiling
temperature / K

the ice lattice

= hydrogen bond

Metallic bonding and physical properties related to bonding type

METALLIC BONDING

The valence electrons in metals become detached from the individual atoms so that metals consist of a close packed lattice of positive ions in a sea of delocalized electrons. A metallic bond is the attraction that two neighbouring positive ions have for the delocalized electrons between them. Metals are malleable, that is, they can be bent and reshaped under pressure. They are also ductile, which means they can be drawn out into a wire.

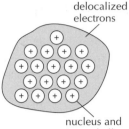

delocalized electrons

nucleus and inner shells

Metals are malleable and ductile because the close-packed layers of positive ions can slide over each other without breaking more bonds than are made.

Impurities added to the metal disturb the lattice and so make the metal less malleable and ductile. This is why alloys are harder than the pure metals they are made from.

TYPE OF BONDING AND PHYSICAL PROPERTIES

Melting and boiling points

When a liquid turns into a gas the attractive forces between the particles are completely broken so boiling point is a good indication of the strength of intermolecular forces. When solids melt the crystal structure is broken down, but there are still some attractive forces between the particles. Melting points are affected by impurities. These weaken the structure and result in lower melting points.

Covalent macromolecular structures have extremely high melting and boiling points. Metals and ionic compounds also tend to have relatively high boiling points due to ionic attractions. Hydrogen bonds are in the order of $\frac{1}{10}$ th the strength of a covalent bond whereas van der Waals' forces are in the order of less than $\frac{1}{100}$ of a covalent bond. The weaker the attractive forces the more volatile the substance.

diamond lattice

NaCl

Diamond (melting point over 4000 °C) All bonds in the macromolecular structure covalent

Sodium chloride (melting point 801 °C) Ions held strongly in ionic lattice

Compound	propane	ethanal	ethanol
M_r	44	44	46
M. pt / °C	−42.2	20.8	78.5
Polarity	non-polar	polar	polar
Bonding type	van der Waals'	dipole:dipole	hydrogen bonding

Solubility

'Like tends to dissolve like'. Polar substances tend to dissolve in polar solvents, such as water, whereas non-polar substances tend to dissolve in non-polar solvents, such as heptane or tetrachloromethane. Organic molecules often contain a polar head and a non-polar carbon chain tail. As the non-polar carbon chain length increases in an homologous series the molecules become less soluble in water. Ethanol itself is a good solvent for other substances as it contains both polar and non-polar ends.

CH_3OH
C_2H_5OH decreasing
C_3H_7OH solubility in water
C_4H_9OH

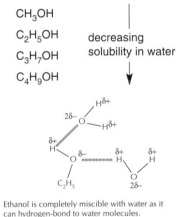

Ethanol is completely miscible with water as it can hydrogen-bond to water molecules.

Conductivity

For conductivity to occur the substance must possess electrons or ions that are free to move. Metals (and graphite) contain delocalized electrons and are excellent conductors. Molten ionic salts also conduct electricity, but are chemically decomposed in the process. Where all the electrons are held in fixed positions, such as diamond or in simple molecules, no electrical conductivity occurs.

When a potential gradient is applied to the metal, the delocalized electrons can move towards the positive end of the gradient carrying charge.

When an ionic compound melts, the ions are free to move to oppositely charged electrodes. Note: in molten ionic compounds it is the ions that carry the charge, not free electrons.

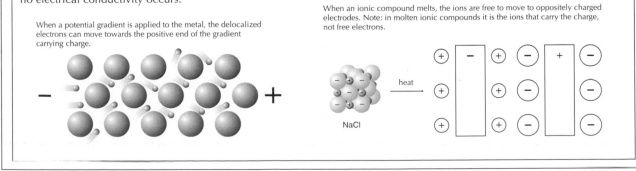

NaCl

heat

OVERLAP OF ATOMIC ORBITALS TO FORM MOLECULAR ORBITALS

Although the Lewis representation is a useful model to represent covalent bonds it does make the false assumption that all the valence electrons are the same. A more advanced model of bonding considers the combination of atomic orbitals to form molecular orbitals.

σ bonds

A σ (sigma) bond is formed when two atomic orbitals on different atoms overlap along a line drawn through the two nuclei. This occurs when two s orbitals overlap, an s orbital overlaps with a p orbital, or when two p orbitals overlap 'head on'.

π bonds

A π (pi) bond is formed when two p orbitals overlap 'sideways on'. The overlap now occurs above and below the line drawn through the two nuclei. A π bond is made up of two regions of electron density.

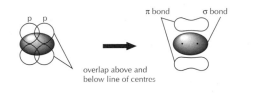

overlap above and below line of centres

σ bonds

HYBRIDIZATION (1)

sp³ hybridization

Methane provides a good example of sp³ hybridization. Methane contains four equal C–H bonds pointing towards the corners of a tetrahedron with bond angles of 109.5°. A free carbon atom has the configuration $1s^2 2s^2 2p^2$. It cannot retain this configuration in methane. Not only are there only two unpaired electrons, but the p orbitals are at 90° to each other and will not give bond angles of 109.5° when they overlap with the s orbitals on the hydrogen atoms.

When the carbon bonds in methane one of its 2s electrons is promoted to a 2p orbital and then the 2s and three 2p orbitals *hybridize* to form four new hybrid orbitals. These four new orbitals arrange themselves to be as mutually repulsive as possible, i.e. tetrahedrally. Four equal σ bonds can then be formed with the hydrogen atoms.

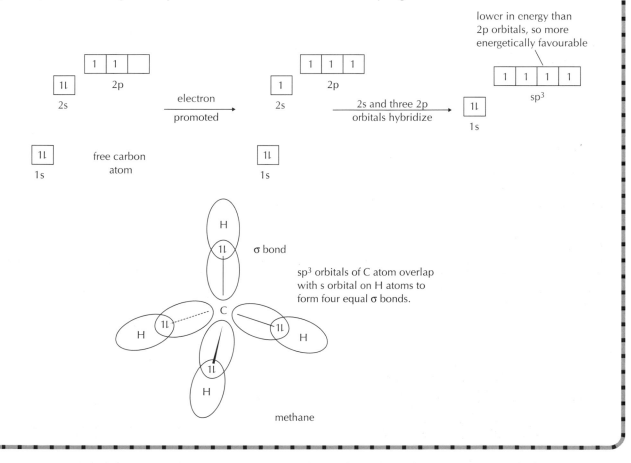

sp³ orbitals of C atom overlap with s orbital on H atoms to form four equal σ bonds.

methane

HYBRIDIZATION (2)

sp² hybridization

sp² hybridization occurs in ethene. After a 2s electron on the carbon atom is promoted the 2s orbital hybridizes with two of the 2p orbitals to form three new planar hybrid orbitals with a bond angle of 120° between them. These can form σ bonds with the hydrogen atoms and also a σ bond between the two carbon atoms. Each carbon atom now has one electron remaining in a 2p orbital. These can overlap to form a π bond. Ethene is thus a planar molecule with a region of electron density above and below the plane.

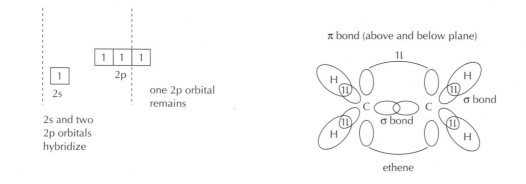

sp hybridization

sp hybridization occurs when the 2s orbital hybridizes with just one of the 2p orbitals to form two new linear sp hybrid orbitals with an angle of 180° between them. The remaining two p orbitals on each carbon atom then overlap to form two π bonds. An example is ethyne.

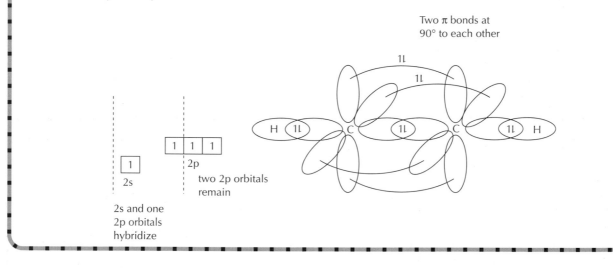

RELATIONSHIP BETWEEN TYPE OF HYBRIDIZATION, LEWIS STRUCTURE, AND MOLECULAR SHAPES

Molecular shapes can be arrived at either by using the VSEPR theory or by knowing the type of hybridization. Hybridization can take place between any s and p orbital in the same energy level and is not just restricted to carbon compounds. If the shape and bond angles are known from using Lewis structures then the type of hybridization can be deduced. Similarly if the type of hybridization is known the shape and bond angles can be deduced.

Hybridization	Regular bond angle	Examples
sp³	109.5°	
sp²	120°	
sp	180°	

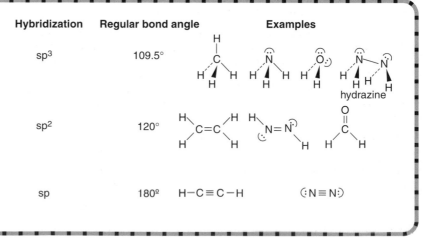

 Delocalization of electrons

RESONANCE STRUCTURES

When writing the Lewis structures for some molecules it is possible to write more than one correct structure. For example, ozone can be written:

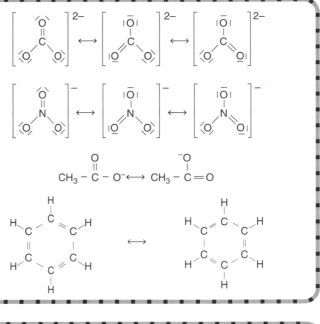

These two structures are known as resonance hybrids. They are extreme forms of the true structure, which lies somewhere between the two. Evidence that this is true comes from bond lengths, as the bond lengths between the oxygen atoms in ozone are both the same and are intermediate between an O=O double bond and an O–O single bond. Resonance structures are usually shown with a double headed arrow between them. Other common compounds which can be written using resonance structures are shown here.

DELOCALIZATION OF ELECTRONS

Resonance structures can also be explained by the delocalization of electrons. For example, in the ethanoate ion the carbon atom and the two oxygen atoms each have a p orbital containing one electron after the σ bonds have been formed. Instead of forming just one double bond between the carbon atom and one of the oxygen atoms the electrons can delocalize over all three atoms. This is energetically more favourable than forming just one double bond. Delocalization can occur whenever alternate double and single bonds occur between carbon atoms. The delocalization energy in benzene is about 150 kJ mol⁻¹, which explains why the benzene ring is so resistant to addition reactions.

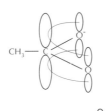

ethanoate ion shown as

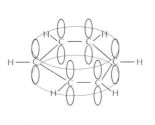

benzene ring shown as

STRUCTURES OF THE ALLOTROPES OF CARBON

Allotropes occur when an element can exist in different crystalline forms. In diamond all the carbon atoms are sp³ hybridized to form a macromolecule in which all the bonds are equally strong. There is no plane of weakness in the molecule, so diamond is exceptionally hard and because all the electrons are localized it does not conduct electricity.

In graphite the carbon atoms are sp² hybridized. This results in three very strong σ bonds between the atoms to give layers of hexagonal rings. The remaining electrons in the p orbitals are then delocalized to form very weak bonds between the layers. The layers can slide over each other so graphite is an excellent lubricant and because the electrons are delocalized it is a good conductor of electricity.

A third allotrope of carbon is buckminsterfullerene. This consists of sixty carbon atoms arranged in hexagons and pentagons to give a geodesic spherical structure similar to a football. Following the initial discovery of buckminsterfullerene a whole family of spherical carbon molecules have been isolated. They are given the collective name 'bucky balls'. In all of them the carbon atoms are sp² hybridized. They also contain delocalized electrons, which gives them the ability to partially conduct electricity.

graphite sp² hybridization

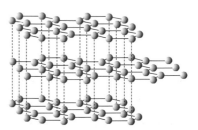

buckminsterfullerene C₆₀ sp² hybridization

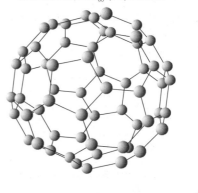

diamond sp³ hybridization

Allotrope	Conductivity / S m⁻¹
graphite	7×10^4
buckminsterfullerene	1.7×10^{-6}
diamond	1×10^{-11}

IB QUESTIONS – BONDING

1. Which compound contains both covalent and ionic bonds?

    A. sodium carbonate, Na_2CO_3

    B. magnesium bromide, $MgBr_2$

    C. dichloromethane, CH_2Cl_2

    D. ethanoic acid, CH_3COOH

2. Which pair of elements is most likely to form a covalently bonded compound?

    A. Li and Cl   **C.** Ca and S

    B. P and O   **D.** Zn and Br

3. Given the following electronegativities,

    H: 2.2   N: 3.0   O: 3.5   F: 4.0

    which bond would be the most polar?

    A. O–H in H_2O **C.** N–O in NO_2

    B. N–F in NF_2 **D.** N–H in NH_3

4. What is the correct Lewis structure for methanal?

    A. H:C:::O:H   **C.** H
                                           C::O:
                                             H

    B.   H   **D.** :C:O:H
         H:C::O: H

5. When CH_4, NH_3, H_2O, are arranged in order of **increasing** bond angle, what is the correct order?

    A. CH_4, NH_3, H_2O **C.** NH_3, CH_4, H_2O

    B. NH_3, H_2O, CH_4 **D.** H_2O, NH_3, CH_4

6. When the H–N–H bond angles in the species NH_2^-, NH_3, NH_4^+ are arranged in order of increasing bond angle (smallest bond angle first), which order is correct?

    A. $NH_2^- < NH_3 < NH_4^+$ **C.** $NH_3 < NH_2^- < NH_4^+$

    B. $NH_4^+ < NH_3 < NH_2^-$ **D.** $NH_2^- < NH_4^+ < NH_3$

7. In which of the following pairs does the second substance have the lower boiling point?

    A. F_2, Cl_2 **C.** C_2H_6, C_3H_8

    B. H_2O, H_2S **D.** CH_3OCH_3, CH_3CH_2OH

8. In which of the following substances would hydrogen bonding be expected to occur?

    I. CH_4

    II. CH_3COOH

    III. CH_3OCH_3

    A. II only   **C.** I and III only

    B. I and III only   **D.** I, II and III

9. Which one of the following statements is correct?

    A. The energy absorbed when liquid ammonia boils is used to overcome the covalent bonds **within** the ammonia molecule.

    B. The energy absorbed when solid phosphorus (P_4) melts is used to overcome the ionic bonds **between** the phosphorus molecules.

    C. The energy absorbed when sodium chloride dissolves in water is used to form ions.

    D. The energy absorbed when copper metal melts is used to overcome the non-directional metallic bonds between the copper atoms.

10. A solid has a melting point of 1440 °C. It conducts heat and electricity. It does not dissolve in water or in organic solvents. The bond between the particles is most likely to be

    A. covalent.   **C.** ionic.

    B. dipole:dipole.   **D.** metallic.

HL

11. What are the types of hybridization of the carbon atoms in the compound

    H_2ClC–CH_2–$COOH$?
           1   2   3

	_1_	_2_	_3_			_1_	_2_	_3_
A.	sp^2	sp^2	sp^2	**C.**		sp^3	sp^3	sp^2
B.	sp^3	sp^2	sp	**D.**		sp^3	sp^3	sp

12. Which molecule or ion does not have a tetrahedral shape?

    A. XeF_4 **C.** BF_4^-

    B. $SiCl_4$ **D.** NH_4^+

13. When the substances below are arranged in order of increasing carbon–carbon bond length (shortest bond first), what is the correct order?

    I. H_2CCH_2 II. H_3CCH_3 III. ⬡

    A. I < II < III   **C.** II < I < III

    B. I < III < II   **D.** III < II < I

14. Which of the following species is considered to involve sp^3 hybridization?

    I. BCl_3 II. CH_4 III. NH_3

    A. I only   **C.** I and III only

    B. II only   **D.** II and III only

15. How many π bonds are present in CO_2?

    A. One   **C.** Three

    B. Two   **D.** Four

16. When the following substances are arranged in order of increasing melting point (lowest melting point first) the correct order is

    A. $CH_3CH_2CH_3$, CH_3COCH_3, $CH_3CH_2CH_2OH$

    B. $CH_3CH_2CH_3$, $CH_3CH_2CH_2OH$, CH_3COCH_3

    C. CH_3COCH_3, $CH_3CH_2CH_2OH$, $CH_3CH_2CH_3$

    D. $CH_3CH_2CH_2OH$, $CH_3CH_2CH_3$, CH_3COCH_3

Changes of state and kinetic theory

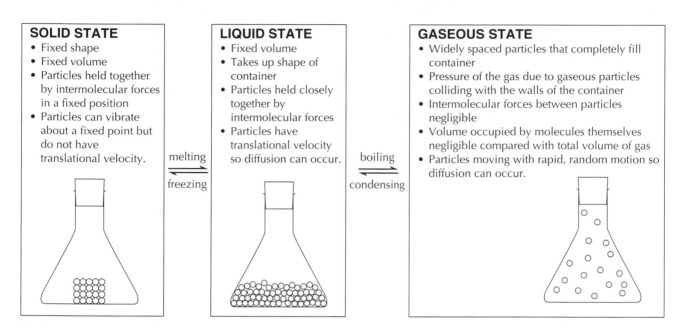

SOLID STATE
- Fixed shape
- Fixed volume
- Particles held together by intermolecular forces in a fixed position
- Particles can vibrate about a fixed point but do not have translational velocity.

melting ⇌ freezing

LIQUID STATE
- Fixed volume
- Takes up shape of container
- Particles held closely together by intermolecular forces
- Particles have translational velocity so diffusion can occur.

boiling ⇌ condensing

GASEOUS STATE
- Widely spaced particles that completely fill container
- Pressure of the gas due to gaseous particles colliding with the walls of the container
- Intermolecular forces between particles negligible
- Volume occupied by molecules themselves negligible compared with total volume of gas
- Particles moving with rapid, random motion so diffusion can occur.

CHANGE OF STATE

The absolute temperature of a substance is proportional to its average kinetic energy ($\frac{1}{2}$ mass × velocity^2). When heat is continually supplied to a substance the following changes take place.

- The solid particles increase in temperature as the kinetic energy increases due to greater vibration of the particles about a fixed position.
- At a certain temperature the vibration is sufficient to overcome the attractive forces holding the lattice together and the solid melts.
- During melting the temperature does not rise as the heat energy (enthalpy of fusion) is needed to overcome these attractive forces.
- Once all the solid has melted the liquid particles move faster and the temperature increases.
- Some particles move faster than others and escape from the surface of the liquid to form a vapour.
- Once the pressure of the vapour is equal to the pressure above the liquid the liquid boils.

- During boiling the temperature remains constant as the heat is used to overcome the intermolecular forces of attraction between the particles (enthalpy of vaporization).
- When all the liquid has turned into a gas the temperature continues to increases as the particles move ever faster.

A cooling curve shows this process in reverse.

COOLING CURVE

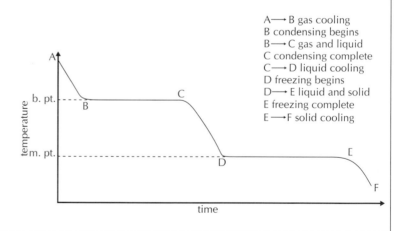

A⟶B gas cooling
B condensing begins
B⟶C gas and liquid
C condensing complete
C⟶D liquid cooling
D freezing begins
D⟶E liquid and solid
E freezing complete
E⟶F solid cooling

MAXWELL–BOLTZMANN DISTRIBUTION

The moving particles in a gas or liquid do not all travel with the same velocity. Some are moving very fast and others much slower. The faster they move the more kinetic energy they possess. The distribution of kinetic energies is shown by a Maxwell–Boltzmann curve. When a liquid evaporates it is the faster moving particles that escape so the average kinetic energy of the remaining particles is lower. This explains why a volatile liquid feels cold as it is evaporating.

At higher temperatures the area under the curve does not change as the total number of particles remains constant. More particles have a very high velocity resulting in an increase in the average kinetic energy, which leads to a broadening of the curve.

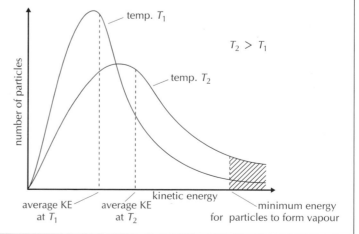

Behaviour of gases

CHANGING THE VARIABLES FOR A FIXED MASS OF GAS

$P \propto \frac{1}{V}$ (or PV = constant)

At constant temperature: as the volume decreases the concentration of the particles increases, resulting in more collisions with the container walls. This increase in pressure is inversely proportional to the volume, i.e. doubling the pressure halves the volume.

$V \propto T$ (or $\frac{V}{T}$ = constant)

At constant pressure: at higher temperatures the particles have a greater average velocity so individual particles will collide with the container walls with greater force. To keep the pressure constant there must be fewer collisions per unit area so the volume of the gas must increase. The increase in volume is directly proportional to the absolute temperature, i.e. doubling the absolute temperature doubles the volume.

$P \propto T$ (or $\frac{P}{T}$ = constant)

At constant volume: increasing the temperature increases the average kinetic energy so the force with which the particles collide with the container walls increases. Hence pressure increases and is directly proportional to the absolute temperature, i.e. doubling the absolute temperature doubles the pressure.

Real gases
An ideal gas exactly obeys the gas laws. As real gases do have some attractive forces between the particles and the particles themselves do occupy some space so they do not exactly obey the laws. If they did they could never condense into liquids. A gas behaves most like an ideal gas at high temperatures and low pressures.

IDEAL GAS EQUATION
The different variables for a gas are all related by the ideal gas equation.

$$PV = nRT$$

P = pressure in Pa ($N\,m^{-2}$)
   (1 atm = 1.013×10^5 Pa)
T = absolute temperature in K
V = volume in m^3
   ($1\,cm^3 = 1 \times 10^{-6}\,m^3$)
n = number of moles
R = gas constant = 8.314 J K^{-1} mol^{-1}

Units
The gas constant can be expressed in different units but it is easier to use SI units.

$$R = \frac{PV}{nT} = \frac{N\,m^{-2} \times m^3}{mol \times K} = N\,m\,mol^{-1}\,K^{-1}$$

$$= J\,K^{-1}\,mol^{-1}$$

Molar volume of a gas
The ideal gas equation depends on the amount of gas (number of moles of gas) but not on the nature of the gas. One mole of any gas will occupy the same volume at the same temperature and pressure. At 273 K and 1.013×10^5 Pa (1 atm) pressure this volume is $2.24 \times 10^{-2}\,m^3$ ($22.4\,dm^3$ or $22\,400\,cm^3$).

When the mass of a particular gas is fixed (nR is constant) a useful expression to convert the pressure, temperature, and volume under one set of conditions (1) to another set of conditions (2) is:

$$\frac{P_1V_1}{T_1} = \frac{P_2V_2}{T_2}$$

In this expression there is no need to convert to SI units as long as the same units for pressure and volume are used on both sides of the equation. However, do not forget that T refers to the absolute temperature and must be in kelvin.

Worked example 1
What volume will be occupied by 0.216 g of carbon dioxide at 21 °C and at a pressure of 1.32 atm?

Step 1. Calculate the number of moles of gas

Amount of carbon dioxide $= \frac{0.216}{44.0} = 4.91 \times 10^{-3}$ mol

Step 2. Express all temperatures as absolute temperatures

21 °C = 294 K

Step 3. Convert all other units to SI units

1.32 atm = $1.32 \times 1.013 \times 10^5 = 1.34 \times 10^5$ Pa

Step 4. Apply ideal gas equation $PV = nRT$

$1.34 \times 10^5 \times V = 4.91 \times 10^{-3} \times 8.314 \times 294$

$V = \frac{4.91 \times 10^{-3} \times 8.314 \times 294}{1.34 \times 10^5} = 8.96 \times 10^{-5}\,m^3$ ($89.6\,cm^3$)

Worked example 2
A gas occupies $127\,cm^3$ at a pressure of 0.830 atm and at 28 °C.

(a) What volume will the same amount of gas occupy at 1.00 atm pressure and at 0 °C?

(b) How many moles of gas are present?

(a) Step 1. Express all temperatures as absolute temperatures

28 °C = 301 K   0 °C = 273 K

Step 2. Apply $\frac{P_1V_1}{T_1} = \frac{P_2V_2}{T_2}$

(Note that there is no need to change the units to SI units.)

$$V_2 = V_1 \times \frac{P_1}{P_2} \times \frac{T_2}{T_1} = 127 \times \frac{0.830}{1.00} \times \frac{273}{301} = 95.6\,cm^3$$

(b) The number of moles can be calculated by using either the molar volume of a gas **(i)** or the ideal gas equation **(ii)**.

(i) 1 mole of any gas occupies $22\,400\,cm^3$ at 273 K, 1 atm
Amount of gas occupying $95.6\,cm^3$ at

273 K, 1 atm $= \frac{95.6}{22\,400} = 4.27 \times 10^{-3}$ mol.

(ii) Convert all units to SI units:

$127\,cm^3 = 1.27 \times 10^{-4}\,m^3$
0.830 atm $= 0.830 \times 1.013 \times 10^5$
   $= 8.41 \times 10^4$ Pa

$$n = \frac{PV}{RT} = \frac{8.41 \times 10^4 \times 1.27 \times 10^{-4}}{8.314 \times 301}$$

$$= 4.27 \times 10^{-3}\,mol$$

IB QUESTIONS – STATES OF MATTER

1. What changes occur when ice at its melting point is converted to liquid water?

   I. movement of the molecules increases

   II. distance between molecules increases

   A. I only
   C. Both I and II
   B. II only
   D. Neither I nor II

2. All of the following are characteristic properties of gases EXCEPT

   A. they can expand without limit.
   B. they diffuse readily.
   C. they are easily compressed.
   D. they have high densities.

3. The boiling points of **four** hydrocarbons are given. Which pair will mix most easily at the temperature specified?

	T_b/K		T_b/K
cyclohexane	354	cyclooctane	421
cycloheptane	392	cyclononane	444

   A. cyclohexane and cycloheptane at 380 K
   B. cycloheptane and cyclooctane at 390 K
   C. cyclooctane and cyclononane at 460 K
   D. cyclononane and cyclohexane at 420 K

4. ^{235}U and ^{238}U are separated by allowing uranium hexafluoride, UF_6, to diffuse through a series of tiny holes. This process works because the molecules of UF_6 that contain ^{235}U and ^{238}U have different

   A. volumes.
   C. chemical properties.
   B. shapes.
   D. speeds.

5. 1.0 g of liquid H_2O has a volume of 1.0 cm^3 but a volume of slightly more than 1 dm^3 when it is converted to vapour at the same temperature. This volume change is best attributed to

   A. an increase in the size of the molecules.
   B. the greater distance between the molecules.
   C. the higher kinetic energy of the molecules.
   D. the presence of more molecules due to the conversion of H_2O to H_2 and O_2.

6. Which of the following best accounts for the observation that gases are easily compressed?

   A. Gas molecules have negligible attractive forces for one another.
   B. The volume occupied by the gas is much greater than that occupied by the molecules.
   C. The average energy of the molecules in a gas is proportional to the absolute temperature of the gas.
   D. The collisions between gas molecules are elastic.

7. Equal volumes of different ideal gases at the same temperature and pressure must contain

   A. the same number of atoms.
   B. one mole of each gas.
   C. the same amount of each gas.
   D. 6.022×10^{23} particles of each gas.

8. All of the following statements are correct about the behaviour of an ideal gas **except**

   A. It expands on heating.
   B. The particles in it are in continuous random motion.
   C. The average energy of the particles doubles when the absolute temperature doubles.
   D. The volume increases as the pressure applied to it increases.

9. A 1.0 dm^3 sample of gas is heated from 27 °C to 54 °C at constant pressure. The volume of the gas will

   A. increase to about 1.1 dm^3
   B. increase to about 2.0 dm^3
   C. decrease to about 0.9 dm^3
   D. decrease to about 0.5 dm^3

10. The temperature (in K) is doubled for a sample of gas in a flexible container while the pressure on it is doubled. The final volume of the gas compared with the initial volume will be:

    A. the same.
    C. four times as large.
    B. twice as large.
    D. half as large.

11. Separate samples of two gases, each containing a pure substance, are found to have the same density under the same conditions of temperature and pressure. Which statement about these two samples **must** be correct?

    A. They have the same volume
    B. They have the same relative molecular mass
    C. There are equal numbers of moles of gas in the two samples
    D. They condense at the same temperature

12. Which expression represents the density of a gas sample of relative molar mass, M_r, at temperature T, and pressure, P?

    A. $\dfrac{PM_r}{T}$
    C. $\dfrac{PM_r}{RT}$

    B. $\dfrac{RT}{PM_r}$
    D. $\dfrac{RM_r}{PT}$

13. A 250 cm^3 sample of an unknown gas has a mass of 1.42 g at 35 °C and 0.85 atmospheres. Which expression gives its molar mass, M_r? (R = 82.05 cm^3 atm K^{-1} mol^{-1})

    A. $\dfrac{1.42 \times 82.05 \times 35}{0.25 \times 0.85}$
    C. $\dfrac{1.42 \times 250 \times 0.85}{82.05 \times 308}$

    B. $\dfrac{1.42 \times 82.05 \times 308}{0.25 \times 0.85}$
    D. $\dfrac{1.42 \times 82.05 \times 308}{250 \times 0.85}$

14. Which graph shows the energy distribution for the particles of a gas when the temperature is increased from T_1 to T_2?

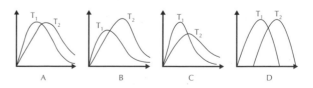

Enthalpy changes

EXOTHERMIC AND ENDOTHERMIC REACTIONS

Energy is defined as the ability to do work, that is, move a force through a distance. It is measured in joules.

Energy = force × distance
(J)   (N × m)

In a chemical reaction energy is required to break the bonds in the reactants, and energy is given out when new bonds are formed in the products. The most important type of energy in chemistry is heat. If the bonds in the products are stronger than the bonds in the reactants then the reaction is said to be **exothermic**, as heat is given out to the surroundings. In **endothermic** reactions heat is absorbed from the surroundings because the bonds in the reactants are stronger than the bonds in the products.

The internal energy stored in the reactants is known as its **enthalpy**, H. The absolute value of the enthalpy of the reactants cannot be known, nor can the enthalpy of the products, but what can be measured is the difference between them, ΔH. By convention ΔH has a negative value for exothermic reactions and a positive value for endothermic reactions. It is normally measured under standard conditions of 1 atm pressure at a temperature of 298 K. The **standard enthalpy change of a reaction** is denoted by $\Delta H^{\ominus}$.

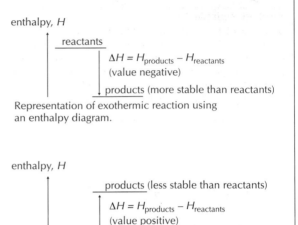

enthalpy, H

reactants

$\Delta H = H_{products} - H_{reactants}$
(value negative)

products (more stable than reactants)

Representation of exothermic reaction using an enthalpy diagram.

enthalpy, H

products (less stable than reactants)

$\Delta H = H_{products} - H_{reactants}$
(value positive)

reactants

Representation of endothermic reaction using an enthalpy diagram.

TEMPERATURE AND HEAT

It is important to be able to distinguish between heat and temperature as the terms are often used loosely.

- Heat is a measure of the total energy in a given amount of substance and therefore depends on the amount of substance present.
- Temperature is a measure of the 'hotness' of a substance. It represents the average kinetic energy of the substance, but is independent of the amount of substance present.

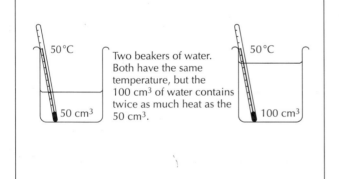

50 °C

Two beakers of water. Both have the same temperature, but the 100 cm³ of water contains twice as much heat as the 50 cm³.

50 cm³

50 °C

100 cm³

CALORIMETRY

The enthalpy change for a reaction can be measured experimentally by using a calorimeter. In a simple calorimeter all the heat evolved in an exothermic reaction is used to raise the temperature of a known mass of water. For endothermic reactions the heat transferred from the water to the reaction can be calculated by measuring the lowering of temperature of a known mass of water.

To compensate for heat lost by the water in exothermic reactions to the surroundings as the reaction proceeds a plot of temperature against time can be drawn. By extrapolating the graph, the temperature rise that would have taken place had the reaction been instantaneous can be calculated.

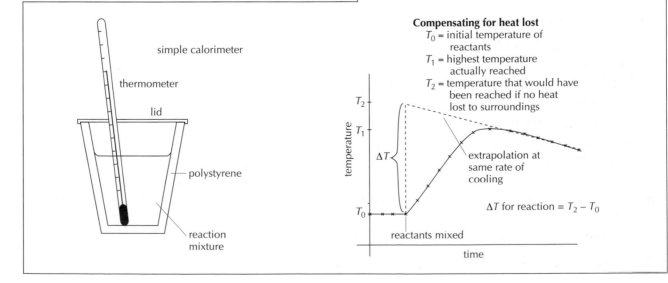

simple calorimeter

thermometer

lid

polystyrene

reaction mixture

Compensating for heat lost
T_0 = initial temperature of reactants
T_1 = highest temperature actually reached
T_2 = temperature that would have been reached if no heat lost to surroundings

T_2
T_1
ΔT
extrapolation at same rate of cooling

T_0

reactants mixed

ΔT for reaction = $T_2 - T_0$

time

ΔH calculations

CALCULATION OF ENTHALPY CHANGES

The heat involved in changing the temperature of any substance can be calculated from the equation:

Heat energy = mass (m) × specific heat capacity (c) × temperature change (ΔT)

The specific heat capacity of water is 4.18 kJ kg^{-1} K^{-1}. That is, it requires one kilojoule of energy to raise the temperature of one kilogram of water by one kelvin.

Enthalpy changes are normally quoted in kJ mol^{-1}, for either a reactant or a product, so it is also necessary to work out the number of moles involved in the reaction which produces the heat change in the water.

Worked example 1

50.0 cm^3 of 1.00 mol dm^{-3} hydrochloric acid solution was added to 50.0 cm^3 of 1.00 mol dm^{-3} sodium hydroxide solution in a polystyrene beaker. The initial temperature of both solutions was 16.7 °C. After stirring and accounting for heat loss the highest temperature reached was 23.5 °C. Calculate the enthalpy change for this reaction.

Step 1. Write equation for reaction

$HCl(aq) + NaOH(aq) \rightarrow NaCl(aq) + H_2O(l)$

Step 2. Calculate molar quantities

Amount of HCl = $\dfrac{50.0}{1000} \times 1.00 = 5.00 \times 10^{-2}$ mol

Amount of NaOH = $\dfrac{50.0}{1000} \times 1.00 = 5.00 \times 10^{-2}$ mol

Therefore the heat evolved will be for 5.00×10^{-2} mol

Step 3. Calculate heat evolved

Total volume of solution = 50.0 + 50.0 = 100 cm^3
Assume the solution has the same density and specific heat capacity as water then
mass of 'water' = 100 g = 0.100 kg
Temperature change = 23.5 − 16.7 = 6.8 °C = 6.8 K
Heat evolved in reaction = 0.100 × 4.18 × 6.8 = 2.84 kJ
= 2.84 kJ (for 5.00×10^{-2} mol)

ΔH for reaction = $-2.84 \times \dfrac{1}{5.00 \times 10^{-2}} = -56.8$ kJ mol^{-1}

(negative value as the reaction is exothermic)

Worked example 2

A student used a simple calorimeter to determine the enthalpy change for the combustion of ethanol.

$C_2H_5OH(l) + 3O_2(g) \rightarrow 2CO_2(g) + 3H_2O(l)$

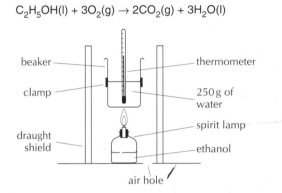

When 0.690 g (0.015 mol) of ethanol was burned it produced a temperature rise of 13.2 K in 250 g of water. Calculate ΔH for the reaction.

Heat evolved by 0.015 mol = $\dfrac{250}{1000} \times 4.18 \times 13.2 = 13.79$ kJ

$\Delta H = -13.79 \times \dfrac{1}{0.015} = -920$ kJ mol^{-1}

Note: the Data Book value is –1371 kJ mol^{-1}. Reasons for the discrepancy include the fact that not all the heat produced is transferred to the water, the water loses some heat to the surroundings, and there is incomplete combustion of the ethanol.

Worked example 3

50.0 cm^3 of 0.200 mol dm^{-3} copper(II) sulfate solution was placed in a polystyrene cup. After two minutes 1.20 g of powdered zinc was added. The temperature was taken every 30 seconds and the following graph obtained. Calculate the enthalpy change for the reaction taking place.

Step 1. Write the equation for the reaction

$Cu^{2+}(aq) + Zn(s) \rightarrow Cu(s) + Zn^{2+}(aq)$

Step 2. Determine the limiting reagent

Amount of Cu^{2+}(aq) = 50.0/1000 × 0.200 = 0.0100 mol

Amount of Zn(s) = 1.20/65.37 = 0.0184 mol

∴ Cu^{2+}(aq) is the limiting reagent

Step 3. Extrapolate the graph (*already done*) to compensate for heat loss and determine ΔT

$\Delta T = 10.4$ °C

Step 4. Calculate the heat evolved in the experiment for 0.0100 mol of reactants

Heat evolved = 50.0/1000 × 4.18 × 10.4 °C = 2.17 kJ

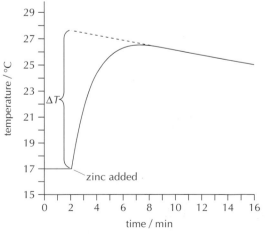

Step 5. Express this as the enthalpy change for the reaction $\Delta H = -2.17 \times \dfrac{1}{0.0100} = -217$ kJ mol^{-1}

Bond enthalpies and Hess' law

BOND ENTHALPIES

Enthalpy changes can also be calculated directly from bond enthalpies.
For a diatomic molecule the bond enthalpy is defined as the enthalpy change
for the process

$$X–Y(g) \rightleftharpoons X(g) + Y(g) \text{ Note the gaseous state.}$$

For bond formation the value is negative and for bond breaking the value is
positive. If the bond enthalpy values are known for all the bonds in the
reactants and products then the overall enthalpy change can be calculated.

Some average bond enthalpies
All values in kJ mol^{-1}

H–H	436	C=C	612	C≡C	837
C–C	348	O=O	496	N≡N	944
C–H	412				
O–H	463				
N–H	388				
N–N	163				

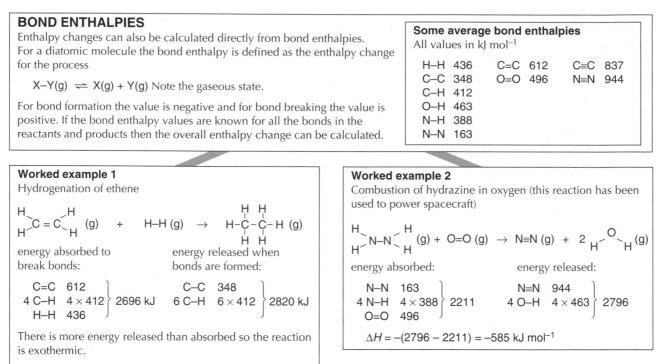

Worked example 1
Hydrogenation of ethene

energy absorbed to
break bonds:

C=C	612
4 C–H	4×412
H–H	436

$\Big\}$ 2696 kJ

energy released when
bonds are formed:

C–C	348
6 C–H	6×412

$\Big\}$ 2820 kJ

There is more energy released than absorbed so the reaction
is exothermic.

$$\Delta H = -(2820 - 2696) = -124 \text{ kJ mol}^{-1}$$

Worked example 2
Combustion of hydrazine in oxygen (this reaction has been
used to power spacecraft)

energy absorbed:

N–N	163
4 N–H	4×388
O=O	496

$\Big\}$ 2211

energy released:

N≡N	944
4 O–H	4×463

$\Big\}$ 2796

$$\Delta H = -(2796 - 2211) = -585 \text{ kJ mol}^{-1}$$

LIMITATIONS OF USING AVERAGE BOND ENTHALPIES

Average bond enthalpies can only be used if all the reactants and products are in the gaseous state. If water were a liquid
product in the above example then even more heat would be evolved since the enthalpy change of vaporization of water would
also need to be included in the calculation.

In the above calculations **average** bond enthalpies have been used. These have been obtained by considering a number of
similar compounds. In practice the energy of a particular bond will vary slightly in different compounds, as it will be affected by
neighbouring atoms. So ΔH values obtained from using average bond enthalpies will not necessarily be very accurate.

HESS' LAW

Hess' law states that the enthalpy change for a reaction depends only on
the difference between the enthalpy of the products and the enthalpy of the
reactants. It is independent of the reaction pathway.

The enthalpy change going
from A to B is the same whether
the reaction proceeds directly
to A or whether it goes
via an intermediate.

$$A \xrightarrow{\Delta H_2} C$$
$$\Delta H_1 \searrow \Delta H_3$$
$$B$$
$$\Delta H_1 = \Delta H_2 + \Delta H_3$$

This law is a statement of the law of conservation of energy. It can be used
to determine enthalpy changes, which cannot be measured directly. For
example, the enthalpy of combustion of both carbon and carbon monoxide
to form carbon dioxide can easily be measured directly, but the combustion
of carbon to carbon monoxide cannot. This can be represented by an
energy cycle.

$$C(s) + \tfrac{1}{2}O_2(g) \xrightarrow{\Delta H_x} CO(g)$$
$$-393 \text{ kJ mol}^{-1} \downarrow O_2(g) \quad \tfrac{1}{2}O_2(g) \; -283 \text{ kJ mol}^{-1}$$
$$CO_2(g)$$

$$-393 = \Delta H_x + (-283)$$
$$\Delta H_x = -393 + 283 = -110 \text{ kJ mol}^{-1}$$

Worked example 3
Calculate the standard enthalpy change when
one mole of methane is formed from its
elements in their standard states. The standard
enthalpies of combustion $\Delta H_c^{\ominus}$ of carbon,
hydrogen, and methane are -393, -286, and
-890 kJ mol^{-1} respectively.

Step 1. Write the equation for the enthalpy
change with the unknown $\Delta H^{\ominus}$ value. Call this
value $\Delta H_x^{\ominus}$.

$$C(s) + 2H_2(g) \xrightarrow{\Delta H_x^{\ominus}} CH_4(g)$$

Step 2. Construct an energy cycle showing the
different routes to the products (in this case the
products of combustion)

$$C(s) + 2H_2(g) \xrightarrow{\Delta H_x^{\ominus}} CH_4(g)$$
$$\downarrow O_2(g) \quad \downarrow O_2(g) \qquad 2O_2(g)$$
$$CO_2(g) + 2H_2O(l)$$

Step 3. Use Hess' law to equate the energy
changes for the two different routes

$$\underbrace{\Delta H_c^{\ominus}(C) + 2\Delta H_c^{\ominus}(H_2)}_{\text{direct route}} = \underbrace{\Delta H_x^{\ominus} + \Delta H_c^{\ominus}(CH_4)}_{\text{route via methane}}$$

Step 4. Rearrange the equation and substitute
the values to give the answer

$$\Delta H_x^{\ominus} = \Delta H_c^{\ominus}(C) + 2\Delta H_c^{\ominus}(H_2) - \Delta H_c^{\ominus}(CH_4)$$
$$= -393 + (2 \times -286) - (-890) \text{ kJ mol}^{-1}$$
$$= -75 \text{ kJ mol}^{-1}$$

Energy cycles

STANDARD ENTHALPY CHANGE OF FORMATION $\Delta H_f^{\ominus}$

The standard enthalpy change of formation of a compound is the enthalpy change when one mole of the compound is formed from its elements in their standard states at 298 K and 1 atm pressure.

From this it follows that $\Delta H_f^{\ominus}$ for an element in its standard state will be zero.

An accurate value for the standard enthalpy change of formation of ethanol can be determined from the following cycle.

$$2C(s) \quad + \quad 3H_2(g) \quad + \quad \tfrac{1}{2}O_2(g) \xrightarrow{\Delta H_f^{\ominus}(C_2H_5OH)} C_2H_5OH(l)$$

$$2 \times \Delta H_f^{\ominus}(CO_2) \Big\downarrow 2O_2(g) \quad 3 \times \Delta H_f^{\ominus}(H_2O) \Big\downarrow 1\tfrac{1}{2}O_2(g) \qquad 3O_2(g)$$

$$2CO_2(g) \quad + \quad 3H_2O(l) \xleftarrow{\Delta H_c^{\ominus}(C_2H_5OH)}$$

By Hess' law: $\Delta H_f^{\ominus}(C_2H_5OH) = 2 \times \Delta H_f^{\ominus}(CO_2) + 3 \times \Delta H_f^{\ominus}(H_2O) - \Delta H_c^{\ominus}(C_2H_5OH)$

Substituting the relevant values $\Delta H_f^{\ominus}(C_2H_5OH) = (2 \times -393.5) + (3 \times -285.8) - (-1371) = -273.4$ kJ mol^{-1}

BORN–HABER CYCLES

Born–Haber cycles are simply energy cycles for the formation of ionic compounds. The enthalpy change of formation of sodium chloride can be considered to occur through a number of separate steps.

$$Na(s) \quad + \quad \tfrac{1}{2}Cl_2 \xrightarrow{\Delta H_f^{\ominus}(NaCl)} Na^+Cl^-(s)$$

$$\Big\downarrow \Delta H_{at}^{\ominus}(Na) \qquad \Big\downarrow \Delta H_{at}^{\ominus}(Cl)$$

$$Na(g) \qquad\qquad Cl(g)$$

$$\Big\downarrow \Delta H_{IE}^{\ominus}(Na) \qquad \Big\downarrow \Delta H_{EA}^{\ominus}(Cl) \quad \Delta H_{latt}^{\ominus}(NaCl)$$

$$Na^+(g) \quad + \quad Cl^-(g)$$

Using Hess' law:

$\Delta H_f^{\ominus}(NaCl) = \Delta H_{at}^{\ominus}(Na) + \Delta H_{IE}^{\ominus}(Na) + \Delta H_{at}^{\ominus}(Cl) + \Delta H_{EA}^{\ominus}(Cl) + \Delta H_{latt}^{\ominus}(NaCl)$

Substituting the relevant values:
$\Delta H_f^{\ominus}(NaCl) = +108 + 494 + 121 - 364 - 771 = -412$ kJ mol^{-1}

Note: it is the large lattice enthalpy that mainly compensates for the endothermic processes and leads to the enthalpy of formation of ionic compounds having a negative value.

LATTICE ENTHALPY $\Delta H_{latt}^{\ominus}$

The lattice enthalpy relates either to the endothermic process of turning a crystalline solid into its gaseous ions or to the exothermic process of turning gaseous ions into a crystalline solid.

$$MX(s) \rightleftharpoons M^+(g) + X^-(g)$$

The sign of the lattice enthalpy indicates whether the lattice is being formed (–) or broken (+).

The size of the lattice enthalpy depends both on the size of the ions and on the charge carried by the ions.

cation size increasing →			anion size increasing →		
LiCl	NaCl	KCl	NaCl	NaBr	NaI
Lattice enthalpy / kJ mol^{-1}					
846	771	701	771	733	684

charge on cation increasing →		charge on anion increasing →	
NaCl	MgCl$_2$	MgCl$_2$	MgO
Lattice enthalpy / kJ mol^{-1}			
771	2493	2493	3889

ENTHALPY OF ATOMIZATION $\Delta H_{at}^{\ominus}$

The standard enthalpy of atomization is the standard enthalpy change when one mole of gaseous atoms is formed from the element in its standard state under standard conditions. For diatomic molecules this is equal to half the bond dissociation enthalpy.

$\tfrac{1}{2}Cl_2(g) \rightarrow Cl(g) \qquad \Delta H_{at}^{\ominus} = +121$ kJ mol^{-1}

ELECTRON AFFINITY $\Delta H_{EA}^{\ominus}$

The electron affinity is the enthalpy change when an electron is added to an isolated atom in the gaseous state, i.e.

$X(g) + e \rightarrow X^-(g)$

Atoms 'want' an extra electron so electron affinity values are negative for the first electron. However, when oxygen forms the O^{2-} ion the overall process is endothermic:

$O(g) + e \rightarrow O^-(g)$	$\Delta H^{\ominus}$	$= -142$ kJ mol^{-1}
$O^-(g) + e \rightarrow O^{2-}(g)$	$\Delta H^{\ominus}$	$= +844$ kJ mol^{-1}
overall $O(g) + 2e \rightarrow O^{2-}(g)$	$\Delta H^{\ominus}$	$= +702$ kJ mol^{-1}

USE OF BORN–HABER CYCLES

Like any energy cycle Born–Haber cycles can be used to find the value of an unknown. They can also be used to assess how ionic a substance is. The lattice enthalpy can be calculated theoretically by considering the charge and size of the constituent ions. It can also be obtained indirectly from the Born–Haber cycle. If there is good agreement between the two values then it is reasonable to assume that there is a high degree of ionic character, e.g. NaCl. However, if there is a big difference between the two values then it is because the compounds possesses a considerable degree of covalent character, e.g. AgCl.

	NaCl	AgCl
Theoretical value / kJ mol^{-1}	766	770
Experimental value / kJ mol^{-1}	771	905

Entropy and free energy

DISORDER

In nature systems naturally tend towards disorder. An increase in disorder can result from:

- mixing different types of particles, e.g. the dissolving of sugar in water
- a change in state where the distance between the particles increases, e.g. liquid water → steam
- the increased movement of particles, e.g. heating a liquid or gas
- increasing the number of particles, e.g.
  $2H_2O_2(l) \rightarrow 2H_2O(l) + O_2(g)$.

The greatest increase in disorder is usually found where the number of particles in the gaseous state increases.

The change in the disorder of a system is known as the entropy change, ΔS. The more disordered the system becomes the more positive the value of ΔS becomes. Systems which become more ordered will have negative ΔS values.

$NH_3(g) + HCl(g) \rightarrow NH_4Cl(s)$ $\Delta S = -284$ J K^{-1} mol^{-1}

(two moles   (one mole
of gas)   of solid)

SPONTANEITY

A reaction is said to be spontaneous if it causes a system to move from a less stable to a more stable state. This will depend both upon the enthalpy change and the entropy change. These two factors can be combined and expressed as the Gibbs energy change ΔG, often known as the 'free energy change'.

The standard free energy change $\Delta G^{\ominus}$ is defined as:

$$\Delta G^{\ominus} = \Delta H^{\ominus} - T\Delta S^{\ominus}$$

Where all the values are measured under standard conditions. For a reaction to be spontaneous it must be able to do work, that is $\Delta G^{\ominus}$ must have a negative value.

POSSIBLE COMBINATIONS FOR FREE ENERGY CHANGES

Some reactions will always be spontaneous. If $\Delta H^{\ominus}$ is negative or zero and $\Delta S^{\ominus}$ is positive then $\Delta G^{\ominus}$ must always have a negative value. Conversely if $\Delta H^{\ominus}$ is positive or zero and $\Delta S^{\ominus}$ is negative then $\Delta G^{\ominus}$ must always be positive and the reaction will never be spontaneous.

For some reactions whether or not they will be spontaneous depends upon the temperature. If $\Delta H^{\ominus}$ is positive or zero and $\Delta S^{\ominus}$ is positive, then $\Delta G^{\ominus}$ will only become negative at high temperatures when the value of $T\Delta S^{\ominus}$ exceeds the value of $\Delta H^{\ominus}$.

Type	$\Delta H^{\ominus}$	$\Delta S^{\ominus}$	$T\Delta S^{\ominus}$	$\Delta H^{\ominus} - T\Delta S^{\ominus}$	$\Delta G^{\ominus}$
1	0	+	+	(0) − (+)	−
2	0	−	−	(0) − (−)	+
3	−	+	+	(−) − (+)	−
4	+	−	−	(+) − (−)	+
5	+	+	+	(+) − (+)	− or +
6	−	−	−	(−) − (−)	+ or −

Type 1. Mixing two gases. $\Delta G^{\ominus}$ is negative so gases will mix of their own accord. Gases do not unmix of their own accord (Type 2) as $\Delta G^{\ominus}$ is positive.

Type 3. $(NH_4)_2Cr_2O_7(s) \rightarrow N_2(g) + Cr_2O_3(s) + 4H_2O(g)$

The decomposition of ammonium dichromate is spontaneous at all temperatures.

Type 4. $N_2(g) + 2H_2(g) \rightarrow N_2H_4(g)$

The formation of hydrazine from its elements will never be spontaneous.

Type 5. $CaCO_3(s) \rightarrow CaO(s) + CO_2(g)$

The decomposition of calcium carbonate is only spontaneous at high temperatures.

Type 6. $C_2H_4(g) + H_2(g) \rightarrow C_2H_6(g)$

Above a certain temperature this reaction will cease to be spontaneous.

Note: the fact that a reaction is spontaneous does not necessarily mean that it will proceed without any input of energy. For example, the combustion of coal is a spontaneous reaction and yet coal is stable in air. It will only burn on its own accord after it has received some initial energy so that some of the molecules have the necessary activation energy for the reaction to occur.

Spontaneity of a reaction

ABSOLUTE ENTROPY VALUES

The standard entropy of a substance is the entropy change per mole that results from heating the substance from 0 K to the standard temperature of 298 K. Unlike enthalpy, absolute values of entropy can be measured. The standard entropy change for a reaction can then be determined by calculating the difference between the entropy of the products and the reactants.

$$\Delta S^{\ominus} = S^{\ominus}(\text{products}) - S^{\ominus}(\text{reactants})$$

e.g. for the formation of ammonia

$$3H_2(g) + N_2(g) \rightleftharpoons 2NH_3(g)$$

the standard entropies of hydrogen, nitrogen, and ammonia are respectively 131, 192, and 192 J K^{-1} mol^{-1}.

Therefore per mole of reaction

$$\Delta S^{\ominus} = 2 \times 192 - [(3 \times 131) + 192] = -201 \text{ J K}^{-1}\text{mol}^{-1}$$

(or per mole of ammonia $\Delta S^{\ominus} = \dfrac{-201}{2} = -101$ J K^{-1} mol^{-1})

DETERMINING THE VALUE OF $\Delta G^{\ominus}$

The precise value of $\Delta G^{\ominus}$ for a reaction can be determined from $\Delta G_f^{\ominus}$ values using an energy cycle, e.g. to find the standard free energy of combustion of methane given the standard free energies of formation of methane, carbon dioxide, water, and oxygen.

$$CH_4(g) + 2O_2 \xrightarrow{\Delta G_x^{\ominus}} CO_2(g) + 2H_2O(l)$$

$$C(s) + 2O_2(g) + 2H_2(g)$$

By Hess' law

$$\Delta G_x^{\ominus} = [\Delta G_x^{\ominus}(CO_2) + 2\Delta G_f^{\ominus}(H_2O)] - [\Delta G_f^{\ominus}(CH_4) + 2\Delta G_f^{\ominus}(O_2)]$$

Substituting the actual values

$$\Delta G_x^{\ominus} = [-394 + 2 \times (-237)] - [-50 + 2 \times 0] = -818 \text{ kJ mol}^{-1}$$

$\Delta G^{\ominus}$ values can also be calculated from using the equation $\Delta G^{\ominus} = \Delta H^{\ominus} - T\Delta S^{\ominus}$. For example, in Type 5 on the previous page the values for $\Delta H^{\ominus}$ and $\Delta S^{\ominus}$ for the thermal decomposition of calcium carbonate are +178 kJ mol^{-1} and +165.3 J K^{-1} mol^{-1} respectively. Note that the units of $\Delta S^{\ominus}$ are different to those of $\Delta H^{\ominus}$.

At 25 °C (298 K) the value for $\Delta G^{\ominus} = 178 - 298 \times \dfrac{165.3}{1000}$

$$= +129 \text{ kJ mol}^{-1}$$

which means that the reaction is not spontaneous.

The reaction will become spontaneous when $T\Delta S^{\ominus} > \Delta H^{\ominus}$.

$$T\Delta S^{\ominus} = \Delta H^{\ominus} \text{ when } T = \dfrac{\Delta H^{\ominus}}{\Delta S^{\ominus}} = \dfrac{178}{165.3/1000} = 1077 \text{ K } (804 \text{ °C})$$

Therefore above 804 °C the reaction will be spontaneous.

Note: this calculation assumes that the entropy value is independent of temperature, which is not strictly true.

IB QUESTIONS – ENERGETICS

1. Which statement about this reaction is correct?

   $$2Fe(s) + 3CO_2(g) \rightarrow Fe_2O_3(s) + 3CO(g) \quad \Delta H^{\ominus} = +26.6 \text{ kJ}$$

   A. 26.6 kJ of energy are released for every mole of Fe reacted

   B. 26.6 kJ of energy are absorbed for every mole of Fe reacted

   C. 53.2 kJ of energy are released for every mole of Fe reacted

   D. 13.3 kJ of energy are absorbed for every mole of Fe reacted

2. When solutions of HCl and NaOH are mixed the temperature increases. The reaction:

   $$H^+(aq) + OH^-(aq) \rightarrow H_2O(l)$$

   A. is endothermic with a positive $\Delta H^{\ominus}$.

   B. is endothermic with a negative $\Delta H^{\ominus}$.

   C. is exothermic with a positive $\Delta H^{\ominus}$.

   D. is exothermic with a negative $\Delta H^{\ominus}$.

3.

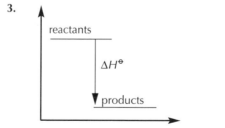

   What can be deduced about the relative stability of the reactants and products and the sign of $\Delta H^{\ominus}$, from the enthalpy level diagram above?

	Relative stability	Sign of $\Delta H^{\ominus}$
A.	Products more stable	−
B.	Products more stable	+
C.	Reactants more stable	−
D.	Reactants more stable	+

4. For the reaction:

$$2C(s) + 2H_2(g) \rightarrow C_2H_4(g) \quad \Delta H_1^{\ominus} = +52.3 \text{ kJ}$$

If $\Delta H_2^{\ominus} = -174.4$ kJ for the reaction:

$$C_2H_2(g) + H_2(g) \rightarrow C_2H_4(g)$$

what can be said about the value of $\Delta H_3^{\ominus}$ for the reaction below?

$$2C(s) + H_2(g) \rightarrow C_2H_2(g)$$

A. $\Delta H_3^{\ominus}$ must be negative.

B. $\Delta H_3^{\ominus}$ must be a positive number smaller than 52.3 kJ.

C. $\Delta H_3^{\ominus}$ must be a positive number larger than 52.3 kJ.

D. No conclusion can be made about $\Delta H_3^{\ominus}$ without the value of H for $H_2(g)$.

5. The enthalpy changes for two different hydrogenation reactions of C_2H_2 are:

$$C_2H_2 + H_2 \rightarrow C_2H_4 \quad \Delta H_1^{\ominus}$$
$$C_2H_2 + 2H_2 \rightarrow C_2H_6 \quad \Delta H_2^{\ominus}$$

Which expression represents the enthalpy change for the reaction below?

$$C_2H_4 + H_2 \rightarrow C_2H_6 \quad \Delta H^{\ominus} = ?$$

A. $\Delta H_1^{\ominus} + \Delta H_2^{\ominus}$

B. $\Delta H_1^{\ominus} - \Delta H_2^{\ominus}$

C. $\Delta H_2^{\ominus} - \Delta H_1^{\ominus}$

D. $-\Delta H_1^{\ominus} - \Delta H_2^{\ominus}$

6. $2KHCO_3(s) \xrightarrow{\Delta H^{\ominus}} K_2CO_3(s) + CO_2(g) + H_2O(l)$

$+2HCl(aq) \searrow \quad \Delta H_2^{\ominus} \swarrow +2HCl(aq)$

$\Delta H_1^{\ominus}$

$2KCl(aq) + 2CO_2(g) + 2H_2O(l)$

This cycle may be used to determine $\Delta H^{\ominus}$ for the decomposition of potassium hydrogencarbonate. Which expression can be used to calculate $\Delta H^{\ominus}$?

A. $\Delta H^{\ominus} = \Delta H_1^{\ominus} + \Delta H_2^{\ominus}$
C. $\Delta H^{\ominus} = \frac{1}{2}\Delta H_1^{\ominus} - \Delta H_2^{\ominus}$

B. $\Delta H^{\ominus} = \Delta H_1^{\ominus} - \Delta H_2^{\ominus}$
D. $\Delta H^{\ominus} = \Delta H_2^{\ominus} - \Delta H_1^{\ominus}$

7. Use the bond energies for H–H (436 kJ mol^{-1}), Br–Br (193 kJ mol^{-1}) and H–Br (366 kJ mol^{-1}) to calculate $\Delta H^{\ominus}$ (in kJ mol^{-1}) for the reaction:

$$H_2(g) + Br_2(g) \rightarrow 2HBr(g)$$

A. 263
C. –103

B. 103
D. –263

8. Which of the changes below occurs with the greatest increase in entropy?

A. $Na_2O(s) + H_2O(l) \rightarrow 2Na^+(aq) + 2OH^-(aq)$

B. $NH_3(g) + HCl(g) \rightarrow NH_4Cl(s)$

C. $H_2(g) + I_2(g) \rightarrow 2HI(g)$

D. $C(s) + CO_2(g) \rightarrow 2CO(g)$

9. How would this reaction at 298 K be described in thermodynamic terms?

$$2H_2O(g) \rightarrow 2H_2(g) + O_2(g)$$

A. Endothermic with a significant increase in entropy

B. Endothermic with a significant decrease in entropy

C. Exothermic with a significant increase in entropy

D. Exothermic with a significant decrease in entropy

10. At 0 °C, the mixture formed when the following reaction reaches equilibrium consists mostly of $N_2O_4(g)$.

$$2NO_2(g) \rightleftharpoons N_2O_4(g)$$

What are the signs of ΔG, ΔH, ΔS at this temperature?

	ΔG	ΔH	ΔS
A.	+	+	+
B.	–	–	–
C.	–	+	+
D.	+	+	–

HL

---

11. Nitroglycerine decomposes violently when it is detonated according to the equation:

$$2C_3H_5(NO_3)_3(l) \rightarrow 3N_2(g) + \frac{1}{2}O_2(g) + 6CO_2(g) + 5H_2O(g)$$

	$\Delta H_f^{\ominus}$ (kJ mol^{-1})
$C_3H_5(NO_3)_3(l)$	–364
$CO_2(g)$	–394
$H_2O(g)$	–242

What is the enthalpy change for the decomposition of 2 moles of nitroglycerine in terms of the $\Delta H_f^{\ominus}$ values above?

A. $5(242) + 6(394) - 2(364)$ kJ

B. $5(-242) - 6(-394) - 2(-364)$ kJ

C. $5(-242) + 6(-394) + 2(364)$ kJ

D. It cannot be determined because $\Delta H_f^{\ominus}$ of oxygen and nitrogen are not given.

12. Which substance has the largest lattice energy?

A. NaF
C. MgO

B. KCl
D. CaS

13. Which factor(s) will cause the lattice enthalpy of ionic compounds to increase in magnitude?

I. an increase in the charge on the ions

II. an increase in the size of ions

A. I only
C. Both I and II

B. II only
D. Neither I nor II

14. The Born–Haber cycle for the formation of potassium chloride includes the steps below:

I. $K(g) \rightarrow K^+(g) + e$ III. $Cl(g) + e \rightarrow Cl^-(g)$

II. $\frac{1}{2}Cl_2(g) \rightarrow Cl(g)$ IV. $K^+(g) + Cl^-(g) \rightarrow KCl(s)$

Which of these steps are exothermic?

A. I and II only
C. I, II and II only

B. III and IV only
D. I, III and IV only

7 KINETICS

Rates of reaction and collision theory

RATE OF REACTION

Chemical kinetics is the study of the factors affecting the rate of a chemical reaction. The rate of a chemical reaction can be defined either as the increase in the concentration of one of the products per unit time or as the decrease in the concentration of one of the reactants per unit time. It is measured in mol dm^{-3} s^{-1}.

The change in concentration can be measured by using any property that differs between the reactants and the products. Common methods include mass or volume changes when a gas is evolved, absorption using a spectrometer when there is a colour change, pH changes when there is a change in acidity, and electrical conductivity when there is a change in the ionic concentrations. A graph of concentration against time is then usually plotted. The rate at any stated point in time is then the gradient of the graph at that time. Rates of reaction usually decrease with time as the reactants are used up.

The reaction of hydrochloric acid with calcium carbonate can be used to illustrate the two typical curves obtained depending on whether the concentration of reactant or product is followed.

$$CaCO_3(s) + 2HCl(aq) \rightarrow CaCl_2(aq) + H_2O(l) + CO_2(g)$$

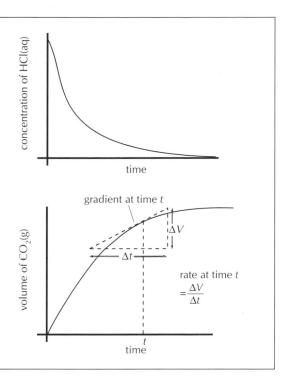

COLLISION THEORY

For a reaction between two particles to occur two conditions must be met. The particles must not only collide so that the reactive parts of the particles come into contact with each other, but they must collide with sufficient energy to bring about the reaction. This minimum amount of energy required is known as the **activation energy**. Any factor that either increases the frequency of the collisions or increases the energy with which they collide will make the reaction go faster.

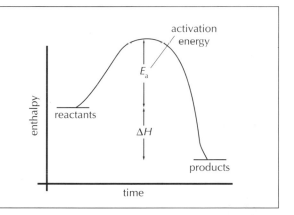

REACTION MECHANISMS

Many reactions do not go in one step. This is particularly true when there are more than two reactant molecules as the chances of a successful collision between three or more particles is extremely small. When there is more than one step then each step will proceed at its own rate. No matter how fast the other steps are the overall rate of the reaction will depend only upon the rate of the slowest step. This slowest step is known as the **rate determining step**.

e.g. consider the reaction between A and B to form A_2B: $2A + B \rightarrow A_2B$. A possible mechanism might be:

$$\text{Step 1} \qquad A + A \xrightarrow{\text{slow}} A\text{--}A \qquad \text{rate determining step}$$

$$\text{Step 2} \qquad A\text{--}A + B \xrightarrow{\text{fast}} A_2B$$

However fast A–A reacts with B the rate of production of A_2B will only depend on how fast A–A is formed.

Factors affecting the rate of reaction

TEMPERATURE

As the temperature increases the particles will move faster so there will be more collisions per second. However, the main reason why an increase in temperature increases the rate is that more of the colliding particles will possess the necessary activation energy resulting in more successful collisions. As a rough rule of thumb an increase of 10 °C doubles the rate of a chemical reaction.

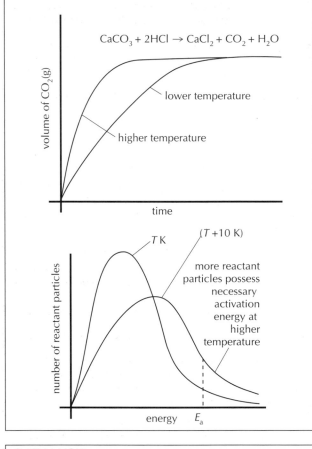

SURFACE AREA

In a solid substance only the particles on the surface can come into contact with a surrounding reactant. If the solid is in powdered form then the surface area increases dramatically and the rate increases correspondingly.

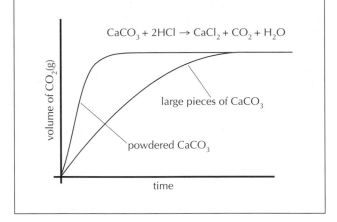

CONCENTRATION

The more concentrated the reactants the more collisions there will be per second per unit volume. As the reactants get used up their concentration decreases. This explains why the rate of most reactions gets slower as the reaction proceeds. (Some exothermic reactions do initially speed up if the heat that is given out more than compensates for the decrease in concentration.)

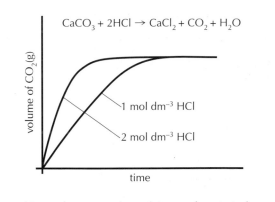

Note: this graph assumes that calcium carbonate is the limiting reagent or that equal amounts (mol) of acid have been added.

CATALYST

Catalysts increase the rate of a chemical reaction without themselves being chemically changed at the end of the reaction. They work essentially by bringing the reactive parts of the reactant particles into close contact with each other. This provides an alternative pathway for the reaction with a lower activation energy. More of the reactants will possess this lower activation energy, so the rate increases.

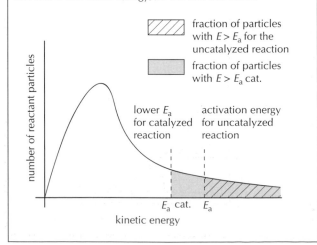

Order of reaction and half-life

RATE EXPRESSIONS

The rate of reaction between two reactants, A and B, can be followed experimentally. The rate will be found to be proportional to the concentration of A raised to some power and also to the concentration of B raised to a power. If square brackets are used to denote concentration this can be written as rate $\propto [A]^x$ and rate $\propto [B]^y$. They can be combined to give the rate expression:

$$\text{rate} = k[A]^x[B]^y$$

where k is the constant of proportionality and is known as the **rate constant**.

x is known as the **order of the reaction** with respect to A.

y is known as the order of the reaction with respect to B.

The overall order of the reaction $= x + y$.

Note: the order of the reaction and the rate expression can only be determined experimentally. They cannot be deduced from the balanced equation for the reaction.

UNITS OF RATE CONSTANT

The units of the rate constant depend on the overall order of the reaction.

First order: rate $= k[A]$

$$k = \frac{\text{rate}}{[A]} = \frac{\text{mol dm}^{-3}\text{ s}^{-1}}{\text{mol dm}^{-3}} = \text{s}^{-1}$$

Second order: rate $= k[A]^2$ or $k = [A][B]$

$$k = \frac{\text{rate}}{[A]^2} = \frac{\text{mol dm}^{-3}\text{ s}^{-1}}{(\text{mol dm}^{-3})^2} = \text{dm}^3\text{ mol}^{-1}\text{ s}^{-1}$$

Third order: rate $= k[A]^2[B]$ or rate $= k[A][B]^2$

$$k = \frac{\text{rate}}{[A]^2[B]} = \frac{\text{mol dm}^{-3}\text{ s}^{-1}}{(\text{mol dm}^{-3})^3} = \text{dm}^6\text{ mol}^{-2}\text{ s}^{-1}$$

GRAPHICAL REPRESENTATIONS OF REACTIONS

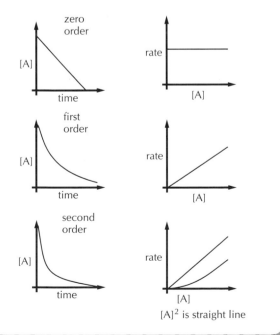

$[A]^2$ is straight line

DERIVING A RATE EXPRESSION BY INSPECTION OF DATA

Experimental data obtained from the reaction between hydrogen and nitrogen monoxide at 1073 K:

$$2H_2(g) + 2NO(g) \rightarrow 2H_2O(g) + N_2(g)$$

Experiment	Initial concentration of H_2(g) / mol dm^{-3}	Initial concentration of NO(g) / mol dm^{-3}	Initial rate of formation of N_2(g) / mol dm^{-3} s^{-1}
1	1×10^{-3}	6×10^{-3}	3×10^{-3}
2	2×10^{-3}	6×10^{-3}	6×10^{-3}
3	6×10^{-3}	1×10^{-3}	0.5×10^{-3}
4	6×10^{-3}	2×10^{-3}	2.0×10^{-3}

From experiments 1 and 2 doubling $[H_2]$ doubles the rate so rate $\propto [H_2]$.

From experiments 3 and 4 doubling [NO] quadruples the rate so rate $\propto [NO]^2$.

Rate expression given by rate $= k[H_2][NO]^2$.

The rate is first order with respect to hydrogen, second order with respect to nitrogen monoxide, and third order overall. The value of k can be found by substituting the values from any one of the four experiments:

$$k = \frac{\text{rate}}{[H_2][NO]^2} = 8.33 \times 10^4\text{ dm}^3\text{ mol}^{-1}\text{ s}^{-1}$$

HALF-LIFE $t_{\frac{1}{2}}$

For a first order reaction the rate of change of concentration of A is equal to $k[A]$. This can be expressed as $\frac{\text{d}[A]}{\text{d}t} = k[A]$.

If this expression is integrated then $kt = \ln[A]_o - \ln[A]$ where $[A]_o$ is the initial concentration and $[A]$ is the concentration at time t. This expression is known as the integrated form of the rate equation.

The half-life is defined as the time taken for the concentration of a reactant to fall to half of its initial value.

At $t_{\frac{1}{2}}$ $[A] = \frac{1}{2}[A]_o$ the integrated rate expression then becomes
$kt_{\frac{1}{2}} = \ln[A]_o - \ln\frac{1}{2}[A]_o = \ln 2$ since $\ln 2 = 0.693$ this simplifies to $t_{\frac{1}{2}} = \frac{0.693}{k}$

From this expression it can be seen that the half-life of a first order reaction is independent of the original concentration of A, i.e. first order reactions have a constant half-life.

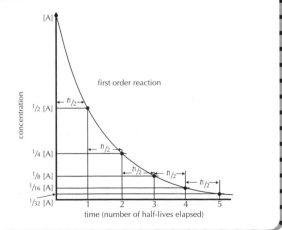

REACTION MECHANISMS

When the separate steps in a chemical reaction are analysed there are essentially only two types of processes. Either a single species can break down into two or more products by what is known as a **unimolecular process**, or two species can collide and interact by a **bimolecular process**.

In a bimolecular process the species collide with the necessary activation energy to give initially an **activated complex**. An activated complex is not a chemical substance which can be isolated, but consists of an association of the reacting particles in which bonds are in the process of being broken and formed. An activated complex breaks down to form either the products or reverts back to the original reactants.

The number of species taking part in any specified step in the reaction is known as the **molecularity**. In most cases the molecularity refers to the slowest step, that is the rate determining step.

In the reaction on the previous page, between nitrogen monoxide and hydrogen, the stoichiometry of the reaction involves two molecules of hydrogen and two molecules of nitrogen monoxide. Any proposed mechanism must be consistent with the rate expression. For third order reactions, such as this, the rate determining step will never be the first step. The proposed mechanism is:

$$NO(g) + NO(g) \xrightarrow{\text{fast}} N_2O_2(g)$$
$$N_2O_2(g) + H_2(g) \xrightarrow{\text{slow}} N_2O(g) + H_2O(g) \quad \text{rate determining step}$$
$$\underline{N_2O(g) + H_2(g) \xrightarrow{\text{fast}} N_2(g) + H_2O(g)}$$

Overall $\quad 2NO(g) + 2H_2(g) \rightarrow N_2(g) + 2H_2O(g)$

If the first step was the slowest step the the rate expression would be rate = $k[NO]^2$ and the rate would be zero order with respect to hydrogen. The rate for the second step depends on $[H_2]$ and $[N_2O_2]$. However, the concentration of N_2O_2 depends on the first step. So the rate expression for the second step becomes rate = $k[H_2][NO]^2$, which is consistent with the experimentally determined rate expression. The molecularity of the reaction is two, as two reacting species are involved in the rate determining step.

ARRHENIUS EQUATION

The rate constant for a reaction is only constant if the temperature remains constant. As the temperature increases the reactants possess more energy and the rate constant increases. The relationship between rate constant and absolute temperature is given by the Arrhenius equation:

$$k = Ae^{(-E_a/RT)}$$

where E_a is the activation energy and R is the gas constant. A is known as the Arrhenius constant and is related to the orientation of the reactants at the point of collision. This equation is often expressed in its logarithmic form:

$$\ln k = \frac{-E_a}{RT} + \ln A$$

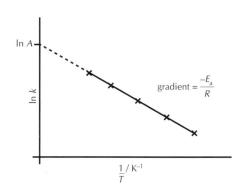

The equation can be used to determine both the Arrhenius constant and the activation energy for the reaction. This can be done either by substitution using simultaneous equations or by plotting $\ln k$ against $\frac{1}{T}$ to give a straight line graph. The gradient of the graph will be equal to $\frac{-E_a}{R}$ from which the activation energy can be calculated. Extrapolating the graph back to the $\ln k$ axis will give an intercept with a value equal to $\ln A$.

CATALYSIS

Catalysts provide an alternative pathway, so that the activation energy required to reach the activated complex is lowered.

Catalysts can be conveniently divided into two types.
- **Homogeneous catalysts** are in the same phase as the reactants and include examples such as concentrated sulfuric acid in the reaction between alcohols and carboxylic acids to form esters.
- **Heterogeneous catalysts** are in a different phase to the reactants. Common examples include metal catalysts in the presence of gases, e.g nickel in the hydrogenation of alkenes and iron in the production of ammonia.

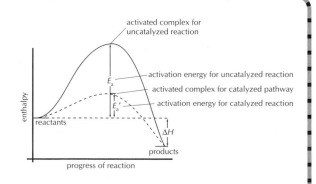

Heterogeneous catalysts tend to work by bringing the reactant particles into close alignment by adsorbing them onto the catalytic surface. In homogeneous catalysis the reactant particles are thought to form reversible complexes with the catalyst. Transition metals and their compounds are often good catalysts as the metals show variable oxidation states. Poisons work by forming irreversible complexes with the catalyst, which cause the active sites to become permanently blocked.

IB QUESTIONS – KINETICS

1. Which graph best represents the change in concentration of products with time for a reaction as it goes to completion?

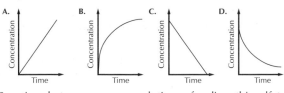

2. Reactions between aqueous solutions of sodium thiosulfate and acid can be followed by timing the appearance of the solid sulfur that is produced. The time required for the appearance of sulfur would be increased by which of the following changes?

   A. Raising the temperature

   B. Diluting the solution

   C. Adding a catalyst

   D. Increasing the concentration of the sodium thiosulfate

3. The reaction between excess calcium carbonate and hydrochloric acid can be followed by measuring the volume of carbon dioxide produced with time. The results or one such reaction are shown below. How does the rate of this reaction change with time and what is the main reason for this change?

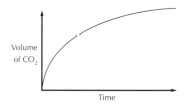

   A. The rate increases with time because the calcium carbonate particles get smaller.

   B. The rate increases with time because the acid becomes more dilute.

   C. The rate decreases with time because the calcium carbonate particles get smaller.

   D. The rate decreases with time because the acid becomes more dilute.

4. The rate of reaction of a strip of magnesium and 50 cm³ of 1 mol dm⁻³ HCl is determined at 25 °C. In which case would **both** new conditions contribute to an increase in the rate of reaction?

   A. Mg powder and 100 cm³ of 1 mol dm⁻³ HCl

   B. Mg powder and 50 cm³ of 0.8 mol dm⁻³ HCl

   C. 100 cm³ of 1 mol dm⁻³ HCl at 30 °C

   D. 50 cm³ of 1.2 mol dm⁻³ HCl at 30 °C

5. Calcium carbonate and hydrochloric acid react according to the equation below:

   $CaCO_3(s) + 2HCl(aq) \rightarrow CO_2(g) + CaCl_2(aq) + H_2O(l)$

   Which conditions will produce the fastest rate of reaction?

   A. 1 mol dm⁻³ HCl and $CaCO_3$ pieces

   B. 2 mol dm⁻³ HCl and $CaCO_3$ pieces

   C. 1 mol dm⁻³ HCl and $CaCO_3$ powder

   D. 2 mol dm⁻³ HCl and $CaCO_3$ powder

6. When ammonia is manufactured commercially a catalyst is used. What is the effect of this catalyst?

   $N_2(g) + 3H_2(g) \rightleftharpoons 2NH_3(g) \quad \Delta H = -92 \text{ kJ}$

   I. To increase the rate of only the forward reaction

   II. To increase the rates of the forward and reverse reactions

   III. To shift the position of equilibrium and increase the yield of ammonia

   A. I only

   B. II only

   C. III only

   D. I and III only

7. The energy diagram for a particular reaction under catalysed and uncatalysed conditions is shown below:

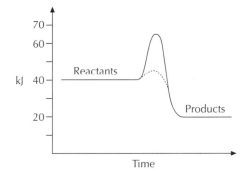

   What are the value and sign of the enthalpy change for the catalysed forward reaction?

   A. −55 kJ   **B.** −30 kJ   **C.** −25 kJ   **D.** −20 kJ

8. Some collisions between reactant molecules do not form products. This is most likely because

   A. the molecules do not collide in the proper ratio.

   B. the molecules do not have enough energy.

   C. the concentration is too low.

   D. the reaction is at equilibrium.

9. The rate-determining step of a reaction is the

   A. fastest step in the reaction.

   B. first step in the reaction.

   C. final step in the reaction.

   D. step with the highest activation energy.

10. Most reactions occur in a series of steps, one of which is the rate determining step. The rate determining step is so called because it is the

    A. first step.   **C.** fastest step.

    B. last step.   **D.** slowest step.

11. Doubling which one of the following will double the rate of a first order reaction?

    A. Concentrations of the reactant

    B. Size of solid particles

    C. Volume of solution in which the reaction is carried out

    D. Activation energy

12. The kinetic data in the table below are for the reaction:

$A + B \rightarrow C$

From these data what are the orders of the reaction with respect to A and B?

[A]	[B]	Initial Rate (mol dm^{-3} sec^{-1})
0.1	0.1	1×10^{-5}
0.1	0.2	4×10^{-5}
0.2	0.1	1×10^{-5}

    A. order of A = 1 order of B = 0

    B. order of A = 0 order of B = 2

    C. order of A = 0 order of B = 4

    D. order of A = 1 order of B = 2

13. The rate information below was obtained for the following reaction:

$2NO_2(g) + F_2(g) \rightarrow 2NO_2F(g)$

[NO$_2$]/mol dm^{-3}	[F$_2$]/mol dm^{-3}	Rate/mol dm^{-3}s^{-1}
0.001	0.005	2×10^{-4}
0.002	0.005	4×10^{-4}
0.002	0.010	8×10^{-4}

What are the orders for NO_2 and F_2?

    A. NO_2 is first order, F_2 is first order

    B. NO_2 is first order, F_2 is second order

    C. NO_2 is second order, F_2 is first order

    D. NO_2 is second order, F_2 is second order

14. Which factor(s) will influence the rate of the reaction shown below?

$NO_2(g) + CO(g) \rightarrow NO(g) + CO_2(g)$

    I. The number of collisions per second

    II. The energy of the collisions

    III. The geometry with which the molecules collide

    A. I only   **C.** I and II only

    B. II only   **D.** I, II, and III

15. The rate constant for a certain reaction has the units **concentration time** $^{-1}$. What is the order of reaction?

    A. 0   **B.** 1   **C.** 2   **D.** 3

16. Which reaction is an example of homogenous catalysis?

    A. $3H_2(g) + N_2(g) \xrightleftharpoons{Fe(s)} 2NH_3(g)$

    B. $2H_2O(aq) \xrightarrow{MnO_2(s)} 2H_2O(l) + O_2(g)$

    C. $2SO_2(g) + O_2(g) \xrightleftharpoons{NO(g)} 2SO_3(g)$

    D. $C_2H_4(g) + H_2(g) \xrightarrow{Ni(s)} C_2H_6(g)$

The equilibrium law

DYNAMIC EQUILIBRIUM

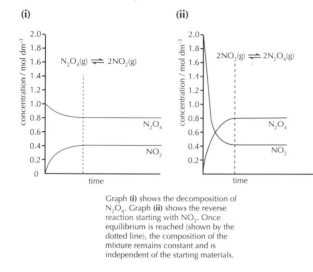

$$A + B \underset{\text{reverse reaction}}{\overset{\text{forward reaction}}{\rightleftharpoons}} C + D$$

Most chemical reactions do not go to completion. Once some products are formed the reverse reaction can take place to reform the reactants. In a closed system the concentrations of all the reactants and products will eventually become constant. Such a system is said to be in a state of **dynamic equilibrium**. The forward and reverse reactions continue to occur, but at equilibrium the rate of the forward reaction is equal to the rate of the reverse reaction.

(i)

(ii)

Graph **(i)** shows the decomposition of N_2O_4. Graph **(ii)** shows the reverse reaction starting with NO_2. Once equilibrium is reached (shown by the dotted line), the composition of the mixture remains constant and is independent of the starting materials.

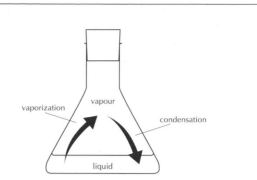

Dynamic equilibrium also occurs when physical changes take place. In a closed flask, containing some water, equilibrium will be reached between the liquid water and the water vapour. The faster moving molecules in the liquid will escape from the surface to become vapour and the slower moving molecules in the vapour will condense back into liquid. Equilibrium will be established when the rate of vaporization equals the rate of condensation.

$$H_2O(l) \rightleftharpoons H_2O(g)$$

CLOSED SYSTEM

A closed system is one in which neither matter nor energy can be lost or gained from the system, that is, the macroscopic properties remain constant. If the system is open some of the products from the reaction could escape and equilibrium would never be reached.

THE EQUILIBRIUM CONSTANT

Consider the following general reversible reaction in which w moles of A react with x moles of B to form y moles of C and z moles of D.

$$wA + xB \rightleftharpoons yC + zD$$

At equilibrium the concentrations of A, B, C, and D can be written as $[A]_{eqm}$, $[B]_{eqm}$, $[C]_{eqm}$, and $[D]_{eqm}$ respectively. The equilibrium law states that for this reaction at a particular temperature

$$K_c = \frac{[C]^y_{eqm} \times [D]^z_{eqm}}{[A]^w_{eqm} \times [B]^x_{eqm}}$$

where K_c is known as the equilibrium constant.

Examples Formation of sulfur trioxide in the Contact process

$$2SO_2(g) + O_2(g) \rightleftharpoons 2SO_3(g)$$

$$K_c = \frac{[SO_3]^2_{eqm}}{[SO_2]^2_{eqm} \times [O_2]_{eqm}}$$

Formation of an ester from ethanol and ethanoic acid

$$C_2H_5OH(l) + CH_3COOH(l) \rightleftharpoons CH_3COOC_2H_5(l) + H_2O(l)$$

$$K_c = \frac{[CH_3COOC_2H_5]_{eqm} \times [H_2O]_{eqm}}{[C_2H_5OH]_{eqm} \times [CH_3COOH]_{eqm}}$$

In both of these examples all the reactants and products are in the same phase. In the first example they are all in the gaseous phase and in the second example they are all in the liquid phase. Such reactions are known as **homogeneous reactions**. Another example of a homogeneous system would be where all the reactants and products are in the aqueous phase.

MAGNITUDE OF THE EQUILIBRIUM CONSTANT

Since the equilibrium expression has the concentration of products on the top and the concentration of reactants on the bottom it follows that the magnitude of the equilibrium constant is related to the position of equilibrium. When the reaction goes nearly to completion $K_c \gg 1$. If the reaction hardly proceeds then $K_c \ll 1$. If the value for K_c lies between about 10^{-2} and 10^2 then both reactants and products will be present in the system in noticeable amounts. The value for K_c in the esterification reaction above is 4 at 100 °C. From this it can be inferred that the concentration of the products present in the equilibrium mixture is roughly twice that of the reactants.

LE CHATELIER'S PRINCIPLE

Provided the temperature remains constant the value for K_c must remain constant. If the concentration of the reactants is increased, or one of the products is removed from the equilibrium mixture then more of the reactants must react in order to keep K_c constant, i.e the position of equilibrium will shift to the right (towards more products). This is the explanation for Le Chatelier's principle, which states that if a system at equilibrium is subjected to a small change the equilibrium tends to shift so as to minimize the effect of the change.

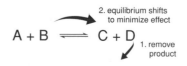

Applications of the equilibrium law

FACTORS AFFECTING THE POSITION OF EQUILIBRIUM

Change in concentration

$$C_2H_5OH(l) + CH_3COOH(l) \rightleftharpoons CH_3COOC_2H_5(l) + H_2O(l)$$

If more ethanoic acid is added the concentration of ethanoic acid increases so that at the point of addition:

$$K_c \neq \frac{[\text{ester}] \times [\text{water}]}{[\text{acid}] \times [\text{alcohol}]}$$

To restore the system so that the equilibrium law is obeyed the equilibrium will move to the right, so that the concentration of ester and water increases and the concentration of the acid and alcohol decreases.

Change in pressure

If there is an overall volume change in a gaseous reaction then increasing the pressure will move the equilibrium towards the side with less volume. This shift reduces the total number of molecules in the equilibrium system and so tends to minimize the pressure.

$$2NO_2(g) \rightleftharpoons N_2O_4(g)$$
brown   colourless
(2 vols)   (1 vol)

If the pressure is increased the mixture will initially go darker as the concentration of NO_2 increases then become lighter as the position of equilibrium is re-established with a greater proportion of N_2O_4.

Change in temperature

In exothermic reactions heat is also a product. Taking the heat away will move the equilibrium towards the right, so more products are formed. The forward reaction in exothermic reactions is therefore increased by lowering the temperature

$$2NO_2(g) \rightleftharpoons N_2O_4(g) \qquad \Delta H^\ominus = -24 \text{ kJ mol}^{-1}$$
brown   colourless

Lowering the temperature will cause the mixture to become lighter as the equilibrium shifts to the right.

For an endothermic reaction the opposite will be true.

Unlike changing the concentration or pressure, a change in temperature will also change the value of K_c. For an exothermic reaction the concentration of the products in the equilibrium mixture decreases as the temperature increases, so the value of K_c will decrease. The opposite will be true for endothermic reactions.

e.g. $H_2(g) + CO_2(g) \rightleftharpoons H_2O(g) + CO(g)$ $\Delta H^\ominus = +41 \text{ kJ mol}^{-1}$

$T\,/\,K$	K_c	
298	1.00×10^{-5}	
500	7.76×10^{-3}	increase
700	1.23×10^{-1}	
900	6.01×10^{-1}	

Adding a catalyst

A catalyst will increase the rate at which equilibrium is reached, as it will speed up both the forward and reverse reactions equally, but it will have no effect on the position of equilibrium and hence on the value of K_c.

APPLICATION OF EQUILIBRIUM AND KINETICS TO INDUSTRIAL PROCESSES

The aim in industry is to produce the highest possible yield of the required product in the shortest time for the least cost (both financial and to the environment) in order to maximize profits.

Haber process

Ammonia is used in the manufacture of fertilizers and in the production of nitric acid.

$$N_2(g) + 3H_2(g) \rightleftharpoons 2NH_3(g) \quad \Delta H^\ominus = -92 \text{ kJ mol}^{-1}$$

The hydrogen is obtained from natural gas and the nitrogen from the fractional distillation of liquid air.

Conditions Four volumes of reactants produce two volumes of product, so a high pressure will be required. Increasing the pressure will also increase the number of particles per unit volume. This increases the rate at which equilibrium is reached. In practice a pressure of about 250 atm is used. Since it is an exothermic reaction a low temperature is required to give a high yield of ammonia. However, lowering the temperature will decrease the rate of reaction and it will take longer to reach equilibrium. What is required is the **optimum temperature** where the best compromise between yield and rate is reached. A temperature of about 450 °C is usually used.
In order to increase the rate at which equilibrium is reached (but not the yield) an iron catalyst is used. It is used in a finely divided form (small pieces) so that the surface area is maximized to increase its efficiency. Even when all these conditions are in place the yield is only about 15%.

Contact process

Sulfuric acid is manufactured by the Contact process. It is the most industrially produced chemical, with over 150 million tonnes being produced world-wide every year. It is used for fertilizers, paints, detergents, and fibres, and as a feedstock for other chemicals.

$$2SO_2(g) + O_2(g) \rightleftharpoons 2SO_3(g) \quad \Delta H^\ominus = -197 \text{ kJ mol}^{-1}$$

The sulfur dioxide is obtained from burning sulfur or sulfide ores, and the oxygen is obtained from the fractional distillation of liquid air.

Conditions Three volumes are converted into two volumes, so a high pressure will favour the production of sulfur trioxide. The reaction is exothermic so an optimum temperature, which is a compromise between yield and rate, is required. In practice a yield of more than 99% is obtained when the pressure is 2 atm at a temperature of 450 °C. The catalyst used is vanadium(V) oxide. Since the yield is so high at 2 atm pressure it is uneconomical and unnecessary to build the converter to withstand higher pressures.

Equilibrium calculations and phase equilibrium

UNITS OF THE EQUILIBRIUM CONSTANT

The units of K_c depend on the powers of the concentrations in the equilibrium expression.
Haber process: units of K_c: $dm^6\ mol^{-2}$

$$K_c = \frac{[NH_3]^2}{[H_2]^3 \times [N_2]} = \frac{concentration^2}{concentration^4} = concentration^{-2} = dm^6\ mol^{-2}$$

If they are the same on the top and bottom then K_c has no units.
Esterification: K_c: no units

$$K_c = \frac{[acid] \times [alcohol]}{[ester] \times [water]} = \frac{concentration^2}{concentration^2}$$

EQUILIBRIUM CALCULATIONS

The equilibrium law can be used either to find the value for the equilibrium constant, or to find the value of an unknown equilibrium concentration.

(a) 23.0 g (0.50 mol) of ethanol was reacted with 60.0 g (1.0 mol) of ethanoic acid and the reaction allowed to reach equilibrium at 373 K. 37.0 g (0.42 mol) of ethyl ethanoate was found to be present in the equilibrium mixture. Calculate K_c to the nearest integer at 373 K.

	$C_2H_5OH(l)$ +	$CH_3COOH(l)$ ⇌	$CH_3COOC_2H_5(l)$ +	$H_2O(l)$
Initial amount / mol	0.50	1.00	–	–
Equilibrium amount / mol	(0.50 – 0.42)	(1.00 – 0.42)	0.42	0.42
Equilibrium concentration / mol dm^{-3}	$\dfrac{(0.50 - 0.42)}{V}$	$\dfrac{(1.00 - 0.42)}{V}$	$\dfrac{0.42}{V}$	$\dfrac{0.42}{V}$
(where V = total volume)				

$$K_c = \frac{[ester] \times [water]}{[alcohol] \times [acid]} = \frac{(0.42/V) \times (0.42/V)}{(0.08/V) \times (0.58/V)} = 4 \text{ (to the nearest integer)}$$

(b) What mass of ester will be formed at equilibrium if 2.0 moles of ethanoic acid and 1.0 moles of ethanol are reacted under the same conditions?

Let x moles of ester be formed and let the total volume be V dm^3.

$$K_c = 4 = \frac{[ester] \times [water]}{[alcohol] \times [ester]} = \frac{x^2/V^2}{(1.0 - x)/V \times (2.0 - x)/V} = \frac{x^2}{(x^2 - 3x + 2)}$$

$$\Rightarrow 3x^2 - 12x + 8 = 0$$

solve by substituting into the quadratic expression $\quad x = \dfrac{-b \pm \sqrt{b^2 - 4ac}}{2a} \Rightarrow x = \dfrac{12 \pm \sqrt{144 - 96}}{6}$

x = 0.845 or ~~3.15~~ (it cannot be 3.15 as only 1.0 mol of ethanol was taken)
Mass of ester = 0.845 × 88.08 = 74.4 g

PHASE EQUILIBRIUM

Dynamic equilibrium between a liquid and its vapour occurs when the rate of vaporization is equal to the rate of condensation. The vapour pressure of a liquid is the pressure exerted by the particles in the vapour phase. It is independent of the surface area of the liquid or of the size of the container, although in a bigger container it may take longer for the equilibrium to become established. The vapour pressure of any liquid does depend both on the strength of the molecular forces holding the liquid particles together and on the temperature.

The stronger the intermolecular forces the lower the vapour pressure at a particular temperature. Vaporization is an endothermic process, as energy is absorbed to break these intermolecular forces. The enthalpy change required to overcome these forces is known as the enthalpy of vaporization. Water is a covalent substance with a low molar mass, but it has strong hydrogen bonding between its molecules. This explains why water has a relatively low vapour pressure and a relatively high enthalpy of vaporization.

As the temperature increases so does the number of particles with sufficient energy to overcome the attractive forces, and the vapour pressure also increases. A liquid boils when its vapour pressure is equal to the external pressure, as this allows bubbles of vapour to form in the body of the liquid. The boiling point of a liquid can be lowered simply by lowering the external pressure. This principle is useful to purify substances which decompose at or near their normal boiling point, by distilling them under reduced pressure. In mountainous regions where the external pressure is low the boiling point of water can be increased by using a pressure cooker.

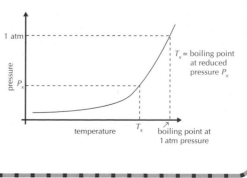

IB QUESTIONS – EQUILIBRIUM

1. Which statement is true about chemical reactions at equilibrium?

   A. The forward and backward reactions proceed at equal rates

   B. The forward and backward reactions have stopped

   C. The concentrations of the reactants and products are equal

   D. The forward reaction is exothermic

2. Chemical equilibrium is referred to as **dynamic** because, at equilibrium, the

   A. equilibrium constant changes.

   B. reactants and products keep reacting.

   C. rates of the forward and backward reactions change.

   D. concentrations of the reactants and products continue to change.

3. What is the equilibrium expression for the reaction:
   $N_2(g) + 3H_2(g) \rightleftharpoons 2NH_3(g)$?

   A. $K_c = \dfrac{[NH_3]}{[N_2][H_2]}$

   B. $K_c = \dfrac{2[NH_3]}{[N_2][H_2]}$

   C. $K_c = \dfrac{2[NH_3]}{3[N_2][H_2]}$

   D. $K_c = \dfrac{[NH_3]^2}{[N_2][H_2]^3}$

4. For a reaction which goes to completion, the equilibrium constant, K_c, is:

   A. >>1   B. <<1   C. = 1   D. = 0

5. The equilibrium constant for the reaction below is 1.0×10^{-14} at 25 °C and 2.1×10^{-14} at 35 °C. What can be concluded from this information?

   $2H_2O(l) \rightleftharpoons H_3O^+(aq) + OH^-(aq)$

   A. $[H_3O^+]$ decreases as the temperature is raised.

   B. $[H_3O^+]$ is greater than $[OH^-]$ at 35 °C.

   C. Water is a stronger electrolyte at 25 °C.

   D. The ionization of water is endothermic.

6. Ethanol is manufactured from ethene using the reaction below:

   $C_2H_4(g) + H_2O(g) \rightleftharpoons C_2H_5OH(g)$ $\Delta H = -46$ kJ

   Which conditions favour the highest yield of ethanol?

   A. High pressure and low temperature

   B. High pressure and high temperature

   C. Low pressure and low temperature

   D. Low pressure and high temperature

7. N_2O_4 and NO_2 produce an equilibrium mixture according to the equation below:

   $N_2O_4(g) \rightleftharpoons 2NO_2(g)$ $\Delta H = 57.2$ kJ mol^{-1}

   An increase in the equilibrium concentration of NO_2 can be produced by increasing which of the factors below?

   I. Pressure   II. Temperature

   A. I only   C. Both I and II

   B. II only   D. Neither I nor II

8. Which change(s) will increase the amount of $SO_3(g)$ at equilibrium?

   $2SO_2(g) + O_2(g) \rightleftharpoons 2SO_3(g)$ $\Delta H^{\ominus} = -197$ kJ

   I.   Increasing the temperature

   II. Decreasing the volume

   III. Adding a catalyst

   A. I only   C. I and III only

   B. II only   D. I, II and III

9. For the reaction $N_2(g) + 3H_2(g) \rightleftharpoons 2NH_3(g)$ $\Delta H = -92$ kJ

   What conditions will produce the highest percentage of NH_3 at equilibrium?

   A. High pressure and high temperature

   B. High pressure and low temperature

   C. Low pressure and high temperature

   D. Low pressure and low temperature

10. The reaction between sulfur dioxide and oxygen occurs according to the equation below:

    $2SO_2(g) + O_2(g) \rightleftharpoons 2SO_3(g)$ $\Delta H^{\ominus} = -197$ kJ

    A higher equilibrium concentration of SO_3 will be produced by all of the following changes in reaction conditions EXCEPT

    A. increasing the pressure.   C. adding a catalyst.

    B. adding more O_2.   D. decreasing the temperature.

11. The reaction between methane and hydrogen sulfide is represented by the equation below:

    $CH_4(g) + 2H_2S(g) \rightleftharpoons CS_2(g) + 4H_2(g)$

    What is the equilibrium expression for this reaction?

    A. $[CS_2][H_2]/[CH_4][H_2S]$

    B. $4[CS_2][H_2]/2[CH_4][H_2S]$

    C. $[CS_2] + 4[H_2]/[CH_4] + 2[H_2S]$

    D. $[CS_2][H_2]^4/[CH_4][H_2S]^2$

12. At 35 °C $K_c = 1.6 \times 10^{-5}$ mol dm^{-3} for the reaction

    $2NOCl(g) \rightleftharpoons 2NO(g) + Cl_2(g)$

    Which relationship must be correct at equilibrium?

    A. $[NO] = [NOCl]$ C. $[NOCl]<[Cl_2]$

    B. $2[NO]=[Cl_2]$ D. $[NO]<[NOCl]$

13. Methanol can be produced from carbon monoxide and hydrogen according to the equation:

    $CO(g) + 2H_2(g) \rightleftharpoons CH_3OH(g)$

    In a certain equilibrium mixture, $[CO] = 0.2$ mol dm^{-3}, $[H_2] = 0.1$ mol dm^{-3}, $[CH_3OH] = 2$ mol dm^{-3}. What is the value of K_c (dm^6 mol^{-2}) for this reaction?

    A. 1×10^{-3} C. 1×10^2

    B. 1×10^{-2} D. 1×10^3

Properties of acids and bases

TYPICAL PROPERTIES OF ACIDS AND BASES

A simple definition of an acid is that it is a substance that produces H^+ ions in aqueous solution. A base is a substance that can neutralize an acid. An alkali is a base that is soluble in water.

The typical reactions of acids are:

1. With indicators.

   Acid–base indicators can be used to determine whether or not a solution is acidic. Common indicators include:

Indicator	Colour in acidic solution	Colour in alkaline solution
litmus	red	blue
phenolphthalein	colourless	pink
methyl orange	red	yellow

2. Neutralization reactions with bases.

   (a) With hydroxides to form a salt and water,

   e.g. $CH_3COOH(aq) + NaOH(aq) \rightarrow NaCH_3COO(aq) + H_2O(l)$

   (b) With metal oxides to form a salt and water,

   e.g. $H_2SO_4(aq) + CuO(s) \rightarrow CuSO_4(aq) + H_2O(l)$

   (c) With ammonia to form a salt.

   e.g. $HCl(aq) + NH_3(aq) \rightarrow NH_4Cl(aq)$

3. With reactive metals (those above copper in the reactivity series) to form a salt and hydrogen, e.g.

   $2HCl(aq) + Mg(s) \rightarrow$
   $MgCl_2(aq) + H_2(g)$

4. With carbonates (soluble or insoluble) to form a salt, carbon dioxide, and water, e.g.

   $2HNO_3(aq) + Na_2CO_3(aq) \rightarrow$
   $2NaNO_3(aq) + CO_2(g) + H_2O(l)$
   $2HCl(aq) + CaCO_3(s) \rightarrow$
   $CaCl_2(aq) + CO_2(g) + H_2O(l)$

5. With hydrogencarbonates to form a salt, carbon dioxide, and water, e.g.

   $HCl(aq) + NaHCO_3(aq) \rightarrow$
   $NaCl(aq) + CO_2(g) + H_2O(aq)$

STRONG AND WEAK ACIDS AND BASES

A strong acid is completely dissociated (ionized) into its ions in aqueous solution. Similarly a strong base is completely dissociated into its ions in aqueous solution. Examples of strong acids and bases include:

Strong acids	Strong bases
hydrochloric acid, HCl	sodium hydroxide, $NaOH$
nitric acid, HNO_3	potassium hydroxide, KOH
sulfuric acid, H_2SO_4	barium hydroxide, $Ba(OH)_2$

Note: because one mole of HCl produces one mole of hydrogen ions it is known as a **monoprotic** acid. Sulfuric acid is known as a **diprotic** acid as one mole of sulfuric acid produces two moles of hydrogen ions.

Weak acids and bases are only slightly dissociated (ionized) into their ions in aqueous solution.

Weak acids	Weak bases
ethanoic acid, CH_3COOH	ammonia, NH_3
'carbonic acid' (CO_2 in water), H_2CO_3	aminoethane, $C_2H_5NH_2$

The difference can be seen in their reactions with water:

Strong acid: $HCl(g) + H_2O(l) \rightarrow H_3O^+(aq) + Cl^-(aq)$
      reaction goes to completion
Weak acid: $CH_3COOH(aq) + H_2O(l) \rightleftharpoons CH_3COO^-(aq) + H_3O^+(aq)$
      equilibrium lies on the left

i.e. a solution of hydrochloric acid consists only of hydrogen ions and chloride ions in water, whereas a solution of ethanoic acid contains mainly undissociated ethanoic acid with only very few hydrogen and ethanoate ions.

Strong base: $KOH(s) \xrightarrow{H_2O(l)} K^+(aq) + OH^-(aq)$
Weak base: $NH_3(g) + H_2O(l) \rightleftharpoons NH_4^+(aq) + OH^-(aq)$
      equilibrium lies on the left

Note: in aqueous solution a hydrogen ion reacts with a water molecule to form the hydronium ion $H_3O^+(aq)$. In IB chemistry both $H^+(aq)$ and $H_3O^+(aq)$ are acceptable to represent a hydrogen ion in aqueous solution.

EXPERIMENTS TO DISTINGUISH BETWEEN STRONG AND WEAK ACIDS AND BASES

1. pH measurement
   Because a strong acid produces a higher concentration of hydrogen ions in solution than a weak acid, with the same concentration, the pH of a strong acid will be lower than a weak acid. Similarly a strong base will have a higher pH in solution than a weak base, with the same concentration. The most accurate way to determine the pH of a solution is to use a pH meter.

   $0.10 \text{ mol dm}^{-3} \text{ } HCl(aq)$ pH = 1.0
   $0.10 \text{ mol dm}^{-3} \text{ } CH_3COOH$ pH = 2.9

2. Conductivity measurement
   Strong acids and strong bases in solution will give much higher readings on a conductivity meter than **equimolar** (equal concentration) solutions of weak acids or bases, because they contain more ions in solution.

The pH scale

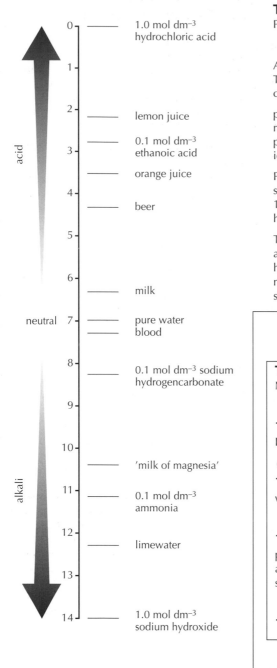

acid

neutral 7

alkali

0	1.0 mol dm^{-3} hydrochloric acid
2	lemon juice
3	0.1 mol dm^{-3} ethanoic acid
	orange juice
4	beer
6	milk
7	pure water / blood
8	0.1 mol dm^{-3} sodium hydrogencarbonate
10	'milk of magnesia'
11	0.1 mol dm^{-3} ammonia
12	limewater
14	1.0 mol dm^{-3} sodium hydroxide

THE pH SCALE

Pure water is very slightly dissociated:

$$H_2O(l) \rightleftharpoons H^+(aq) + OH^-(aq)$$

At 25 °C the equilibrium constant for this reaction is 1×10^{-14} mol^2 dm^{-6}. The concentration of the hydrogen ions (which is the same as the concentration of the hydroxide ions) equals 1×10^{-7} mol dm^{-3}.

pH (which stands for **p**ower of **H**ydrogen) is defined as being equal to minus the logarithm to the base ten of the hydrogen ion concentration. In practice this means that it is equal to the power of ten of the hydrogen ion concentration with the sign reversed. The pH of pure water is thus 7.

Pure water is neutral, so the pH of any neutral solution is 7. If the solution is acidic the hydrogen ion concentration will be greater than 10^{-7} mol dm^{-3} and the pH will decrease. Similarly alkaline solutions will have a pH greater than 7.

The pH scale runs from 0 to 14. Because it depends on the power of ten a change in one unit in the pH corresponds to a tenfold change in the hydrogen ion concentration. A 0.1 mol dm^{-3} solution of a strong monoprotic acid will have a pH of 1, a 0.001 mol dm^{-3} solution of the same acid will have a pH of 3.

THE LOG$_{10}$ SCALE AND p-SCALE

Normal scale – the distance between each number is equal

–5	–4	–3	–2	–1	0	1	2	3	4	5

Log$_{10}$ scale – the distances between powers of ten are equal

0.00001	0.0001	0.001	0.01	0.1	1	10	100	1000	10 000	100 000

which can be written

10^{-5}	10^{-4}	10^{-3}	10^{-2}	10^{-1}	10^{0}	10^{1}	10^{2}	10^{3}	10^{4}	10^{5}

p-scale – sometimes used by chemists to express equilibrium constants and concentration. It is equal to minus the power of ten in the logarithmic scale so the scale becomes:

5	4	3	2	1	0	–1	–2	–3	–4	–5

DETERMINATION OF pH

The pH of a solution can be determined by using a pH meter or by using 'universal' indicator, which contains a mixture of indicators which give a range of colours at different pH values.

pH	[H$^+$]/ mol dm^{-3}	[OH$^-$]/ mol dm^{-3}	Description	Colour of universal indicator
0	1	1×10^{-14}	very acidic	red
4	1×10^{-4}	1×10^{-10}	acidic	orange
7	1×10^{-7}	1×10^{-7}	neutral	green
10	1×10^{-10}	1×10^{-4}	basic	blue
14	1×10^{-14}	1	very basic	purple

STRONG, CONCENTRATED, AND CORROSIVE

In English the words strong and concentrated are often used interchangeably. In chemistry they have very precise meanings:

- **strong**: completely dissociated into ions
- **concentrated**: a high number of moles of solute per litre (dm^3) of solution
- **corrosive**: chemically reactive.

Similarly weak and dilute also have very different chemical meanings:

- **weak**: only slightly dissociated into ions
- **dilute**: a low number of moles of solute per litre of solution.

BRØNSTED–LOWRY ACIDS AND BASES

An acid was originally distinguished by its sour taste. The ionic theory defines an acid in terms of a substance giving a hydrogen ion concentration in aqueous solution greater than 10^{-7} mol dm^{-3}.

A Brønsted–Lowry acid is a substance that can *donate* a proton. A Brønsted–Lowry base is a substance that can *accept* a proton.

Consider the reaction between hydrogen chloride gas and water.

$$\underset{\text{acid}}{HCl(g)} + \underset{\text{base}}{H_2O(l)} \rightleftharpoons \underset{\text{acid}}{H_3O^+(aq)} + \underset{\text{base}}{Cl^-(aq)}$$

Under this definition both HCl and H_3O^+ are acids as both can donate a proton. Similarly both H_2O and Cl^- are bases as both can accept a proton. Cl^- is said to be the **conjugate base** of HCl and H_2O is the conjugate base of H_3O^+. The conjugate base of an acid is the species remaining after the acid has lost a proton. Every base also has a conjugate acid, which is the species formed after the base has accepted a proton. In the reaction with hydrogen chloride water is behaving as a base. Water can also behave as an acid.

$$\underset{\text{base}}{NH_3(g)} + \underset{\text{acid}}{H_2O(l)} \rightleftharpoons \underset{\text{acid}}{NH_4^+(aq)} + \underset{\text{base}}{OH^-(aq)}$$

Substances such as water, which can act both as an acid and as a base, are described as **amphiprotic**.

Strong acids have weak conjugate bases, whereas weak acids have strong conjugate bases.

acid strength ↓

Acid	pKa at 298 K	Conjugate base
ethanoic acid, CH_3COOH	4.76	ethanoate ion, CH_3COO^-
phenol, C_6H_5OH	10.0	phenoxide ion, $C_6H_5O^-$
water, H_2O	14.0	hydroxide ion, OH^-
ethanol, C_2H_5OH	approx 16	ethoxide ion, $C_2H_5O^-$

base strength ↓

LEWIS ACIDS AND BASES

Brønsted–Lowry bases must contain a non-bonding pair of electrons to accept the proton. The Lewis definition takes this further and describes bases as substances which can *donate* a pair of electrons, and acids as substances which can *accept* a pair of electrons. In the process a co-ordinate (both electrons provided by one species) covalent bond is formed between the base and the acid.

The Lewis theory is all-embracing, so the term Lewis acid is usually reserved for substances which are not also Brønsted–Lowry acids. Many Lewis acids do not even contain hydrogen.

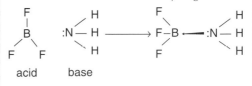

BF_3 is a good Lewis acid as there are only six electrons around the central boron atom which leaves room for two more. Other common Lewis acids are $AlCl_3$, and also transition metal ions, which accept a pairs of electrons from their surrounding ligands.

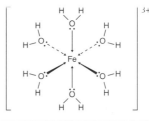

SALT HYDROLYSIS

Sodium chloride is neutral in aqueous solution. It is the salt of a strong acid and a strong base. Salts made from a weak acid and a strong base, such as sodium ethanoate, are alkaline in solution. This is because the ethanoate ions will combine with hydrogen ions from water to form mainly undissociated ethanoic acid, leaving excess hydroxide ions in solution.

$$NaCH_3COO(aq) \rightarrow Na^+(aq) + CH_3COO^-(aq)$$
$$+$$
$$H_2O(l) \rightleftharpoons OH^-(aq) + H^+(aq)$$
strong base so completely dissociated $CH_3COOH\ (aq)$

Similarly salts derived from a strong acid and a weak base will be acidic in solution.

$$NH_4Cl(aq) \rightarrow NH_4^+(aq) + Cl^-(aq)$$ strong acid
$$+$$ so completely
$$H_2O(l) \rightleftharpoons OH^-(aq) + H^+(aq)$$ dissociated
$$NH_3(aq) + H_2O(l)$$

The acidity of salts also depends on the size and charge of the cation. Aluminium chloride reacts vigorously with water to give an acidic solution.

$$AlCl_3(s) + 3H_2O(l) \rightarrow Al(OH)_3(s) + 3HCl(aq)$$

The +3 charge is spread over a very small ion, which gives the Al^{3+} ion a very high charge density. The lone pair of one

of the six water molecules surrounding the ion will be strongly attracted to the ion and the water molecule will lose a hydrogen ion in the process. This process will continue until aluminium hydroxide is formed. The equilibrium can be moved further to the right by adding $OH^-(aq)$ ions or back to the left by adding $H^+(aq)$ ions, which exemplifies the amphoteric nature of aluminium hydroxide. Similar examples of hydrolysis in aqueous solution occur with other small highly charged ions such as Fe^{3+}.

$$[Fe(H_2O)_6]^{3+} \underset{H^+}{\overset{-H^+}{\rightleftharpoons}} [Fe(H_2O)_5OH]^{2+} \underset{H^+}{\overset{-H^+}{\rightleftharpoons}}$$

$$[Fe(H_2O)_4(OH)_2]^+ \underset{H^+}{\overset{-H^+}{\rightleftharpoons}} Fe(H_2O)_3(OH)_3$$
$$OH^- \| H^+$$
$$[Fe(H_2O)_2(OH)_4]^-$$

Even $MgCl_2$ is slightly acidic in aqueous solution for the same reason.

	Charge	Ionic radius / nm	Aqueous solution
Na^+	+1	0.098	neutral
Mg^{2+}	+2	0.065	acidic
Al^{3+}	+3	0.045	acidic

THE IONIC PRODUCT OF WATER

Pure water is very slightly ionized:

$$H_2O(l) \rightleftharpoons H^+(aq) + OH^-(aq) \qquad \Delta H^\ominus = +57.3 \text{ kJ mol}^{-1}$$

$$K_c = \frac{[H^+(aq)] \times [OH^-(aq)]}{[H_2O(l)]}$$

Since the equilibrium lies far to the left the concentration of water can be regarded as constant so

$K_w = [H^+(aq)] \times [OH^-(aq)] = 1.00 \times 10^{-14} \text{ mol}^2 \text{ dm}^{-6}$ at 298 K, where K_w is known as the ionic product of water.

The dissociation of water into its ions is an endothermic process, so the value of K_w will increase as the temperature is increased.

Variation of K_w with temperature

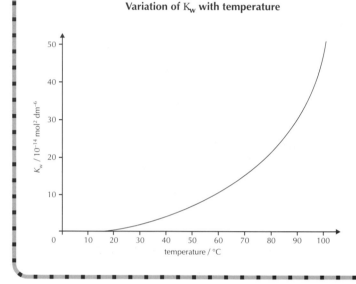

For pure water $[H^+(aq)] = [OH^-(aq)]$
$= 1.00 \times 10^{-7} \text{ mol dm}^{-3}$ at 298 K

From the graph the value for $K_w = 1.00 \times 10^{-13}$ at 334 K (61 °C)

At this temperature $[H^+(aq)] = \sqrt{1.00 \times 10^{-13}}$
$= 3.16 \times 10^{-7} \text{ mol dm}^{-3}$

pH, pOH, AND pK_w

The pH of a solution depends only on the hydrogen ion concentration and is independent of the volume of solution.

$$pH = -\log_{10}[H^+]$$

Strong acids

For strong monoprotic acids $[H^+]$ will be equal to the concentration of the acid,

e.g. for 0.100 mol dm^{-3} HCl

$[H^+] = 0.100 \text{ mol dm}^{-3}$
$pH = -\log_{10} 0.100 = 1.0$

For a strong diprotic acid the hydrogen ion concentration will be equal to twice the acid concentration,

e.g. for 0.025 mol dm^{-3} H$_2$SO$_4$

$[H^+] = 2 \times 0.025 = 0.050 \text{ mol dm}^{-3}$
$pH = 1.3$

Note: it has been assumed that sulfuric acid is a strong acid. In reality the HSO$_4^-$ ion is only partially dissociated in aqueous solution.

pOH for strong bases

For a strong base the hydrogen ion concentration can be calculated using the ionic product of water,

e.g. for 1.00 × 10^{-3} mol dm^{-3} NaOH

$[OH^-] = 1.00 \times 10^{-3} \text{ mol dm}^{-3}$
$[H^+] \times [OH^-] = 1.00 \times 10^{-14}$

$\Rightarrow [H^+] = \dfrac{1.00 \times 10^{-14}}{1.00 \times 10^{-3}} = 1.00 \times 10^{-11} \text{ mol dm}^{-3}$

$pH = -\log_{10} 1.00 \times 10^{-11} = 11.0$

The pH of alkaline solutions can also be calculated by using pOH and the relationship between pOH, pH, and pK_w.

$[H^+] \times [OH^-] = K_w \quad pOH = -\log_{10}[OH^-] \quad pK_w = -\log_{10} K_w = 14$

$$pH + pOH = 14$$

e.g for 4.00 × 10^{-2} mol dm^{-3} Ba(OH)$_2$

$[OH^-] = 2 \times (4.00 \times 10^{-2}) = 8.00 \times 10^{-2} \text{ mol dm}^{-3}$

$pOH = 1.10$

$\Rightarrow pH = 14 - 1.10 = 12.9$

Calculations with weak acids and bases

WEAK ACIDS

The dissociation of a weak acid HA in water can be written:

$$HA(aq) \rightleftharpoons H^+(aq) + A^-(aq)$$

The equilibrium expression for this reaction is:

$$K_a = \frac{[H^+] \times [A^-]}{[HA]}$$ where K_a is known as the acid dissociation constant

For example, to calculate the pH of 0.10 mol dm^{-3} CH_3COOH given that $K_a = 1.8 \times 10^{-5}$ mol dm^{-3} at 298 K:

$$CH_3COOH(aq) \rightleftharpoons CH_3COO^-(aq) + H^+(aq)$$

Initial concentration / mol dm^{-3}
    0.10   –   –

Equilibrium concentration / mol dm^{-3}
  (0.10 – x) x x

$$K_a = \frac{[CH_3COO^-] \times [H^+]}{[CH_3COOH]} = \frac{x^2}{(0.10 - x)}$$

$$= 1.8 \times 10^{-5} \text{ mol dm}^{-3}$$

$$\Rightarrow x^2 + (1.8 \times 10^{-5}x) - 1.8 \times 10^{-6} = 0$$

by solving the quadratic equation
$$x = 1.33 \times 10^{-3} \text{ mol dm}^{-3}$$
$$pH = -\log_{10} 1.33 \times 10^{-3} = 2.88$$

If the acids are quite weak the equilibrium concentration of the acid can be assumed to be the same as its initial concentration. Provided the assumption is stated it is usual to simplify the expression in calculations to avoid a quadratic equation. In the above example:

$$K_a = \frac{[CH_3COO^-] \times [H^+]}{[CH_3COOH]} \approx \frac{[H^+]^2}{0.10}$$

$$= 1.8 \times 10^{-5} \text{ mol dm}^{-3}$$

$$\Rightarrow [H^+] = \sqrt{1.8 \times 10^{-6}} = 1.34 \times 10^{-3} \text{ mol dm}^{-3}$$
$$pH = 2.87$$

Examples of other weak acid calculations

1. The pH of a 0.020 mol dm^{-3} solution of a weak acid is 3.9. Find the K_a of the acid.

$$K_a = \frac{[H^+]^2}{(0.020 - [H^+])} \approx \frac{10^{-3.9} \times 10^{-3.9}}{0.020}$$

$$= 7.92 \times 10^{-7} \text{ mol dm}^{-3}$$

2. An acid whose K_a is 4.1×10^{-6} mol dm^{-3} has a pH of 4.5. Find the concentration of the acid.

$$[HA] = \frac{[H^+]^2}{K_a} = \frac{10^{-4.5} \times 10^{-4.5}}{4.1 \times 10^{-6}}$$

$$= 2.44 \times 10^{-4} \text{ mol dm}^{-3}$$

WEAK BASES

The reaction of a weak base can be written:

$$B(aq) + H_2O(l) \rightleftharpoons BH^+(aq) + OH^-(aq)$$

Since the concentration of water is constant:

$$K_b = \frac{[BH^+] \times [OH^-]}{[B]}$$

where K_b is the base dissociation constant

If one considers the reverse reaction of BH^+ acting as an acid to give B and H^+ then:

$$K_a = \frac{[B] \times [H^+]}{[BH^+]}$$

then
$$K_a \times K_b = \frac{[B] \times [H^+]}{[BH^+]} \times \frac{[BH^+] \times [OH^-]}{[B]}$$

$$= [H^+] \times [OH^-] = K_w$$

since $pK_a = -\log_{10} K_a$; $pK_b = -\log_{10} K_b$ and $pK_w = -\log_{10} K_w = 14$ this can also be expressed as:

$$pK_a + pK_b = 14$$

Examples of calculations

1. The K_b value for ammonia is 1.8×10^{-5} mol dm^{-3}. Find the pH of a 1.00×10^{-2} mol dm^{-3} solution.

Since $[NH_4^+] = [OH^-]$ then

$$K_b = \frac{[OH^-]^2}{[NH_3]} \approx \frac{[OH^-]^2}{1.00 \times 10^{-2}} = 1.8 \times 10^{-5}$$

$$[OH^-] = \sqrt{1.8 \times 10^{-7}} = 4.24 \times 10^{-4} \text{ mol dm}^{-3}$$

$$\Rightarrow pOH = -\log_{10} 4.24 \times 10^{-4} = 3.37$$

$$\Rightarrow pH = 14 - 3.37 = 10.6$$

2. The pH of a 3.00×10^{-2} mol dm^{-3} solution of weak base is 10.0. Calculate the pK_b value of the base.

pH = 10.0 so pOH = 4.0

$$K_b = \frac{10^{-4} \times 10^{-4}}{3.00 \times 10^{-2}} = 3.33 \times 10^{-7} \text{ mol dm}^{-3}$$

$$\Rightarrow pK_b = 6.48$$

3. The IB Data Booklet value for the pK_a of methylamine (aminomethane) is 10.64. Calculate the concentration of a solution of methylamine with a pH of 10.8.

$$pK_b = 14 - 10.64 = 3.36; \quad pOH = 3.2$$

$$[CH_3NH_2] = \frac{[OH^-]^2}{K_b} = \frac{10^{-3.2} \times 10^{-3.2}}{10^{-3.36}}$$

$$= 9.13 \times 10^{-4} \text{ mol dm}^{-3}$$

Buffers

BUFFER SOLUTIONS

A buffer solution resists changes in pH when small amounts of acid or alkali are added to it.

An acidic buffer solution can be made by mixing a weak acid together with the salt of that acid and a strong base. An example is a solution of ethanoic acid and sodium ethanoate. The weak acid is only slightly dissociated in solution, but the salt is fully dissociated into its ions, so the concentration of ethanoate ions is high.

$$NaCH_3COO(aq) \rightarrow Na^+(aq) + CH_3COO^-(aq)$$
$$CH_3COOH(aq) \rightleftharpoons CH_3COO^-(aq) + H^+(aq)$$

If an acid is added the extra H^+ ions coming from the acid are removed as they combine with ethanoate ions to form undissociated ethanoic acid, so the concentration of H^+ ions remains unaltered.

$$CH_3COO^-(aq) + H^+(aq) \rightleftharpoons CH_3COOH(aq)$$

If an alkali is added the hydroxide ions from the alkali are removed by their reaction with the undissociated acid to form water, so again the H^+ ion concentration stays constant.

$$CH_3COOH(aq) + OH^-(aq) \rightarrow CH_3COO^-(aq) + H_2O(l)$$

In practice acidic buffers are often made by taking a solution of a strong base and adding excess weak acid to it, so that the solution contains the salt and the unreacted weak acid.

$$NaOH(aq) + CH_3COOH(aq) \rightarrow NaCH_3COO(aq) + H_2O(l) + CH_3COOH(aq)$$

limiting reagent   salt   excess weak acid

buffer solution

An alkali buffer with a fixed pH greater than 7 can be made from a weak base together with the salt of that base with a strong acid. An example is ammonia with ammonium chloride.

$$NH_4Cl(aq) \rightarrow NH_4^+(aq) + Cl^-(aq)$$
$$NH_3(aq) + H_2O \rightleftharpoons NH_4^+(aq) + OH^-(aq)$$

If H^+ ions are added they will combine with OH^- ions to form water and more of the ammonia will dissociate to replace them. If more OH^- ions are added they will combine with ammonium ions to form undissociated ammonia. In both cases the hydroxide ion concentration and the hydrogen ion concentration remain constant.

BLOOD

An important buffer is blood which only functions correctly within a very narrow pH range. Blood is a complex buffering system, which is responsible for carrying oxygen around the body. One of the components of the system is that the oxygen adds on reversibly to the haemoglobin in the blood.

$$HHb + O_2 \rightleftharpoons H^+ + HbO_2^-$$

If the pH increases ($[H^+]$ falls) the equilibrium will move to the right and the oxygen will tend to be bound to the haemoglobin more tightly. If the pH decreases ($[H^+]$ increases) the oxygen will tend to be displaced from the haemoglobin. Both of these processes are potentially life threatening.

BUFFER CALCULATIONS

The equilibrium expression for weak acids also applies to acidic buffer solutions,

e.g. ethanoic acid/sodium ethanoate solution.

$$K_a = \frac{[H^+] \times [CH_3COO^-]}{[CH_3COOH]}$$

The essential difference is that now the concentrations of the two ions from the acid will not be equal.

Since the sodium ethanoate is completely dissociated the concentration of the ethanoate ions in solution will be almost the same as the concentration of the sodium ethanoate, as very little will come from the acid.

If logarithms are taken and the equation is rearranged then:

$$pH = pK_a + \log_{10}\frac{[CH_3COO^-]}{[CH_3COOH]}$$

Two facts can be deduced from this expression. Firstly the pH of the buffer does not change on dilution, as the concentration of the ethanoate ions and the acid will be affected equally. Secondly the buffer will be most efficient

when $[CH_3COO^-] = [CH_3COOH]$. At this point, which equates to the half equivalence point when ethanoic acid is titrated with sodium hydroxide, the pH of the solution will equal the pK_a value of the acid.

Calculate the pH of a buffer containing 0.200 mol of sodium ethanoate in 500 cm³ of 0.100 mol dm⁻³ ethanoic acid (given that K_a for ethanoic acid = 1.8×10^{-5} mol dm⁻³).

$[CH_3COO^-]$ = 0.400 mol dm⁻³; $[CH_3COOH]$ = 0.100 mol dm⁻³

$$K_a \approx \frac{[H^+] \times 0.400}{0.100} = 1.8 \times 10^{-5} \text{ mol dm}^{-3}$$

$[H^+] = 4.5 \times 10^{-6}$ mol dm⁻³

pH = 5.35

Calculate what mass of sodium propanoate must be dissolved in 1.00 dm³ of 1.00 mol dm³ propanoic acid (pK_a = 4.87) to give a buffer solution with a pH of 4.5.

$$[C_2H_5COO^-] = \frac{K_a \times [C_2H_5COOH]}{[H^+]} = \frac{10^{-4.87} \times 1.00}{10^{-4.5}}$$

$$= 0.427 \text{ mol dm}^{-3}$$

Mass of NaC_2H_5COO required = $0.427 \times 96.07 = 41.0$ g

Titration curves and indicators

STRONG ACID – STRONG BASE TITRATION

The change in pH during an acid–base titration can be followed using a pH meter. Consider starting with 50 cm³ of 1.0 mol dm⁻³ hydrochloric acid. Since $[H^+(aq)] = 1.0$ mol dm⁻³ the initial pH will be 0. After 49 cm³ of 1.0 mol dm⁻³ NaOH have been added there will be 1.0 cm³ of the original 1.0 mol dm⁻³ hydrochloric acid left in 99 cm³ of solution. At this point $[H^+(aq)] \approx 1.0 \times 10^{-2}$ mol dm⁻³ so the pH = 2.

When 50 cm³ of the NaOH solution has been added the solution will be neutral and the pH will be 7. This is indicated by the point of inflexion, which is known as the equivalence point. It can be seen that there is a very large change in pH around the equivalence point. Almost all of the common acid–base indicators change colour (reach their end point) within this pH region. This means that it does not matter which indicator is used.

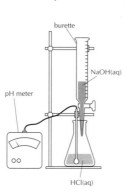

burette

NaOH(aq)

pH meter

HCl(aq)

This curve shows what happens when 1.0 mol dm⁻³ sodium hydroxide is added to 50 cm³ of 1.0 mol dm⁻³ hydrochloric acid

$$NaOH(aq) + HCl(aq) \longrightarrow NaCl(aq) + H_2O(l)$$

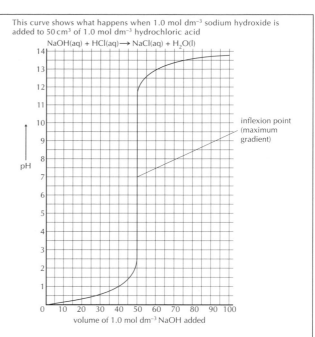

inflexion point (maximum gradient)

pH

volume of 1.0 mol dm⁻³ NaOH added

WEAK ACID – STRONG BASE TITRATION

Consider titrating 50.0 cm³ of 1.0 mol dm⁻³ CH₃COOH with 1.0 mol dm⁻³ NaOH.

$K_a = 1.8 \times 10^{-5}$. Making the usual assumptions the initial $[H^+] = \sqrt{K_a \times [CH_3COOH]}$ and pH = 2.37.

When 49.0 cm³ of the 1.0 mol dm⁻³ NaOH has been added $[CH_3COO^-] \approx 0.05$ mol dm⁻³ and $[CH_3COOH] \approx 1.0 \times 10^{-2}$ mol dm⁻³.

$$[H^+] = \frac{K_a \times [CH_3COOH]}{[CH_3COO^-]} \approx \frac{1.8 \times 10^{-5} \times 1 \times 10^{-3}}{0.05}$$

$$= 3.6 \times 10^{-7} \text{ mol dm}^{-3} \text{ and pH} = 6.44$$

After the equivalence point the graph will follow the same pattern as the strong acid – strong base curve as more sodium hydroxide is simply being added to the solution.

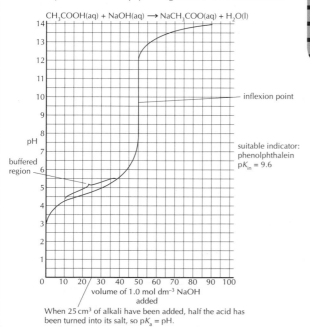

$$CH_3COOH(aq) + NaOH(aq) \longrightarrow NaCH_3COO(aq) + H_2O(l)$$

pH

buffered region

inflexion point

suitable indicator: phenolphthalein $pK_{in} = 9.6$

volume of 1.0 mol dm⁻³ NaOH added

When 25 cm³ of alkali have been added, half the acid has been turned into its salt, so $pK_a = pH$.

INDICATORS

An indicator is a weak acid (or base) in which the dissociated form is a different colour to the undissociated form.

$$HIn(aq) \rightleftharpoons H^+(aq) + In^-(aq)$$

colour A   colour B
(colour in acid solution)   (colour in alkali solution)

$K_{in} = [H^+] \times \dfrac{[In^-]}{[HIn]}$ Assuming the colour changes when

$[In^-] \approx [HIn]$ then the end point of the indicator will be when $[H^+] \approx K_{in}$, i.e. when pH $\approx pK_{in}$. Different indicators have different K_{in} values and so change colour within different pH ranges.

Indicator	pK_{in}	pH range	Use
methyl orange	3.7	3.1–4.4	titrations with strong acids
phenolphthalein	9.6	8.3–10.0	titrations with strong bases

Similar arguments can be used to explain the shapes of pH curves for strong acid – weak base, and weak acid – weak base titrations. Since there is no sharp inflexion point titrations involving weak acids with weak bases should not be used in analytical chemistry.

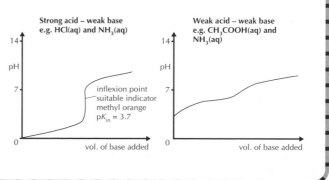

Strong acid – weak base e.g. HCl(aq) and NH₃(aq)

pH

inflexion point suitable indicator methyl orange $pK_{in} = 3.7$

vol. of base added

Weak acid – weak base e.g. CH₃COOH(aq) and NH₃(aq)

pH

vol. of base added

IB QUESTIONS – ACIDS AND BASES

1. Which statement about hydrochloric acid is false?

   A. It can react with copper to give hydrogen

   B. It can react with sodium carbonate to give carbon dioxide

   C. It can react with ammonia to give ammonium chloride

   D. It can react with copper oxide to give water

2. 1.00 cm^3 of a solution has a pH of 3. 100 cm^3 of the same solution will have pH of:

   A. 1   C. 5

   B. 3   D. Impossible to calculate from the data given.

3. Which statement(s) is/are true about separate solutions of a strong acid and a weak acid both with the same concentration?

   I. They both have the same pH.

   II. They both have the same electrical conductivity.

   A. I and II   B. I only   C. II only   D. Neither I nor II

4. Identify the correct statement about 25 cm^3 of a solution of 0.1 mol dm^{-3} ethanoic acid CH$_3$COOH.

   A. It will contain more hydrogen ions than 25 cm^3 of 0.1 mol dm^{-3} hydochloric acid.

   B. It will have a pH greater than 7.

   C. It will react exactly with 25 cm^3 of 0.1 mol dm^{-3} sodium hydroxide.

   D. It is completely dissociated into ethanoate and hydrogen ions in solution.

5. NH$_3$(aq),   HCl(aq),   NaOH(aq),   CH$_3$COOH(aq)

   When 1.0 mol dm^{-3} solutions of the substances above are arranged in order of **decreasing** pH the order is:

   A. NaOH(aq), NH$_3$(aq), CH$_3$COOH(aq), HCl(aq)

   B. NH$_3$(aq), NaOH(aq), HCl(aq), CH$_3$COOH(aq)

   C. CH$_3$COOH(aq), HCl(aq), NaOH(aq), NH$_3$(aq)

   D. HCl(aq), CH$_3$COOH(aq), NH$_3$(aq), NaOH(aq)

6. A solution with a pH of 8.5 would be described as:

   A. very basic.   C. slightly acidic.

   B. slightly basic.   D. very acidic.

7. Which statement is true about two solutions one with a pH of 3 and the other with a pH of 6?

   A. The solution with a pH of 3 is twice as acidic as the solution with a pH of 6

   B. The solution with a pH of 6 is twice as acidic as the solution with a pH of 3

   C. The hydrogen ion concentration in the solution with a pH of 6 is one thousand times greater than that in the solution with a pH of 3

   D. The hydrogen ion concentration in the solution with a pH of 3 is one thousand times greater than that in the solution with a pH of 6

8. Which of the following reagents could not be added together to make a buffer solution?

   A. NaOH(aq) and CH$_3$COOH(aq)

   B. NaCH$_3$COO(aq) and CH$_3$COOH(aq)

   C. NaOH(aq) and NaCH$_3$COO(aq)

   D. NH$_4$Cl(aq) and NH$_3$(aq)

9. When 1.0 cm^3 of a weak acid solution is added to 100 cm^3 of a buffer solution

   A. The volume of the resulting mixture will be 100 cm^3.

   B. There will be almost no change in the pH of the solution.

   C. The pH of the solution will increase noticeably.

   D. The pH of the solution will decrease noticeably.

10. During the titration of a known volume of a strong acid with a strong base

    A. there is a steady increase in pH.

    B. there is a sharp increase in pH around the end point.

    C. there is a steady decrease in pH.

    D. there is a sharp decrease in pH around the end point.

HL ──────────────────────────────

11. Which of the following is not a conjugate acid–base pair?

    A. HNO$_3$/NO$_3^-$ C. NH$_3$/NH$_2^-$

    B. H$_2$SO$_4$/HSO$_4^-$ D. H$_3$O$^+$/OH$^-$

12. Which species cannot act as a Lewis acid?

    A. NH$_3$ B. BF$_3$ C. Fe^{2+} D. AlCl$_3$

13. Three acids, HA, HB, and HC have the following K_a values

    K_a(HA) = 1×10^{-5} K_a(HB) = 2×10^{-5} K_a(HC) = 1×10^{-6}

    What is the correct order of increasing acid strength (weakest first)?

    A. HA, HB, HC   C. HC, HA, HB

    B. HC, HB, HA   D. HB, HA, HC

14. What is the pH of a buffer solution in which the concentration of the acid HX and the salt NaX are both 0.1 mol dm^{-3} ($K_a = 1 \times 10^{-5}$)?

    A. 3   B. 4   C. 5   D. 6

15. Which salt does not form an acidic solution in water?

    A. MgCl$_2$ B. Na$_2$CO$_3$ C. FeCl$_3$ D. NH$_4$NO$_3$

16. An indicator changes colour in the pH range 8.3–10.0. This indicator should be used when titrating a known volume of:

    A. a strong acid with a weak base

    B. a weak acid with a weak base

    C. a weak base with a strong acid

    D. a weak acid with a strong base.

Redox reactions (1)

DEFINITIONS OF OXIDATION AND REDUCTION

Oxidation used to be narrowly defined as the addition of oxygen to a substance. For example, when magnesium is burned in air the magnesium is oxidized to magnesium oxide.

$$2Mg(s) + O_2(g) \rightarrow 2MgO(s)$$

The electronic configuration of magnesium is 2.8.2. During the oxidation process it loses two electrons to form the Mg^{2+} ion with the electronic configuration of 2.8. **Oxidation** is now defined as the *loss of one or more electrons from a substance*. This is a much broader definition, as it does not necessarily involve oxygen. Bromide ions, for example, are oxidized by chlorine to form bromine.

$$2Br^-(aq) + Cl_2(aq) \rightarrow Br_2(aq) + 2Cl^-(aq)$$

If a substance loses electrons then something else must be gaining electrons. *The gain of one or more electrons* is called **reduction**. In the first example oxygen is reduced as it is gaining two electrons from magnesium to form the oxide ion O^{2-}. Similarly, in the second example chlorine is reduced as each chlorine atom gains one electron from a bromide ion to form a chloride ion.

Since the processes involve the transfer of electrons oxidation and reduction must occur simultaneously. Such reactions are known as **redox reactions**. In order to distinguish between the two processes half-equations are often used:

$2Mg(s) \rightarrow 2Mg^{2+}(s) + 4e$	— OXIDATION —	$2Br^-(aq) \rightarrow Br_2(aq) + 2e$
$O_2(g) + 4e \rightarrow 2O^{2-}(s)$	— REDUCTION —	$Cl_2(aq) + 2e \rightarrow 2Cl^-(aq)$
$2Mg(s) + O_2(g) \rightarrow 2MgO(s)$	OVERALL REDOX EQUATION	$2Br^-(aq) + Cl_2(aq) \rightarrow Br_2(aq) + 2Cl^-(aq)$

Understanding that magnesium must lose electrons and oxygen must gain electrons when magnesium oxide MgO is formed from its elements is a good way to remember the definitions of oxidation and reduction. Some students prefer to use the mnemonic OILRIG: **O**xidation **I**s the **L**oss of electrons, **R**eduction **I**s the **G**ain of electrons.

RULES FOR DETERMINING OXIDATION NUMBERS

It is not always easy to see how electrons have been transferred in redox processes. Oxidation numbers can be a useful tool to identify which species have been oxidized and which reduced. Oxidation numbers are assigned according to a set of rules:

1. In an ionic compound between two elements the oxidation number of each element is equal to the charge carried by the ion, e.g.

    $$Na^+Cl^- \qquad Ca^{2+}Cl^-_2$$
    $$(Na = +1; Cl = -1) \quad (Ca = +2; Cl = -1)$$

2. For covalent compounds assume the compound is ionic with the more electronegative element forming the negative ion, e.g.

    $$CCl_4 \qquad NH_3$$
    $$(C = +4; Cl = -1) \quad (N = -3; H = +1)$$

3. The algebraic sum of all the oxidation numbers in a compound = zero, e.g.

    $$CCl_4 [(+4) + 4 \times (-1) = 0];$$
    $$H_2SO_4 [2 \times (+1) + (+6) + 4 \times (-2) = 0]$$

4. The algebraic sum of all the oxidation numbers in an ion = the charge on the ion, e.g.

    $$SO_4^{2-} [(+6) + 4 \times (-2) = -2];$$
    $$MnO_4^- [(+7) + 4 \times (-2) = -1]; NH_4^+ [(-3) + 4 \times (+1) = +1]$$

5. Elements not combined with other elements have an oxidation number of zero, e.g. O_2; P_4; S_8.

6. Oxygen when combined always has an oxidation number of –2 except in peroxides (e.g. H_2O_2) when it is –1.

7. Hydrogen when combined always has an oxidation number of +1 except in certain metal hydrides (e.g. NaH) when it is –1.

Many elements can show different oxidation numbers in different compounds, e.g. nitrogen in:

NH_3	N_2H_4	N_2	N_2O	NO	NO_2	NO_3^-
(–3)	(–2)	(0)	(+1)	(+2)	(+4)	(+5)

When elements show more than one oxidation state the oxidation number is represented by using Roman numerals when naming the compound,

e.g. $FeCl_2$ iron(II) chloride; $FeCl_3$ iron(III) chloride
$K_2Cr_2O_7$ potassium dichromate(VI); $KMnO_4$ potassium manganate(VII)
Cu_2O copper(I) oxide; CuO copper(II) oxide.

Redox reactions (2)

OXIDATION AND REDUCTION IN TERMS OF OXIDATION NUMBERS

When an element is oxidized its oxidation number *increases*,

e.g. $Mg(s) \rightarrow Mg^{2+}(aq) + 2e$

    (0)   (+2)

When an element is reduced its oxidation number *decreases*,

e.g. $SO_4^{2-}(aq) + 2H^+(aq) + 2e \rightarrow SO_3^{2-}(aq) + H_2O(l)$

   (+6)   (+4)

The change in the oxidation number will be equal to the number of electrons involved in the half-equation.

Using oxidation numbers makes it easy to identify whether or not a reaction is a redox reaction.

Redox reactions (change in oxidation numbers)

$CuO(s) + H_2(g) \rightarrow Cu(s) + H_2O(l)$

 (+2)   (0)   (0)   (+1)

$5Fe^{2+}(aq) + MnO_4^-(aq) + 8H^+(aq) \rightarrow Fe^{3+}(aq) + Mn^{2+}(aq) + 4H_2O(l)$

 (+2)   (+7)   (+3)   (+2)

Not redox reactions (no change in oxidation numbers)

precipitation $Ag^+(aq) + Cl^-(aq) \rightarrow AgCl(s)$

                (+1)   (−1)   (+1) (−1)

neutralization $HCl(aq) + NaOH(aq) \rightarrow NaCl(aq) + H_2O(l)$

              (+1)(−1) (+1)(−2)(+1) (+1)(−1) (+1)(−2)

Note: reactions where an element is uncombined on one side of the equation and combined on the other side *must* be redox reactions since there must be a change in oxidation number,

e.g. $Mg(s) + 2HCl(aq) \rightarrow MgCl_2(aq) + H_2(g)$

OXIDIZING AGENTS AND REDUCING AGENTS

A substance that readily oxidizes other substances is known as an **oxidizing agent**. Oxidizing agents are thus substances that readily accept electrons. Usually they contain elements that are in their highest oxidation state,

e.g. O_2, Cl_2, F_2, SO_3 (SO_4^{2-} in solution), MnO_4^-, and $Cr_2O_7^{2-}$.

Reducing agents readily donate electrons and include H_2, Na, C, CO, and SO_2 (SO_3^{2-} in solution),

e.g. $Cr_2O_7^{2-}(aq)$ + $3SO_3^{2-}(aq) + 8H^+(aq) \rightarrow 2Cr^{3+}(aq) + 3SO_4^{2-}(aq) + 4H_2O(l)$

        (+6)   (+4)   (+3)   (+6)

      (orange)   (green)

(oxidizing agent)   (reducing agent)

        $Cl_2(aq)$ + $2Br^-(aq)$ $\rightarrow$ $2Cl^-(aq)$ + $Br_2(aq)$

          (0)   (−1)   (−1)   (0)

(oxidizing agent)   (reducing agent)

Reactivity series

REACTIVITY

Lithium, sodium, and potassium all react with cold water to give similar products but the reactivity increases down the group.

$$2M(s) + 2H_2O(l) \rightarrow 2M^+(aq) + 2OH^-(aq) + H_2(g) \quad (M = Li, Na, or K)$$

Slightly less reactive metals react with steam and will give hydrogen with dilute acids, e.g.

$$Mg(s) + 2H_2O(g) \rightarrow Mg(OH)_2 + H_2(g)$$

$$Mg(s) + 2HCl(aq) \rightarrow Mg^{2+}(aq) + 2Cl^-(aq) + H_2(g)$$

In all of these reactions the metal is losing electrons – that is, it is being oxidized and in the process it is acting as a reducing agent. A reactivity series of reducing agents can be deduced by considering the reactivity of metals with water and acids, and the reactions of metals with the ions of other metals.

Reactivity series of reducing agents

$$K(s) \rightleftharpoons e + K^+(aq)$$
$$Na(s) \rightleftharpoons e + Na^+(aq)$$
$$Li(s) \rightleftharpoons e + Li^+(aq)$$
$$Ca(s) \rightleftharpoons 2e + Ca^{2+}(aq)$$
$$Mg(s) \rightleftharpoons 2e + Mg^{2+}(aq)$$
$$Al(s) \rightleftharpoons 3e + Al^{3+}(aq)$$
$$Zn(s) \rightleftharpoons 2e + Zn^{2+}(aq)$$
$$Fe(s) \rightleftharpoons 2e + Fe^{2+}(aq)$$
$$Pb(s) \rightleftharpoons 2e + Pb^{2+}(aq)$$
$$\tfrac{1}{2}H_2(g) \rightleftharpoons e + H^+(aq)$$
$$Cu(s) \rightleftharpoons 2e + Cu^{2+}(aq)$$
$$Ag(s) \rightleftharpoons e + Ag^+(aq)$$

Increasing reactivity

The more readily the metal loses its outer electrons the more reactive it is. Metals higher in the series can displace metal ions lower in the series from solution, e.g. zinc can react with copper ions to form zinc ions and precipitate copper metal.

$$Zn(s) + Cu^{2+}(aq) \rightarrow Zn^{2+}(aq) + Cu(s)$$

$$Zn(s) \rightarrow Zn^{2+}(aq) + 2e \qquad \text{Zn loses electrons in preference to Cu}$$
$$Cu^{2+}(aq) + 2e \rightarrow Cu(s) \qquad \text{Cu}^{2+} \text{ gains electrons in preference to Zn}^{2+}$$

This also explains why only metals above hydrogen can react with acids (displace hydrogen ions) to produce hydrogen gas, e.g.

$$Zn(s) + 2H^+(aq) \rightarrow Zn^{2+}(aq) + H_2(g)$$

The series can be extended for oxidizing agents. The most reactive oxidizing agent will be the species that gains electrons the most readily. For example, in group 7

$$I^-(aq) = e + \tfrac{1}{2}I_2(aq)$$
$$Br^-(aq) = e + \tfrac{1}{2}Br_2(aq)$$
$$Cl^-(aq) = e + \tfrac{1}{2}Cl_2(aq)$$
$$F^-(aq) = e + \tfrac{1}{2}F_2(aq)$$

increasing oxidizing ability

Oxidizing agents lower in the series gain electrons from species higher in the series, e.g.

$$Cl_2(aq) + 2Br^-(aq) \rightarrow 2Cl^-(aq) + Br_2(aq)$$

SIMPLE VOLTAIC CELLS

A half-cell is simply a metal in contact with an aqueous solution of its own ions. A voltaic cell consists of two different half-cells, connected together to enable the electrons transferred during the redox reaction to produce energy in the form of electricity. The cells are connected by an external wire and by a salt bridge, which allows the free movement of ions.

A good example of a voltaic cell is a zinc half-cell connected to a copper half-cell. Because zinc is higher in the reactivity series the electrons will flow from the zinc half-cell towards the copper half-cell. To complete the circuit and to keep the half-cells electrically neutral, ions will flow through the salt bridge. The voltage produced by a voltaic cell depends on the relative difference between the two metals in the reactivity series. Thus the voltage from a Mg(s)/Mg^{2+}(aq) half-cell connected to a Cu(s)/Cu^{2+}(aq) half-cell will be greater than that obtained from a Zn(s)/Zn^{2+}(aq) half-cell connected to a Fe(s)/Fe^{2+}(aq) half-cell.

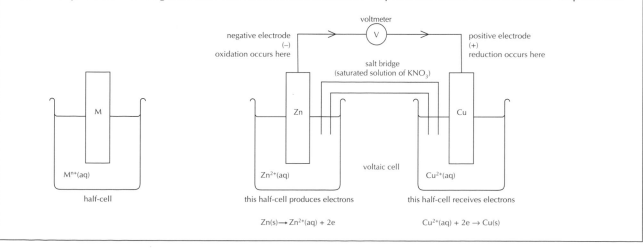

Electrolysis (1)

ELECTROLYTIC CELL

In a voltaic cell electricity is produced by the spontaneous redox reaction taking place. Electrolytic cells are used to make non-spontaneous redox reactions occur by providing energy in the form of electricity from an external source. In an electrolytic cell electricity is passed through an **electrolyte** and electrical energy is converted into chemical energy. An electrolyte is a substance which does not conduct electricity when solid, but does conduct electricity when molten or in aqueous solution and is chemically decomposed in the process. A simple example of an electrolytic cell is the electrolysis of molten sodium chloride.

During the electrolysis: sodium is formed at the negative electrode (cathode)
$$Na^+(l) + e \rightarrow Na(l) \text{ reduction}$$

chlorine is formed at the positive electrode (anode)
$$2Cl^-(l) \rightarrow Cl_2(g) + 2e \text{ oxidation}$$

Electrolysis is an important industrial process used to obtain reactive metals, such as sodium, from their common ores. It can also be used to coat one metal with a thin layer of another metal. This process is known as **electroplating**.

For example, in copper plating the negative electrode (cathode) is made from the metal to be copper plated. It is placed into a solution of copper(II) sulfate together with a positive electrode (anode) made from a piece of copper. As electricity is passed through the solution the copper anode dissolves in the solution to form $Cu^{2+}(aq)$ ions and the $Cu^{2+}(aq)$ ions in solution are deposited onto the cathode. By making the anode of impure copper and the cathode from a small piece of pure copper this process can also be used to purify impure copper. This is an important industrial process as one of the main uses of copper is for electrical wiring, where purity is important since impure copper has a much higher electrical resistance.

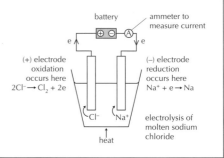

FACTORS AFFECTING THE DISCHARGE OF IONS DURING ELECTROLYSIS

During the electrolysis of molten salts there are only usually two ions present, so the cation will be discharged at the negative electrode (cathode) and the anion at the positive electrode (anode). However, for aqueous electrolytes there will also be hydrogen ions and hydroxide ions from the water present. There are three main factors which influence which ions will be discharged at their respective electrodes.

1. Position in the electrochemical series

The lower the metal ion is in the electrochemical series the more readily it will gain electrons (be reduced) to form the metal at the cathode. Thus in the electrolysis of a solution of sodium hydroxide, hydrogen will be evolved at the negative electrode in preference to sodium, whereas in a solution of copper(II) sulfate, copper will be deposited at the negative electrode in preference to hydrogen.

For negative ions the order of discharge follows $OH^- > Cl^- > SO_4^{2-}$.

2. Concentration

If one of the ions is much more concentrated than another ion then it may be preferentially discharged. For example, when electricity is passed through an aqueous solution of sodium chloride both oxygen and chlorine are evolved at the positive electrode. For dilute solutions mainly oxygen is evolved, but for concentrated solutions of sodium chloride more chlorine than oxygen is evolved.

3. The nature of the electrode

It is normally safe to assume that the electrode is inert, i.e. does not play any part in the reaction. However, if copper electrodes are used during the electrolysis of a solution of copper sulfate then the positive electrode is itself oxidized to release electrons and form copper(II) ions. Since copper is simultaneously deposited at the negative electrode the concentration of the solution will remain constant throughout the electrolysis.

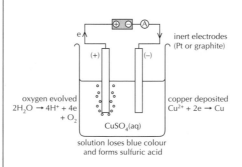

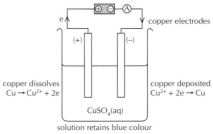

Electrolysis (2) and standard electrode potentials

The amount of substance deposited will depend on:

1. The number of electrons flowing through the system, i.e. the charge passed. This in turn depends on the current and the time for which it flows. If the current is doubled then twice as many electrons pass through the system and twice as much product will be formed. Similarly if the time is doubled twice as many electrons will pass through the system and twice as much product will be formed.

charge = current × time
(1 coulomb = 1 ampere × 1 second)

2. The charge on the ion. To form one mole of sodium in the electrolysis of molten sodium chloride requires one mole of electrons to flow through the cell. However, the formation of one mole of lead during the electrolysis of molten lead(II) bromide requires two moles of electrons.

$$Na^+(l) + e \rightarrow Na(l)$$
$$Pb^{2+}(l) + 2e \rightarrow Pb(l)$$

If cells are connected in series…

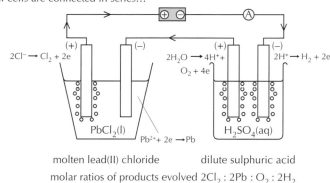

$2Cl^- \rightarrow Cl_2 + 2e$ (+) (−) (+) $2H_2O \rightarrow 4H^+ + O_2 + 4e$ (−) $2H^+ \rightarrow H_2 + 2e$

$PbCl_2(l)$ $Pb^{2+} + 2e \rightarrow Pb$ $H_2SO_4(aq)$

molten lead(II) chloride   dilute sulphuric acid

molar ratios of products evolved $2Cl_2 : 2Pb : O_2 : 2H_2$

If cells are connected in series then the same amount of electricity will pass through both cells and the relative amounts of products obtained can be determined.

STANDARD ELECTRODE POTENTIALS

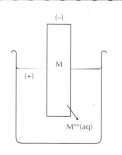

There are two opposing tendencies in a half-cell. The metal may dissolve in the solution of its own ions to leave the metal with a negative potential compared with the solution, or the metal ions may deposit on the metal, which will give the metal a positive potential compared with the solution. It is impossible to measure this potential, as any attempt to do so interferes with the system being investigated. However, the electrode potential of one half-cell can be compared against another half-cell. The hydrogen half-cell is normally used as the standard. Under standard conditions of 1 atm pressure, 298 K, and 1.0 mol dm^{-3}

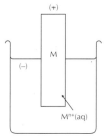

hydrogen ion concentration the standard electrode potential of the hydrogen electrode is assigned a value of zero volts.

When the half-cell contains a metal above hydrogen in the reactivity series electrons flow from the half-cell to the hydrogen electrode, and the electrode potential is given a negative value. If the half-cell contains a metal below hydrogen in the reactivity series electrons flow from the hydrogen electrode to the half-cell, and the electrode potential has a positive value. The standard electrode potentials are arranged in increasing order to form the electrochemical series.

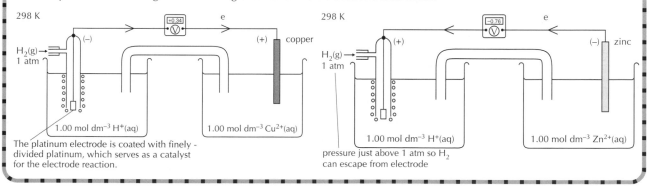

298 K   +0.34   e   (−)   (+) copper
$H_2(g)$ 1 atm
1.00 mol dm^{-3} H$^+$(aq)   1.00 mol dm^{-3} Cu^{2+}(aq)

The platinum electrode is coated with finely-divided platinum, which serves as a catalyst for the electrode reaction.

298 K   −0.76   e   (+)   (−) zinc
$H_2(g)$ 1 atm
1.00 mol dm^{-3} H$^+$(aq)   1.00 mol dm^{-3} Zn^{2+}(aq)

pressure just above 1 atm so H$_2$ can escape from electrode

ELECTROCHEMICAL SERIES

(more complete series can be found in the IB Data Booklet)

Couple	$E^{\ominus}$ / V
$K(s)/K^+(aq)$	−2.92
$Ca(s)/Ca^{2+}(aq)$	−2.87
$Na(s)/Na^+(aq)$	−2.71
$Mg(s)/Mg^{2+}(aq)$	−2.36
$Al(s)/Al^{3+}(aq)$	−1.66
$Zn(s)/Zn^{2+}(aq)$	−0.76
$Fe(s)/Fe^{2+}(aq)$	−0.44
$\frac{1}{2}H_2(g)/H^+(aq)$	0.00
$Cu(s)/Cu^{2+}(aq)$	+0.34
$I^-(aq)/\frac{1}{2}I_2(aq)$	+0.62
$Ag(s)/Ag^+(aq)$	+0.80
$Br^-(aq)/\frac{1}{2}Br_2(aq)$	+1.09
$Cl^-(aq)/\frac{1}{2}Cl_2(aq)$	+1.36
$F^-(aq)/\frac{1}{2}F_2(aq)$	+2.87

SHORTHAND NOTATION FOR A CELL

To save drawing out the whole cell a shorthand notation has been adopted. A half-cell is denoted by a / between the metal and its ions, and two vertical lines are used to denote the salt bridge between the two half cells,

e.g. $Cu(s)/Cu^{2+}(aq) \, || \, H^+(aq)/H_2(g)$ and
$Zn(s)/Zn^{2+}(aq) \, || \, H^+(aq)/H_2(g)$

The standard electromotive force (emf) of any cell $E^{\ominus}_{cell}$ is simply the difference between the standard electrode potentials of the two half-cells,

e.g. $Cu(s)/Cu^{2+}(aq) \, || \, Zn(s)/Zn^{2+}(aq)$
    $E^{\ominus} + 0.34 \, V$ $-0.76 \, V$
        $E^{\ominus}_{cell} = 1.10 \, V$

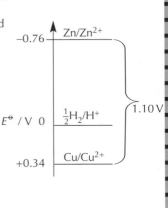

ELECTRON FLOW AND SPONTANEOUS REACTIONS

By using standard electrode potentials it is easy to determine what will happen when two half-cells are connected together. The electrons will always flow from the more negative half-cell to the more positive half-cell, e.g. consider an iron half-cell connected to a magnesium half-cell:

$E^{\ominus}$ $Fe(s)/Fe^{2+}(aq)$ $||$ $Mg^{2+}(aq)/Mg(s)$
        $-0.44 \, V$ $-2.36 \, V$
    more positive   more negative
    $Fe^{2+} + 2e \rightarrow Fe$ $Mg \rightarrow Mg^{2+} + 2e$

Spontaneous reaction: $Mg(s) + Fe^{2+}(aq) \rightarrow$
        $Mg^{2+}(aq) + Fe(s)$ $E^{\ominus}_{cell} = 1.92 \, V$

Positive $E^{\ominus}_{cell}$ values give negative $\Delta G^{\ominus}$ values as the reaction can provide electrical energy, i.e. do work, and the reaction is spontaneous. The reverse reaction ($Fe(s) + Mg^{2+}(aq) \rightarrow Fe^{2+}(aq) + Mg(s)$) has a negative $E^{\ominus}_{cell}$ value which gives a positive value for $\Delta G^{\ominus}$ and the reaction will not be spontaneous, so can only proceed if an external voltage greater than 1.92 V is applied to force the reaction in the opposite direction. Note: just because a reaction has a positive $E^{\ominus}_{cell}$ value does not mean that it will necessarily proceed, as there may be a large activation energy which needs to be overcome.

REDOX EQUATIONS

Standard electrode potentials can be extended to cover any half-equation. The values can then be used to determine whether a particular redox reaction is spontaneous, e.g. $Cr_2O_7^{2-}(aq)/Cr^{3+}(aq)$, $E^{\ominus} = +1.33 \, V$, but this only takes place in acid solution, so hydrogen ions and water are required to balance the half-equation. Consider the reaction between dichromate(VI) ions and sulfite ions SO_3^{2-}. Using standard electrode potentials the cell becomes:

$E^{\ominus}$ $Cr_2O_7^{2-}(aq)/Cr^{3+}(aq)$ $||$ $SO_3^{2-}(aq)/SO_4^{2-}(aq)$
        $+1.33 \, V$ $+0.17 \, V$
    more positive   more negative
$Cr_2O_7^{2-}$ will gain electrons SO_3^{2-} will lose electrons
$Cr_2O_7^{2-} + 14H^+ + 6e \rightarrow$ $SO_3^{2-} + H_2O \rightarrow$
    $2Cr^{3+} + 7H_2O$ $SO_4^{2-} + 2H^+ + 2e$

Since the number of electrons transferred must be the same for both equations the overall equation is obtained by multiplying the sulfite equation by three, adding the two equations together, and simplifying the water and hydrogen ion components which appear on both sides.

$Cr_2O_7^{2-}(aq) + 8H^+(aq) + 3SO_3^{2-}(aq) \rightarrow$
$2Cr^{3+}(aq) + 3SO_4^{2-}(aq) + 4H_2O(l)$ $E^{\ominus}_{cell} = 1.16 \, V$

CONVENTION FOR WRITING CELLS

By convention in a cell diagram the half-cell undergoing oxidation is placed on the left of the diagram and the half-cell undergoing reduction on the right of the diagram. The two aqueous solutions are then placed either side of the salt bridge e.g.

$Zn(s)/Zn^{2+}(aq) \, || \, Cu^{2+}(aq)/Cu(s)$

Some text books then state that:

$E^{\ominus}_{cell} = E^{\ominus}_{right\ hand\ side} - E^{\ominus}_{left\ hand\ side}$

This can cause much confusion. Essentially a cell consists of two half-cells connected by a salt bridge and it makes no difference which is placed on the left or on the right. Provided you remember that the electron flow is always *from* the half-cell with the more negative electrode potential *to* the half-cell with the more positive electrode potential the convention can be safely ignored.

IB QUESTIONS – OXIDATION AND REDUCTION

1. $5Fe^{2+}(aq) + MnO_4^-(aq) + 8H^+(aq) \rightarrow 5Fe^{3+}(aq) + Mn^{2+}(aq) + 4H_2O(l)$

   In the equation above:

   A. $Fe^{2+}(aq)$ is the oxidizing agent

   B. $H^+(aq)$ ions are reduced

   C. $Fe^{2+}(aq)$ ions are oxidized

   D. $MnO_4^-(aq)$ is the reducing agent

2. The oxidation numbers of nitrogen in NH_3, HNO_3, and NO_2 are, respectively

   A. –3, –5, +4   **C.** –3, +5, –4

   B. +3, +5, +4   **D.** –3, +5, +4

3. Which one of the following reactions is **not** a redox reaction?

   A. $Ag^+(aq) + Cl^-(aq) \rightarrow AgCl(s)$

   B. $2Na(s) + Cl_2(g) \rightarrow 2NaCl(s)$

   C. $Mg(s) + 2HCl(aq) \rightarrow MgCl_2(aq) + H_2(g)$

   D. $Cu^{2+}(aq) + Zn(s) \rightarrow Cu(s) + Zn^{2+}(aq)$

4. Which substance does not have the correct formula?

   A. iron(III) sulfate $Fe_2(SO_4)_3$

   B. iron(II) oxide Fe_2O

   C. copper(I) sulfate Cu_2SO_4

   D. copper(II) nitrate $Cu(NO_3)_2$

5. For which conversion is an oxidising agent required?

   A. $2H^+(aq) \rightarrow H_2(g)$ **C.** $SO_3(g) \rightarrow SO_4^{2-}(aq)$

   B. $2Br^-(aq) \rightarrow Br_2(aq)$ **D.** $MnO_2(s) \rightarrow Mn^{2+}(aq)$

6. $Mn(s) + Hg^{2+}(aq) \rightarrow Mn^{2+}(aq) + Hg(l)$

   $Ni^{2+}(aq) + Mn(s) \rightarrow Ni(s) + Mn^{2+}(aq)$

   $Hg^{2+}(aq) + Ni(s) \rightarrow Hg(l) + Ni^{2+}(aq)$

   From the above reactions the **increasing** reactivity of the three metals is:

   A. Hg, Ni, Mn   **C.** Ni, Mn, Hg

   B. Mn, Hg, Ni   **D.** Mn, Ni, Hg

7. When an $Fe(s)/Fe^{2+}(aq)$ half-cell is connected to a $Cu(s)/Cu^{2+}(aq)$ half-cell by a salt bridge and a current allowed to flow between them

   A. the electrons will flow from the copper to the iron.

   B. the salt bridge allows the flow of ions to complete the circuit.

   C. the salt bridge allows the flow of electrons to complete the circuit.

   D. the salt bridge can be made of copper or iron.

8. During the electrolysis of molten sodium chloride using platinum electrodes

   A. sodium is formed at the negative electrode

   B. chlorine is formed at the negative electrode

   C. sodium is formed at the positive electrode

   D. oxygen is formed at the postitive electrode

9. Which statement is true?

   A. Lead chloride is ionic so solid lead chloride will conduct electricity.

   B. When a molten ionic compound conducts electricity free electrons pass through the liquid.

   C. When liquid mercury conducts electricity mercury ions move towards the negative electrode.

   D. During the electrolysis of a molten salt reduction will always occur at the negative electrode.

10. When a metal is electroplated with copper by electrolysis which of the following statement(s) is/are true?

    I. The metal to be plated with copper is the negative electrode.

    II. The electrolyte is a solution of aqueous copper(II) ions, $Cu^{2+}(aq)$.

    III. Copper(II) ions are oxidised during the process.

    A. I, II, and III   **C.** II only

    B. I and II only   **D.** II and III only

HL

---

11. Ethanol can be oxidised to ethanal by an acidic solution of dichromate(VI) ions.

    $\_C_2H_5OH(aq) + \_H^+(aq) + \_Cr_2O_7^{2-}(aq) \rightarrow$
    $\_CH_3CHO(aq) + \_Cr^{3+}(aq) + \_H_2O(l)$

    The sum of all the co-efficients in the balanced equation is:

    A. 24   **B.** 26   **C.** 28   **D.** 30

12. Which statement(s) is/are true about the standard hydrogen electrode?

    I. The electrode potential is assigned a value of 0.00 volts.

    II. The hydrogen ion concentration is $0.100 \ mol \ dm^{-3}$.

    III. The temperature is 273 K.

    A. I, II and III   **C.** I only

    B. I and II only   **D.** II and III only

Use the following information to answer questions 13 and 14.

$Sn^{2+}(aq) + 2e \rightleftharpoons Sn(s)$	$E^\ominus = -0.14 \ V$
$Sn^{4+}(aq) + 2e \rightleftharpoons Sn^{2+}(aq)$	$E^\ominus = +0.15 \ V$
$Fe^{2+}(aq) + 2e \rightleftharpoons Fe(s)$	$E^\ominus = -0.44 \ V$
$Fe^{3+}(aq) + e \rightleftharpoons Fe^{2+}(aq)$	$E^\ominus = +0.77 \ V$

13. Under standard conditions which statement is correct?

    A. $Sn^{2+}(aq)$ can reduce $Fe^{3+}(aq)$.

    B. $Fe(s)$ can oxidise $Sn^{2+}(aq)$.

    C. $Sn(s)$ can reduce $Fe(s)$.

    D. $Fe^{3+}(aq)$ can reduce $Sn^{4+}(aq)$.

14. When a half-cell of $Fe^{2+}(aq)/Fe^{3+}(aq)$ is connected by a salt bridge to a half-cell of $Sn^{2+}(aq)/Sn^{4+}(aq)$ under standard conditions and a current allowed to flow in an external circuit the total e.m.f. of the spontaneous reaction will be:

    A. +0.92 V   **B.** –0.92 V   **C.** +0.62 V   **D.** –0.62 V

Functional groups and homologous series

NAMING ORGANIC COMPOUNDS

Organic chemistry is concerned with the compounds of carbon. Since there are more compounds of carbon known than all the other elements put together, it is helpful to have a systematic way of naming them.

1. Identify the longest carbon chain.
   - 1 carbon = **meth-**
   - 2 carbons = **eth-**
   - 3 carbons = **prop-**
   - 4 carbons = **but-**
   - 5 carbons = **pent-**
   - 6 carbons = **hex-**
   - 7 carbons = **hept-**
   - 8 carbons = **oct-**

2. Identify the type of bonding in the chain or ring
   All single bonds in the carbon chain = **-an-**
   One double bond in the carbon chain = **-en-**
   One triple bond in the carbon chain = **-yn-**

3. Identify the functional group joined to the chain or ring. This may come at the beginning or at the end of the name, e.g.

   alkane: only hydrogen (-H) joined to chain = **-e**
   alcohol: –OH = **-ol**
   amine: –NH$_2$ = **amino-**

   halogenoalkane: -X: **chloro-, bromo, or iodo-**

   aldehyde: $\overset{O}{\overset{\|}{-C}}$–H (on the end of the chain) = **-al**

   ketone: $\overset{O}{\overset{\|}{-C}}$ – (not on the end of the chain) = **-one**

   carboxylic acid: $\overset{O}{\overset{\|}{-C}}$–OH = **-oic acid**

   ester: $\overset{O}{\overset{\|}{-C}}$–OR: = **-oate**

   amide: $\overset{O}{\overset{\|}{-C}}$–NH$_2$ = **-amide**

4. Numbers are used to give the positions of groups or bonds along the chain.

HOMOLOGOUS SERIES

The alkanes form a series of compounds all with the general formula C_nH_{2n+2}, e.g.

   methane CH_4

   ethane C_2H_6

   propane C_3H_8

   butane C_4H_{10}

If one of the hydrogen atoms is removed what is left is known as an alkyl radical R – (e.g methyl CH_3–; ethyl C_2H_5–). When other atoms or groups are attached to an alkyl radical they can form a different series of compounds. These atoms or groups attached are known as functional groups and the series formed are all homologous series.

Homologous series have the same general formula with the neighbouring members of the series differing by –CH$_2$; for example the general formula of alcohols is $C_nH_{2n+1}OH$. The chemical properties of the individual members of an homologous series are similar and they show a gradual change in physical properties.

SOME COMMON FUNCTIONAL GROUPS

Formula	Name	Examples		
R–H	alkane	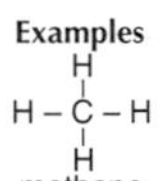 methane	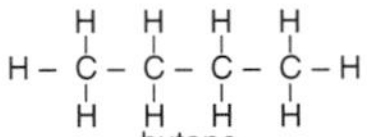 butane	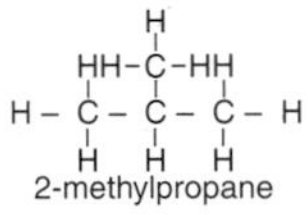 2-methylpropane
R–OH	alcohol	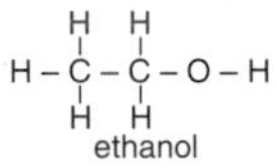 ethanol	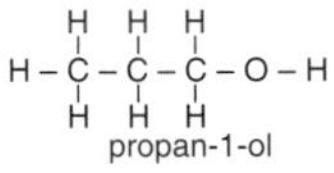 propan-1-ol	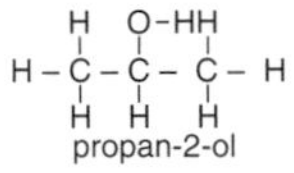 propan-2-ol
R–NH$_2$	amine	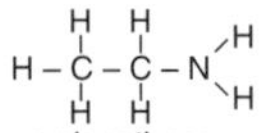 aminoethane	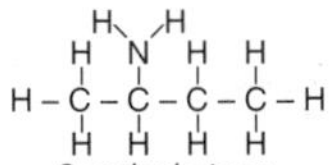	2-aminobutane
R–X (X = F, Cl, Br, or I)	halogenoalkane	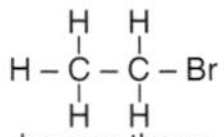 bromoethane	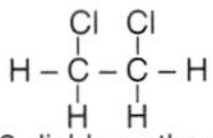 1,2-dichloroethane	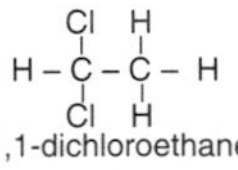 1,1-dichloroethane
R–C̈–H (O)	aldehyde	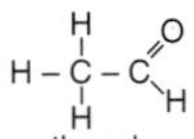 ethanal	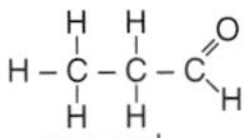 propanal	
R–C̈–R′ (O) ketone (R′ may be the same as or different to R)	ketone	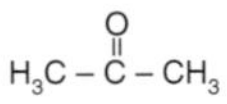 propanone	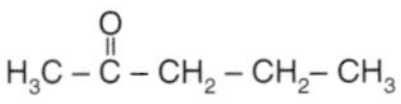 pentan-2-one	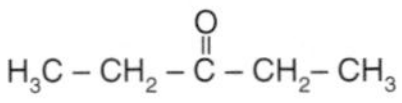 pentan-3-one
R–C̈–OH (O)	carboxylic acid	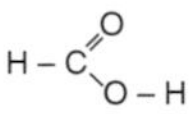 methanoic acid	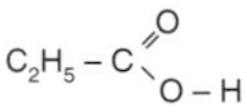 propanoic acid	
[R–C̈–OR′] (O)	ester	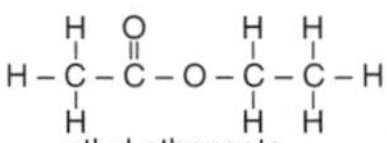 ethyl ethanoate	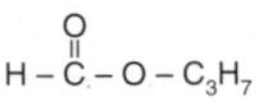 propyl methanoate	
[R–C̈–NH$_2$] (O)	amide	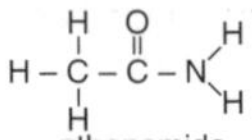 ethanamide		

Properties of different functional groups

BOILING POINTS

As the carbon chain gets longer the mass of the molecules increases and the van der Waals' forces of attraction increase. A plot of boiling point against number of carbon atoms shows a sharp increase at first, as the percentage increase in mass is high, but as successive $-CH_2-$ groups are added the rate of increase in boiling point decreases.

When branching occurs the molecules become more spherical in shape, which reduces the contact surface area between them and lowers the boiling point.

Other homologous series show similar trends but the actual temperatures at which the compounds boil will depend on the types of attractive forces between the molecules. The volatility of the compounds also follows the same pattern. The lower members of the alkanes are all gases as the attractive forces are weak and the next few members are volatile liquids. Methanol, the first member of the alcohols is a liquid at room temperature, due to the presence of hydrogen bonding. Methanol is classed as volatile as its boiling point is 64.5 °C but when there are four or more carbon atoms in the chain the boiling points exceed 100 °C and the higher alcohols have low volatility.

Compound	Formula	M_r	Functional group	Strongest type of attraction	B. pt / °C
butane	C_4H_{10}	58	alkane	van der Waals'	−0.5
butene	C_4H_8	56	alkene	van der Waals'	−6.2
butyne	C_4H_6	54	alkyne	van der Waals'	8.1
methyl methanoate	$HCOOCH_3$	60	ester	dipole:dipole	31.5
propanal	CH_3CH_2CHO	58	aldehyde	dipole:dipole	48.8
propanone	CH_3COCH_3	58	ketone	dipole:dipole	56.2
aminopropane	$CH_3CH_2CH_2NH_2$	59	amine	hydrogen bonding	48.6
propan-1-ol	$CH_3CH_2CH_2OH$	60	alcohol	hydrogen bonding	97.2
ethanoic acid	CH_3COOH	60	carboxylic acid	hydrogen bonding	118

b. pt 36.3 °C

b. pt 27.9 °C

b. pt 9.5 °C

SOLUBILITY IN WATER

Whether or not an organic compound will be soluble in water depends on the polarity of the functional group and on the chain length. The lower members of alcohols, amines, aldehydes, ketones, and carboxylic acids are all water soluble. However, as the length of the non-polar hydrocarbon chain increases the solubility in water decreases. For example, ethanol and water mix in all proportions, but hexan-1-ol is only slightly soluble in water. Compounds with non-polar functional groups, such as alkanes, and alkenes, do not dissolve in water but are soluble in other non-polar solvents. Propan-1-ol is a good solvent because it contains both polar and non-polar groups and can to some extent dissolve both polar and non-polar substances.

ACIDITY AND BASICITY

Most organic compounds do not show acidic or basic properties, as they do not readily either donate or accept protons. However, there are two significant exceptions.

Carboxylic acids are weak acids, e.g.

$$CH_3COOH(aq) \rightleftharpoons CH_3COO^-(aq) + H^+(aq)$$

This is because the negative charge on the acid anion can be spread out over three atoms and does not so readily attract H^+ ions to return to the undissociated acid.

Amines are weak bases, e.g.

$$C_2H_5NH_2(aq) + H_2O(l) \rightleftharpoons C_2H_5NH_3^+(aq) + OH^-(aq)$$

This is easy to understand if amines are thought of as simply substituted compounds of ammonia. They still contain the non-bonding pair of electrons present in ammonia and can readily accept a proton. The reaction between amines and dilute hydrochloric acid can be used to make the higher amines soluble in water, since the substituted ammonium salt that is formed is ionic and very soluble.

Alkanes and structural isomers

STRUCTURES OF HYDROCARBONS

When drawing organic structures all the hydrogen atoms must be shown. However, unless specifically asked, Lewis structures showing all the valence electrons are not necessary. The bonding must be clearly indicated. Structures may be shown using lines as bonds or in their shortened form, e.g. $CH_3CH_2CH_2CH_2CH_3$ or $CH_3-(CH_2)_3-CH_3$ but the molecular formula C_5H_{12} will not suffice.

Isomers of alkanes

Each carbon atom contains four single bonds. There is only one possible structure for each of methane, ethane, and propane however two structures of butane are possible.

butane

2-methylpropane

These are examples of structural isomers. **Structural isomers** have the same molecular formula but a different structural formula. They normally have similar chemical properties but their physical properties may be slightly different. There are three structural isomers of pentane.

pentane
(b. pt 36.3 °C)

2-methylbutane
(b. pt 27.9 °C)

2,2-dimethylpropane
(b. pt 9.5 °C)

Structures of alkenes

Ethene and propene only have one possible structure each but butene has three structural isomers.

ethene

propene

but-1-ene

but-2-ene

2-methylpropene

FUNCTIONAL GROUP ISOMERS

Structural isomers can also exist within other homologous series, for example propan-1-ol and propan-2-ol.

propan-1-ol

propan-2-ol

A particular type of structural isomerism is functional group isomerism. **Functional group isomers** have the same molecular formula but contain a different functional group.

carboxylic acid

aldehyde

ester

ketone

molecular formula
$C_2H_4O_2$

molecular formula
C_3H_6O

Both the chemical and physical properties of functional group isomers are likely to be very different.

CHEMICAL PROPERTIES OF ALKANES

Because of the relatively strong C–C and C–H bonds alkanes tend to be quite unreactive. The main reaction of alkanes is combustion. Alkanes are hydrocarbons – compounds that contain carbon and hydrogen only. All hydrocarbons burn in a plentiful supply of oxygen to give carbon dioxide and water. The general equation for the combustion of any hydrocarbon is:

$$C_xH_y + (x+\tfrac{y}{4})O_2 \rightarrow xCO_2 + \tfrac{y}{2}H_2O$$

Although the C–C and C–H bonds are strong the C=O and O–H bonds in the products are even stronger, so the reaction is very exothermic and much use is made of the alkanes as fuels,

e.g methane (natural gas)

$$CH_4(g) + 2O_2(g) \rightarrow CO_2(g) + 2H_2O(l) \quad \Delta H^\ominus = -890.4 \text{ kJ mol}^{-1}$$

octane (gasoline or petrol)

$$C_8H_{18}(l) + 12\tfrac{1}{2}O_2(g) \rightarrow 8CO_2(g) + 9H_2O(l) \quad \Delta H^\ominus = -5512 \text{ kJ mol}^{-1}$$

The use of hydrocarbons as fuels causes considerable environmental problems. The carbon dioxide emitted absorbs infrared radiation reflected by the Earth's surface and so contributes to global warming. The temperature inside the internal combustion engine is high enough for nitrogen and oxygen to combine, which leads to the formation of oxides of nitrogen and ultimately acid rain. If incomplete combustion (oxidation) of the hydrocarbon occurs, as it does in the internal combustion engine, then carbon monoxide and carbon itself are produced. Carbon monoxide is poisonous as it prevents the blood from carrying oxygen. The carbon can form particulates, which may interfere with the respiratory system.

Alkenes

ADDITION REACTIONS

The bond enthalpy of the C=C double bond in alkenes has a value of 612 kJ mol^{-1}. This is less than twice the average value of 348 kJ mol^{-1} for the C–C single bond and accounts for the relative reactivity of alkenes compared to alkanes. The most important reactions of alkenes are addition reactions. Reactive molecules are able to add across the double bond. The double bond is said to be **unsaturated** and the product, in which each carbon atom is bonded by four single bonds, is said to be **saturated**.

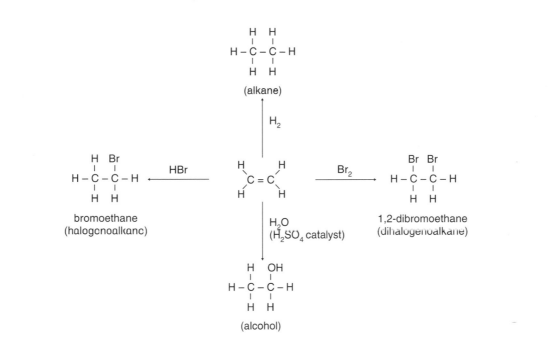

Addition reactions include the addition of hydrogen, bromine, hydrogen halides, and water.

USES OF ADDITION REACTIONS

1. Bromination
Pure bromine is a red liquid but it has a distinctive yellow/orange colour in solution. When a solution of bromine is added to an alkene the product is colourless. This decolorization of bromine solution provides a useful test to indicate the presence of an alkene group.

2. Hydration
Ethene is an important product formed during the cracking of oil. Although ethanol can be made from the fermentation of starch and sugars, much industrial ethanol is formed from the addition of water to ethene.

3. Hydrogenation
The addition of hydrogen to unsaturated vegetable oils is used industrially to make margarine. Hydrogenation reduces the number of double bonds in the polyunsaturated vegetable oils present in the margarine, which causes it to become a solid at room temperature.

ADDITION POLYMERIZATION

Under certain conditions ethene can also undergo addition reactions with itself to form a long chain polymer containing many thousands (typically 40 000 to 800 000) of carbon atoms.

These addition reactions can be extended to other substituted alkenes to give a wide variety of different addition polymers.

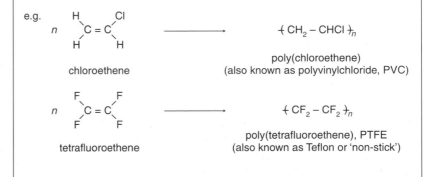

Alcohols

CONDENSATION REACTIONS

Alcohols can react with carboxylic acids to form esters. In this process the alcohol molecule **condenses** with the carboxylic acid molecule to eliminate water,

e.g.

ethanoic acid   ethanol   ethyl ethanoate

Most esters have a distinctive, pleasant, fruity smell and are used both as natural and artificial flavouring agents in food. For example, ethyl methanoate $HCOOCH_2CH_3$ is added to chocolate to give it the characteristic flavour of 'rum truffle'. Esters are also used as solvents in perfumes and as plasticizers (substances used to modify the properties of polymers by making them more flexible).

OXIDATION OF ETHANOL

Ethanol can be readily oxidized by warming with an acidified solution of potassium dichromate(VI). During the process the orange dichromate(VI) ion $Cr_2O_7^{2-}$ is reduced from an oxidation state of +6 to the green Cr^{3+} ion. Use is made of this in simple breathalyser tests, where a motorist who is suspected of having exceeded the alcohol limit blows into a bag containing crystals of potassium dichromate(VI).

ethanol ('wine') $\xrightarrow{Cr_2O_7^{2-}/H^+}$ ethanal $\xrightarrow{Cr_2O_7^{2-}/H^+}$ ethanoic acid ('vinegar')

Ethanol is initially oxidized to ethanal. The ethanal is then oxidized further to ethanoic acid.

Unlike ethanol (b. pt 78.5 °C) and ethanoic acid (b. pt 118 °C) ethanal (b. pt 20.8 °C) does not have hydrogen bonding between its molecules, and so has a lower boiling point. To stop the reaction at the aldehyde stage the ethanal can be distilled from the reaction mixture as soon as it is formed. If the complete oxidation to ethanoic acid is required, then the mixture can be heated under reflux so that none of the ethanal can escape.

OXIDATION OF ALCOHOLS

Ethanol is a primary alcohol, that is the carbon atom bonded to the –OH group is bonded to two hydrogen atoms and one alkyl group. The oxidation reactions of alcohols can be used to distinguish between primary, secondary, and tertiary alcohols.

All **primary alcohols** are oxidized by acidified potassium dichromate(VI), first to aldehydes then to carboxylic acids.

primary alcohol $\xrightarrow{Cr_2O_7^{2-}/H^+}$ aldehyde $\xrightarrow{Cr_2O_7^{2-}/H^+}$ carboxylic acid

Secondary alcohol are oxidized to ketones, which cannot undergo further oxidation.

Tertiary alcohols cannot be oxidized by acidified dichromate(VI) ions as they have

secondary alcohol $\xrightarrow{Cr_2O_7^{2-}/H^+}$ ketone   tertiary alcohol

no hydrogen atoms attached directly to the carbon atom containing the –OH group. It is not true to say that tertiary alcohols can never be oxidized, as they burn readily, but when this happens the carbon chain is destroyed.

DEHYDRATION OF ALCOHOLS

Alcohols can be made by the addition of water to alkenes. The reverse process in which water is eliminated can also occur. This process is known as **dehydration** and takes place when alcohols are refluxed with concentrated sulfuric acid. Sulfuric acid is a good dehydrating agent, but it is also an oxidizing agent and can react with the product, so phosphoric acid H_3PO_4 is frequently used instead.

alcohol $\xrightarrow[\text{reflux}]{\text{conc. } H_2SO_4 \text{ or } H_3PO_4}$ alkene $+$ H_2O

Condensation polymerization and amino acids

CONDENSATION POLYMERIZATION

Condensation involves the reaction between two molecules to eliminate a smaller molecule, such as water or hydrogen chloride. If each of the reacting molecules contain *two* functional groups that can undergo condensation, then the condensation can continue to form a polymer.

An example of a polyester is polyethene terephthalate (known as Terylene in the UK and as Dacron in the USA) used for textiles, which is made from benzene-1,4-dicarboxylic acid and ethane-1,2-diol.

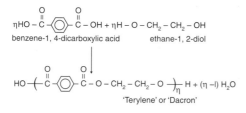

benzene-1, 4-dicarboxylic acid   ethane-1, 2-diol

'Terylene' or 'Dacron'

Amines can also condense with carboxylic acids to form an amide link (also known as a peptide bond). One of the best known examples of a polyamide is nylon.

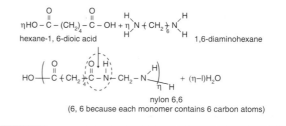

hexane-1, 6-dioic acid   1,6-diaminohexane

nylon 6,6
(6, 6 because each monomer contains 6 carbon atoms)

AMINO ACIDS AND POLYPEPTIDES

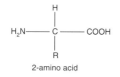

2-amino acid

Amino acids contain both an amine functional group and a carboxylic acid functional group. When they are both attached to the same carbon atom they are known as 2-amino acids (or α-amino acids).

There are about twenty 2-amino acids that occur naturally. They are the basic 'building blocks' of proteins in the body, which consist of long chains of polypeptides formed by condensation reactions between the amino acids.

When two amino acids condense they form a **dipeptide and water**. A **tripeptide** is made up from three amino acids. If the three amino acids are all different then there are six possible ways in which they can combine.

Possible tripeptides formed from alanine $H_2N–CHCH_3–COOH$, glycine $H_2N–CH_2–COOH$, and cysteine $H_2N–CH(CH_2SH)–COOH$:

$H_2N – Ala – Gly – Cys – COOH$

$H_2N – Ala – Cys – Gly – COOH$

$HOOC – Ala – Gly – Cys – NH_2$

$HOOC – Ala – Cys – Gly – NH_2$

$H_2N – Gly – Ala – Cys – COOH$

$HOOC – Gly – Ala – Cys – NH_2$

OPTICAL ISOMERISM

All 2-amino acids (except glycine $H_2N–CH_2COOH$) show one property in common with all compounds that contain an **asymmetric** or **chiral** carbon atom – that is, one that contains four *different* groups attached to it. They can exist as mirror images of each other and the two different isomers are optically active with plane-polarized light.

Normal light consists of electromagnetic radiation which vibrates in all planes. When it is passed through a polarizing filter the waves only vibrate in one plane and the light is said to be plane polarized. The two different mirror images, known as enantiomers, both rotate the plane of plane-polarized light. One of the enantiomers rotates it to the left and the other rotates it by the same amount to the right. Apart from their behaviour towards plane-polarized light enantiomers have identical physical properties. Their chemical properties are identical too, except when they interact with other optically active substances. This is often the case in the body where the different enantiomers can have completely different physiological effects. For example, one of the enantiomers of the amino acid asparagine $H_2N–CH(CH_2CONH_2)–COOH$ tastes bitter whereas the other enantiomer tastes sweet.

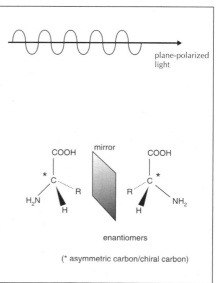

plane-polarized light

enantiomers

(* asymmetric carbon/chiral carbon)

ⓗⓛ Substitution reactions

SUBSTITUTION REACTIONS OF ALKANES

Alkanes can react with chlorine (or other halogens) in the presence of ultraviolet light to form hydrogen chloride and a substituted alkane, e.g.

When chemical bonds break they may break **heterolytically** or **homolytically**. In heterolytic fission both of the shared electrons go to one of the atoms resulting in a negative and a positive ion. In homolytic fission each of the two atoms forming the bond retains one of the shared electrons resulting in the formation of two **free radicals**. Free radicals are reactive species, which contain an unpaired electron.

The bond between two halogen atoms is weaker than the C–H or C–C bond and can break homolytically in the presence of ultraviolet light. When this happens the halogen radical that is formed has sufficient energy to react with alkanes, and substitute the halogen atom for a hydrogen atom to form the halogenoalkane.

SUBSTITUTION REACTIONS OF HALOGENOALKANES

Because of the greater electronegativity of the halogen atom compared with the carbon atom halogenoalkanes have a polar bond. Reagents that have a non-bonding pair of electrons are attracted to the carbon atom in halogenoalkanes and a substitution reaction occurs. Such reagents are called **nucleophiles**

A double-headed curly arrow represents the movement of a pair of electrons. It shows where they come from and where they move to.

Typical nucleophiles are CN^-, OH^-, and NH_3. Nucleophilic substitution reactions are useful in organic synthesis. For example, when substitution occurs with cyanide ions the resulting nitrile formed can be hydrolysed by dilute acid to give a carboxylic acid. In this reaction the number of carbon atoms in the carbon chain has been increased by one.

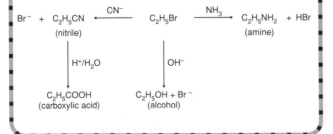

SUBSTITUTION REACTIONS OF BENZENE

The Kekulé structure (cyclohexa-1,3,5-triene) of benzene consists of three double bonds, which suggests that benzene will readily undergo addition reactions. In fact it will only undergo addition reactions with difficulty, and more commonly undergoes substitution reactions, e.g.

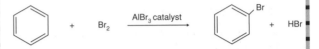

There is both physical and chemical evidence to support the fact that benzene does not contain three separate double bonds but exists as a delocalized structure.

1. The C–C bond lengths are all the same and have a value of 0.139 nm, which lies between the values for C–C (0.154 nm) and C=C (0.134 nm).
2. The enthalpy of hydrogenation of benzene (–210 kJ mol^{-1}) is not equal to 3 × the enthalpy of hydrogenation of cyclohexene (3 × –120 kJ mol^{-1}). The difference of 150 kJ mol^{-1} is the extra energy associated with the delocalization energy.
3. The high resolution ^{1}H NMR spectrum of benzene shows a single peak. This would not be the case if benzene had three alternate double bonds.
4. Only one isomer exists for 1,2- disubstituted benzene compounds. If there were alternate double bonds two isomers would exist.

The actual bonding in benzene is best described by the delocalization of the π electrons as described in Topic 14. For this reason benzene is often represented by a hexagon with a circle in the middle of it.

🆗 Nucleophilic substitution

MECHANISM OF NUCLEOPHILIC SUBSTITUTION

Primary halogenoalkanes (one alkyl group attached to the carbon atom bonded to the halogen)

e.g. the reaction between bromoethane and warm dilute sodium hydroxide solution.

$$C_2H_5Br + OH^- \rightarrow C_2H_5OH + Br^-$$

The experimentally determined rate expression is:

rate = $k[C_2H_5Br][OH^-]$

The proposed mechanism involves the formation of a transition state which involves both of the reactants.

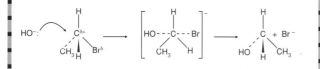

Because the molecularity of this single-step mechanism is two it is known as an S_N2 mechanism (bimolecular nucleophilic substitution).

Tertiary halogenoalkanes (three alkyl groups attached to the carbon atom bonded to the halogen)

e.g. the reaction between 2-bromo-2-methylpropane and warm dilute sodium hydroxide solution.

$$C(CH_3)_3Br + OH^- \longrightarrow C(CH_3)_3OH + Br^-$$

The experimentally determined rate expression for this reaction is: rate = $k[C(CH_3)_3Br]$

A two-step mechanism is proposed that is consistent with this rate expression.

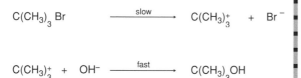

In this reaction it is the first step that is the rate determining step. The molecularity of this step is one and the mechanism is known as S_N1 (unimolecular nucleophilic substitution).

The mechanism for the hydrolysis of secondary halogenoalkanes (e.g 2-bromopropane $CH_3CHBrCH_3$) is more complicated as they can proceed by either S_N1 or S_N2 pathways or a combination of both.

FACTORS AFFECTING THE RATE OF NUCLEOPHILIC SUBSTITUTION

1. The nature of the halogen atom

   For both S_N1 and S_N2 reactions the iodoalkanes react faster than bromoalkanes, which in turn react faster than chloroalkanes. This is due to the relative bond energies, as the C–I bond is much weaker than the C–Cl bond and therefore breaks more readily.

   Bond enthalpy / kJ mol^{-1}

C–I	238
C–Br	276
C–Cl	338

2. Tertiary halogenoalkanes react faster than secondary halogenoalkanes, which in turn react faster than primary halogenoalkanes. The S_N1 route, which involves the formation of an intermediate carbocation, is faster than the S_N2 route, which involves a transition state with a relatively high activation energy.

(HL) Determination of structure – mass spectrometry

The classic way to determine the structure of an organic compound was to determine both its empirical formula and relative molar mass experimentally, then deduce the nature of the functional groups from its chemical reactivity. It can still be useful to determine the empirical formula of a substance by burning a known mass of it in excess oxygen, and finding the mass of the oxides formed from the different elements. However, modern well-equipped laboratories now employ a variety of instrumental techniques, which if used in combination are able to unambiguously determine the exact structural formula.

MASS SPECTROMETRY

The principles of mass spectrometry have already been explained in Topic 12. However, in addition to giving the precise molar mass of the substance mass spectrometry gives considerable information about the actual structure of a compound from the fragmentation patterns.

When a sample is introduced into the machine the vaporized sample becomes ionized to form the molecular ion $M^+(g)$. Inside the mass spectrometer some of the molecular ions break down to give fragments, which are also deflected by the external magnetic field and which then show up as peaks on the detector. By looking at the difference in mass from the parent peak it is often possible to identify particular fragments,

e.g. $(M_r - 15)^+$ = loss of CH_3

$(M_r - 29)^+$ = loss of C_2H_5 or CHO

$(M_r - 31)^+$ = loss of CH_3O

$(M_r - 45)^+$ = loss of $COOH$

This can provide a useful way of distinguishing between structural isomers.

The mass spectra of propan-1-ol and propan-2-ol both show a peak at 60 due to the molecular ion $C_3H_8O^+$. However, the mass spectrum of propan-1-ol shows a strong peak at 31 due to the loss of $-C_2H_5$, which is absent in the mass spectrum of propan-2-ol. There is a strong peak at 45 in the spectrum of propan-2-ol as it contains two different methyl groups, which can fragment.

Mass spectrum of propan-1-ol $CH_3 - CH_2 - CH_2 - OH$

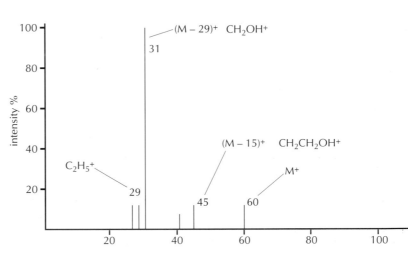

Mass spectrum of propan-2-ol

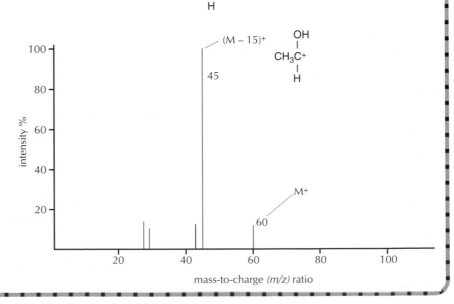

Determination of structure – IR and ^{1}H NMR spectroscopy

INFRARED SPECTROSCOPY

When molecules absorb energy in the infrared region of the spectrum they vibrate (i.e. the bonds stretch and bend). The precise value of the energy they absorb depends on the particular bond, and to a lesser extent on the other groups attached to the two atoms forming the bond. When infrared radiation is passed through a sample the spectrum shows characteristic absorptions, which confirm that a particular bond is present in the molecule. Some absorptions, e.g. those due to C–H, are not particularly useful as they are shown by most organic compounds. However, others give a very clear indication of a particular group.

In addition to these particular absorptions infrared spectra also possess a 'fingerprint' region. This is a characteristic pattern between about 1400 and 400 cm^{-1}, which is specific to a particular compound. It is often possible to identify an unknown sample by comparing the fingerprint region with a library of spectra of known samples.

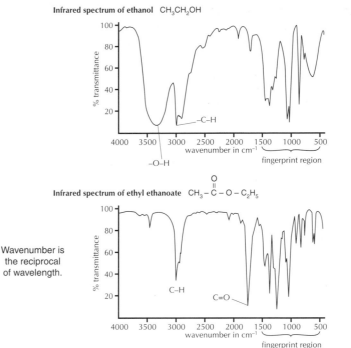

Infrared spectrum of ethanol CH_3CH_2OH

A simplified correlation chart

Bond	Wavenumber / cm^{-1}
C–O	1000–1300
C=C	1610–1680
C=O	1680–1750
C≡C	2070–2250
O–H (in carboxylic acids)	2500–3300
C–H	2840–3095
O–H (in alcohols)	3230–3500

Wavenumber is the reciprocal of wavelength.

Infrared spectrum of ethyl ethanoate $CH_3 - \overset{\displaystyle O}{\overset{\|}{C}} - O - C_2H_5$

^{1}H NMR SPECTROSCOPY

Whereas infrared spectroscopy gives information about the types of bonds in a molecule, ^{1}H NMR spectroscopy provides information on the chemical environment of all the hydrogen atoms in the molecule. The nuclei of hydrogen atoms possess spin and can exist in two possible states of equal energy. If a strong magnetic field is applied the spin states may align themselves either with the magnetic field, or against it, and there is a small energy difference between them. The nuclei can absorb energy when transferring from the lower to the higher spin state. This energy is very small and occurs in the radio region of the spectrum. The precise energy difference depends on the chemical environment of the hydrogen atoms.

The position in the ^{1}H NMR spectrum where the absorption occurs for each hydrogen atom in the molecule is known as the chemical shift, and is measured in parts per million (ppm). The area under each peak corresponds to the number of hydrogen atoms in that particular environment.

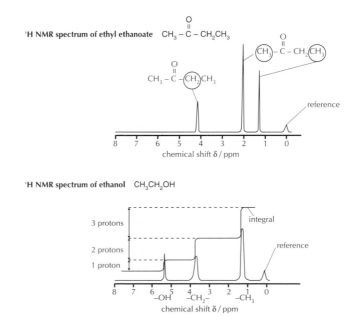

^{1}H NMR spectrum of ethyl ethanoate $CH_3 - \overset{\displaystyle O}{\overset{\|}{C}} - CH_2CH_3$

A simplified correlation chart

Type of proton	Chemical shift / ppm
R–CH$_3$	0.9
R$_2$–CH$_2$	1.3
R$_3$CH	2.0
CH$_3$COR	2.0
R–C(=O)–CH$_3$	2.1
R–OH	4.5
⬡–H	7.3
R–C(=O)–H	9.7

^{1}H NMR spectrum of ethanol CH_3CH_2OH

The additional trace integrates the area under each peak. The height of each section is proportional to the number of hydrogen atoms in each chemical environment.

IB QUESTIONS – ORGANIC CHEMISTRY

1. Which of the following is/are true about alkanes?

   I. They form an homologous series with the general formula C_nH_{2n-2}

   II. They all have identical physical properties

   III. They all have similar chemical properties

   A. I,II and III   **C.** I and III only

   B. II and III only   **D.** III only

2. Which of the following two compounds both belong to the same homologous series?

   A. CH_3COOH and $HCOOCH_3$

   B. CH_3OH and C_2H_5OH

   C. C_2H_4 and C_2H_6

   D. C_2H_5Cl and $C_2H_4Cl_2$

3. How many different isomers of C_5H_{12} exist?

   A. 1   **B.** 2   **C.** 3   **D.** 4

4. Give the correct name for:

   $$H_3C-\underset{\underset{\displaystyle CH_3}{\underset{\displaystyle |}{\underset{\displaystyle CH_2}{|}}}}{\overset{\overset{\displaystyle CH_3}{|}}{C}}-CH_3$$

   A. 2-methyl-2-ethylpropane   **C.** 2,2-dimethylbutane

   B. hexane   **D.** 2-methylpentane

5. Which compound can exhibit optical isomerism?

   A. $HOOC-\overset{\overset{\displaystyle H}{|}}{\underset{\underset{\displaystyle H}{|}}{C}}-NH_2$ **C.** $HOO-\overset{\overset{\displaystyle H}{|}}{\underset{\underset{\displaystyle CH_3}{|}}{C}}-OH$

   B. $HOOC-\overset{\overset{\displaystyle H}{|}}{\underset{\underset{\displaystyle CH_3}{|}}{C}}-COOH$ **D.** $H_2N-\overset{\overset{\displaystyle H}{|}}{\underset{\underset{\displaystyle H}{|}}{C}}-NH_2$

6. Which compound is an ester?

   A. CH_3COOH **C.** C_2H_5CHO

   B. $CH_3OC_2H_5$ **D.** $HCOOCH_3$

7. When ethanol is partially oxidized by an acidified solution of potassium dichromate(VI), the product that can be obtained by distillation as soon as it is formed is:

   A. ethanal   **C.** ethanoic acid

   B. ethene   **D.** ethane-1,2-diol

8. Identify which two compounds can react together to form a condensation polymer.

   A. C_2H_4 and Br_2

   B. C_2H_5OH and CH_3COOH

   C. $H_2N(CH_2)_6NH_2$ and CH_3COOH

   D. $HOOC(CH_2)_4COOH$ and $HOCH_2CH_2OH$

9. Which cannot undergo an addition reaction with ethene?

   A. oxygen   **C.** water

   B. hydrogen   **D.** hydrogen bromide

10. The compound $CH_3CH_2CH_2\overset{\overset{\displaystyle O}{||}}{C}-O-\overset{\overset{\displaystyle CH_3}{|}}{C}H-CH_3$ can be synthesized most readily from:

    A. propanoic acid and butan-2-ol

    B. butanoic acid and propan-1-ol

    C. butanoic acid and propan-2-ol

    D. propanoic acid and propan-2-ol

HL

---

11. Which compound will give three distinct peaks in its ^1H NMR spectrum in addition to the peak due to the TMS reference sample?

    A. $C_2H_5OCH_3$ **C.** $CH_3CH_2CH_2OH$

    B. CH_3COCH_3 **D.** $CH_3CH_2CH_3$

12. The compound that will not show a fragment with a mass of 45 in its mass spectrum is:

    A. $H_3C-\overset{\overset{\displaystyle O}{||}}{C}-O-CH_3$ **C.** $HO-CH_2-CH_2-\overset{\overset{\displaystyle O}{||}}{C}-H$

    B. $CH_3-CH_2-\overset{\overset{\displaystyle O}{||}}{C}-OH$ **D.** $H_3C-O-CH_2-\overset{\overset{\displaystyle O}{||}}{C}-H$

13. Which statement(s) is/are true about the reactions of halogenoalkanes with warm dilute sodium hydroxide solution

    I. CH_3I reacts faster than CH_3F

    II. $(CH_3)_3CBr$ reacts faster than CH_3Br

    III. $(CH_3)_3CBr$ and $(CH_3)_3CCl$ both react by S_N1 mechanisms.

    A. I and II only   **C.** II and III only

    B. I, II and III   **D.** I only

14. A ketone containing four carbon atoms will be formed from the reaction of an acidified solution of potassium dichromate(IV) solution with:

    A. $CH_3CH_2CH_2CH_2OH$ **C.** $CH_3CH_2CH_2CHO$

    B. $(CH_3)_3COH$ **D.** $CH_3CH_2CH(OH)CH_3$

15. Which statement about benzene is false?

    A. The bonds between all the carbon atoms are equal.

    B. It undergoes addition reactions rather than substitution reactions

    C. The delocalisation of π electrons contributes to the stability of the ring

    D. It can be hydrogenated to form cyclohexane.

16. Identify which statement(s) about the reaction of butane with chlorine is/are true.

    I. The reaction involves the heterolytic fission of the Cl–Cl bond in chlorine molecules.

    II. The reaction is described as an addition reaction.

    III. One of the products is hydrogen chloride.

    A. I and II only   **C.** I, II and III

    B. I and III only   **D.** III only

Determination of structure – mass spectrometry

The classic way to determine the structure of an organic compound was to determine both its empirical formula and relative molar mass experimentally, then deduce the nature of the functional groups from its chemical reactivity. It can still be useful to determine the empirical formula of a substance by burning a known mass of it in excess oxygen, and finding the mass of the oxides formed from the different elements. However, modern well-equipped laboratories now employ a variety of instrumental techniques, which if used in combination are able to unambiguously determine the exact structural formula.

MASS SPECTROMETRY

The principles of mass spectrometry have already been explained in Topic 12. However, in addition to giving the precise molar mass of the substance mass spectrometry gives considerable information about the actual structure of a compound from the fragmentation patterns.

When a sample is introduced into the machine the vaporized sample becomes ionized to form the molecular ion $M^+(g)$. Inside the mass spectrometer some of the molecular ions break down to give fragments, which are also deflected by the external magnetic field and which then show up as peaks on the detector. By looking at the difference in mass from the parent peak it is often possible to identify particular fragments,

e.g. $(M_r – 15)^+ =$ loss of CH_3

$(M_r – 29)^+ =$ loss of C_2H_5 or CHO **Mass spectrum of propan-1-ol** $CH_3 – CH_2 – CH_2–OH$

$(M_r – 31)^+ =$ loss of CH_3O

$(M_r – 45)^+ =$ loss of $COOH$

This can provide a useful way of distinguishing between structural isomers.

The mass spectra of propan-1-ol and propan-2-ol both show a peak at 60 due to the molecular ion $C_3H_8O^+$. However, the mass spectrum of propan-1-ol shows a strong peak at 31 due to the loss of $–C_2H_5$, which is absent in the mass spectrum of propan-2-ol. There is a strong peak at 45 in the spectrum of propan-2-ol as it contains two different methyl groups, which can fragment.

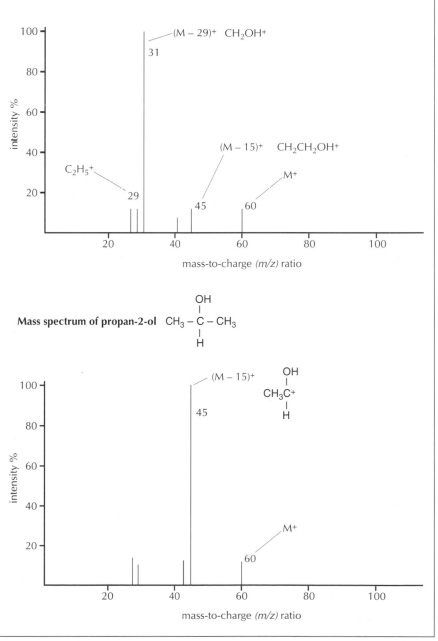

Mass spectrum of propan-2-ol

Determination of structure – IR and ^{1}H NMR

INFRARED SPECTROSCOPY

When molecules absorb energy in the infrared region of the spectrum they vibrate (i.e. the bonds stretch and bend). The precise value of the energy they absorb depends on the particular bond, and to a lesser extent on the other groups attached to the two atoms forming the bond. When infrared radiation is passed through a sample the spectrum shows characteristic absorptions, which confirm that a particular bond is present in the molecule. Some absorptions, e.g. those due to C–H, are not particularly useful as they are shown by most organic compounds. However, others give a very clear indication of a particular group.

In addition to these particular absorptions infrared spectra also possess a 'fingerprint' region. This is a characteristic pattern between about 1400–400 cm^{-1}, which is specific to a particular compound. It is often possible to identify an unknown sample by comparing the fingerprint region with a library of spectra of known samples.

A simplified correlation chart

Bond	Wavenumber / cm^{-1}
C–O	1000–1300
C=C	1610–1680
C=O	1680–1750
C≡C	2070–2250
O–H (in carboxylic acids)	2500–3300
C–H	2840–3095
O–H (in alcohols)	3230–3500

Wavenumber is the reciprocal of wavelength.

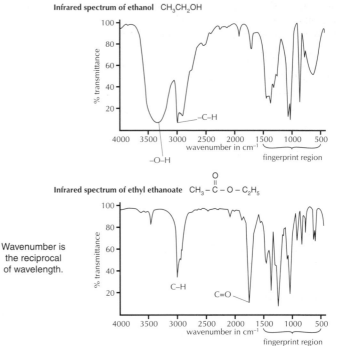

^{1}H NMR SPECTROSCOPY

Whereas infrared spectroscopy gives information about the types of bonds in a molecule, ^{1}H NMR spectroscopy provides information on the chemical environment of all the hydrogen atoms in the molecule. The nuclei of hydrogen atoms possess spin and can exist in two possible states of equal energy. If a strong magnetic field is applied the spin states may align themselves either with the magnetic field, or against it, and there is a small energy difference between them. The nuclei can absorb energy when transferring from the lower to the higher spin state. This energy is very small and occurs in the radio region of the spectrum. The precise energy difference depends on the chemical environment of the hydrogen atoms.

The position in the ^{1}H NMR spectrum where the absorption occurs for each hydrogen atom in the molecule is known as the chemical shift, and is measured in parts per million (ppm). The area under each peak corresponds to the number of hydrogen atoms in that particular environment.

A simplified correlation chart

Type of proton	Chemical shift / ppm
R–CH$_3$	0.9
R$_2$–CH$_2$	1.3
R$_3$CH	2.0
CH$_3$COR	2.0
R–C–CH$_3$	2.1
R–OH	4.5
⬡–H	7.3
R–C–H	9.7

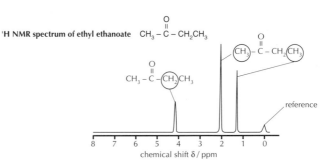

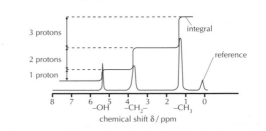

The additional trace integrates the area under each peak. The height of each section is proportional to the number of hydrogen atoms in each chemical environment.

Order of reaction and half-life

RATE EXPRESSIONS

The rate of reaction between two reactants, A and B, can be followed experimentally. The rate will be found to be proportional to the concentration of A raised to some power and also to the concentration of B raised to a power. If square brackets are used to denote concentration this can be written as rate $\propto [A]^x$ and rate $\propto [B]^y$. They can be combined to give the rate expression:

rate $= k[A]^x[B]^y$

where k is the constant of proportionality and is known as the **rate constant**.

x is known as the **order of the reaction** with respect to A.

y is known as the order of the reaction with respect to B.

The overall order of the reaction $= x + y$.

Note: the order of the reaction and the rate expression can only be determined experimentally. They cannot be deduced from the balanced equation for the reaction.

UNITS OF RATE CONSTANT

The units of the rate constant depend on the overall order of the reaction.

First order: rate $= k[A]$

$$k = \frac{\text{rate}}{[A]} = \frac{\text{mol dm}^{-3}\,\text{s}^{-1}}{\text{mol dm}^{-3}} = \text{s}^{-1}$$

Second order: rate $= k[A]^2$ or $k = [A][B]$

$$k = \frac{\text{rate}}{[A]^2} = \frac{\text{mol dm}^{-3}\,\text{s}^{-1}}{(\text{mol dm}^{-3})^2} = \text{dm}^3\,\text{mol}^{-1}\,\text{s}^{-1}$$

Third order: rate $= k[A]^2[B]$ or rate $= k[A][B]^2$

$$k = \frac{\text{rate}}{[A]^2[B]} = \frac{\text{mol dm}^{-3}\,\text{s}^{-1}}{(\text{mol dm}^{-3})^3} = \text{dm}^6\,\text{mol}^{-2}\,\text{s}^{-1}$$

GRAPHICAL REPRESENTATIONS OF REACTIONS

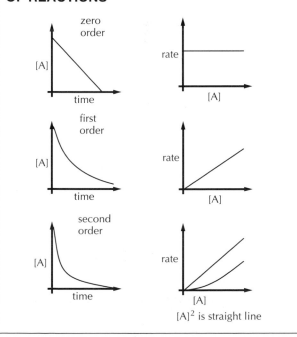

$[A]^2$ is straight line

DERIVING A RATE EXPRESSION BY INSPECTION OF DATA

Experimental data obtained from the reaction between hydrogen and nitrogen monoxide at 1073 K:

$$2H_2(g) + 2NO(g) \rightarrow 2H_2O(g) + N_2(g)$$

Experiment	Initial concentration of $H_2(g)$ / mol dm^{-3}	Initial concentration of NO(g) / mol dm^{-3}	Initial rate of formation of $N_2(g)$ / mol dm^{-3} s^{-1}
1	1×10^{-3}	6×10^{-3}	3×10^{-3}
2	2×10^{-3}	6×10^{-3}	6×10^{-3}
3	6×10^{-3}	1×10^{-3}	0.5×10^{-3}
4	6×10^{-3}	2×10^{-3}	2.0×10^{-3}

From experiments 1 and 2 doubling $[H_2]$ doubles the rate so rate $\propto [H_2]$.

From experiments 3 and 4 doubling [NO] quadruples the rate so rate $\propto [NO]^2$.

Rate expression given by rate $= k[H_2][NO]^2$.

The rate is first order with respect to hydrogen, second order with respect to nitrogen monoxide, and third order overall. The value of k can be found by substituting the values from any one fo the four experiments:

$$k = \frac{\text{rate}}{[H_2][NO]^2} = 8.33 \times 10^4 \text{ dm}^3\,\text{mol}^{-1}\,\text{s}^{-1}$$

HALF-LIFE $t_{\frac{1}{2}}$

For a first order reaction the rate of change of concentration of A is equal to $k[A]$. This can be expressed as $\frac{d[A]}{dt} = k[A]$.

If this expression is integrated then $kt = \ln [A]_o - \ln [A]$ where $[A]_o$ is the initial concentration and $[A]$ is the concentration at time t. This expression is known as the integrated form of the rate equation.

The half-life is defined as the time taken for the concentration of a reactant to fall to half of its initial value.

At $t_{\frac{1}{2}}[A] = \frac{1}{2}[A]_o$ the integrated rate expression then becomes

$kt_{\frac{1}{2}} = \ln [A]_o - \ln \frac{1}{2}[A]_o = \ln 2$ since $\ln 2 = 0.693$ this
   simplifies to $t_{\frac{1}{2}} = \frac{0.693}{k}$

From this expression it can be seen that the half-life of a first order reaction is independent of the original concentration of A, i.e. first order reactions have a constant half-life.

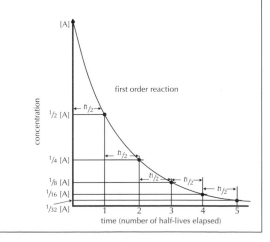

Reaction mechanisms and the reactions and structure of benzene

REACTION MECHANISMS

When the separate steps in a chemical reaction are analysed there are essentially only two types of processes. Either a single species can break down into two or more products by what is known as a **unimolecular process**, or two species can collide and interact by a **bimolecular process**.

In a bimolecular process the species collide with the necessary activation energy to give initially an **activated complex**. An activated complex is not a chemical substance which can be isolated, but consists of an association of the reacting particles in which bonds are in the process of being broken and formed. An activated complex breaks down to form either the products or reverts back to the original reactants.

The number of species taking part in any specified step in the reaction is known as the **molecularity**. In most cases the molecularity refers to the slowest step, that is the rate determining step.

In the reaction on the previous page, between nitrogen monoxide and hydrogen, the stoichiometry of the reaction involves two molecules of hydrogen and two molecules of nitrogen monoxide. Any proposed mechanism must be consistent with the rate expression.

For third order reactions, such as this, the rate determining step will never be the first step. The proposed mechanism is:

$$NO(g) + NO(g) \xrightarrow{fast} N_2O_2(g)$$

$$N_2O_2(g) + H_2(g) \xrightarrow{slow} N_2O(g) + H_2O(g) \text{ rate determining step}$$

$$N_2O(g) + H_2(g) \xrightarrow{fast} N_2(g) + H_2O(g)$$

Overall $\quad 2NO(g) + 2H_2(g) \rightarrow N_2(g) + 2H_2O(g)$

If the first step was the slowest step the the rate expression would be rate = $k[NO]^2$ and the rate would be zero order with respect to hydrogen. The rate for the second step depends on $[H_2]$ and $[N_2O_2]$. However, the concentration of N_2O_2 depends on the first step. So the rate expression for the second step becomes rate = $k[H_2][NO]^2$, which is consistent with the experimentally determined rate expression. The molecularity of the reaction is two, as two reacting species are involved in the rate determining step.

THE REACTIONS AND STRUCTURE OF BENZENE

The Kekulé structure (cyclohexa-1,3,5-triene) of benzene consists of three double bonds, which suggests that benzene will readily undergo addition reactions. In fact it will only undergo addition reactions with difficulty, and more commonly undergoes substitution reactions.

e.g.

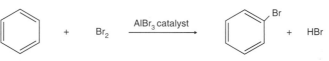

There is both physical and chemical evidence to support the theory that benzene does not contain three separate double bonds but exists as a resonance hybid between the two structures known as the resonance forms. The two forms can be shown as

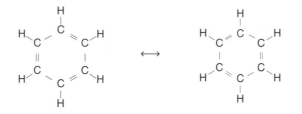

1. The C–C bond lengths are all the same and have a value of 0.139 nm, which lies between the values for C–C (0.154 nm) and C=C (0.134 nm).

2. The enthalpy of hydrogenation of benzene (–210 kJ mol⁻¹) is not equal to 3 × the enthalpy of hydrogenation of cyclohexene (–120 kJ mol⁻¹). The difference of 150 kJ mol⁻¹ is the extra energy associated with the formation of the resonance hybrid.

3. The ^1H NMR spectrum of benzene shows a single peak. This would not be the case if benzene had three alternate double bonds.

4. Only one isomer exists for 1,2-disubstituted benzene compounds. If there were alternate double bonds two isomers would exist.

Because the true structure lies between the two resonance forms benzene is often represented by a hexagonal ring with a circle in the middle of it.

Nucleophilic substitution reactions

SUBSTITUTION REACTIONS OF HALOGENOALKANES

Because of the greater electronegativity of the halogen atom compared with the carbon atom halogenoalkanes have a polar bond. Reagents that have a non-bonding pair of electrons are attracted to the carbon atom, in halogenoalkanes and a substitution reaction occurs. Such reagents are called **nucleophiles**.

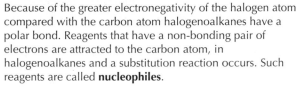

A double-headed curly arrow represents the movement of a pair of electrons. It shows where they come from and where they move to.

Typical nucleophiles are CN^-, OH^-, and NH_3. Nucleophilic substitution reactions are useful in organic synthesis. For example, when substitution occurs with cyanide ions the resulting nitrile formed can be hydrolysed by dilute acid to give a carboxylic acid. In this reaction the number of carbon atoms in the carbon chain has been increased by one.

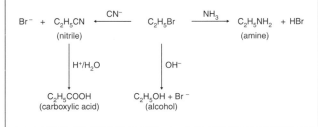

FACTORS AFFECTING THE RATE OF NUCLEOPHILIC SUBSTITUTION

1. The nature of the halogen atom

   For both S_N1 and S_N2 reactions the iodoalkanes react faster than bromoalkanes, which in turn react faster than chloroalkanes. This is due to the relative bond energies, as the C–I bond is much weaker than the C–Cl bond and therefore breaks more readily.

   Bond enthalpy / kJ mol^{-1}

C–I	238
C–Br	276
C–Cl	338

2. Tertiary halogenoalkanes react faster than secondary halogenoalkanes, which in turn react faster than primary halogenoalkanes. The S_N1 route, which involves the formation of an intermediate carbocation, is faster than the S_N2 route, which involves a transition state with a relatively high activation energy.

MECHANISM OF NUCLEOPHILIC SUBSTITUTION

Primary halogenoalkanes (one alkyl group attached to the carbon atom bonded to the halogen)

e.g. the reaction between bromoethane and warm dilute sodium hydroxide solution.

$$C_2H_5Br + OH^- \rightarrow C_2H_5OH + Br^-$$

The experimentally determined rate expression is:
$$\text{rate} = k[C_2H_5Br][OH^-]$$

The proposed mechanism involves the formation of a transition state which involves both of the reactants.

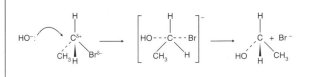

Because the molecularity of this single-step mechanism is two it is known as an S_N2 mechanism (bimolecular nucleophilic substitution).

Tertiary halogenoalkanes (three alkyl groups attached to the carbon atom bonded to the halogen)

e.g. the reaction between 2-bromo-2-methylpropane and warm dilute sodium hydroxide solution.

$$C(CH_3)_3Br + OH^- \longrightarrow C(CH_3)_3OH + Br^-$$

The experimentally determined rate expression for this reaction is: $\text{rate} = k[C(CH_3)_3Br]$

A two-step mechanism is proposed that is consistent with this rate expression.

$$C(CH_3)_3Br \xrightarrow{\text{slow}} C(CH_3)_3^+ + Br^-$$

$$C(CH_3)_3^+ + OH^- \xrightarrow{\text{fast}} C(CH_3)_3OH$$

In this reaction it is the first step that is the rate determining step. The molecularity of this step is one and the mechanism is known as S_N1 (unimolecular nucleophilic substitution).

The mechanism for the hydrolysis of secondary halogenoalkanes (e.g 2-bromopropane $CH_3CHBrCH_3$) is more complicated as they can proceed by either S_N1 or S_N2 pathways or a combination of both.

Calculations involving pH and pOH

THE IONIC PRODUCT OF WATER

Pure water is very slightly ionized:

$$H_2O(l) \rightleftharpoons H^+(aq) + OH^-(aq) \qquad \Delta H^\ominus = +57.3 \text{ kJ mol}^{-1}$$

$$K_c = \frac{[H^+(aq)] \times [OH^-(aq)]}{[H_2O(l)]}$$

Since the equilibrium lies far to the left the concentration of water can be regarded as constant so

$K_w = [H^+(aq)] \times [OH^-(aq)] = 1.00 \times 10^{-14} \text{ mol}^2 \text{ dm}^{-6}$ at 298 K, where K_w is known as the ionic product of water.

The dissociation of water into its ions is an endothermic process, so the value of K_w will increase as the temperature is increased.

Variation of K_w with temperature

For pure water $[H^+(aq)] = [OH^-(aq)]$
$= 1.00 \times 10^{-7}$ mol dm^{-3} at 298 K

From the graph the value for $K_w = 1.00 \times 10^{-13}$ at 334 K (61 °C)

At this temperature $[H^+(aq)] = \sqrt{1.00 \times 10^{-13}}$
$= 3.16 \times 10^{-7}$ mol dm^{-3}

pH, pOH, AND pK_w

The pH of a solution depends only on the hydrogen ion concentration and is independent of the volume of solution.

$$pH = -\log_{10}[H^+]$$

Strong acids

For strong monoprotic acids $[H^+]$ will be equal to the concentration of the acid,

e.g. for 0.100 mol dm^{-3} HCl

$[H^+] = 0.100$ mol dm^{-3}
$pH = -\log_{10} 0.100 = 1.0$

For a strong diprotic acid the hydrogen ion concentration will be equal to twice the acid concentration,

e.g. for 0.025 mol dm^{-3} H$_2$SO$_4$

$[H^+] = 2 \times 0.025 = 0.050$ mol dm^{-3}
$pH = 1.3$

Note: it has been assumed that sulfuric acid is a strong acid. In reality the HSO$_4^-$ ion is only partially dissociated in aqueous solution.

pOH for strong bases

For a strong base the hydrogen ion concentration can be calculated using the ionic product of water,

e.g. for 1.00 × 10^{-3} mol dm^{-3} NaOH

$[OH^-] = 1.00 \times 10^{-3}$ mol dm^{-3}
$[H^+] \times [OH^-] = 1.00 \times 10^{-14}$

$$\Rightarrow [H^+] = \frac{1.00 \times 10^{-14}}{1.00 \times 10^{-3}} = 1.00 \times 10^{-11} \text{ mol dm}^{-3}$$

$pH = -\log_{10} 1.00 \times 10^{-11} = 11.0$

The pH of alkaline solutions can also be calculated by using pOH and the relationship between pOH, pH, and pK_w.

$$[H^+] \times [OH^-] = K_w \quad pOH = -\log_{10}[OH^-] \quad pK_w = -\log_{10} K_w = 14$$
$$pH + pOH = 14$$

e.g for 4.00 × 10^{-2} mol dm^{-3} Ba(OH)$_2$

$[OH^-] = 2 \times (4.00 \times 10^{-2}) = 8.00 \times 10^{-2}$ mol dm^{-3}
$pOH = 1.10$
$\Rightarrow pH = 14 - 1.10 = 12.9$

Calculations involving weak acids, weak bases, and buffer solutions

WEAK ACIDS

The dissociation of a weak acid HA in water can be written:

$$HA(aq) \rightleftharpoons H^+(aq) + A^-(aq)$$

The equilibrium expression for this reaction is:

$$K_a = \frac{[H^+] \times [A^-]}{[HA]}$$ where K_a is known as the acid dissociation constant.

If the acids are quite weak the equilibrium concentration of the acid can be assumed to be the same as its initial concentration.

For example, to calculate the pH of 0.10 mol dm^{-3} CH$_3$COOH given that $K_a = 1.8 \times 10^{-5}$ mol dm^{-3} at 298 K:

$$CH_3COOH(aq) \rightleftharpoons CH_3COO^-(aq) + H^+(aq)$$

$$K_a = \frac{[CH_3COO^-] \times [H^+]}{[CH_3COOH]} \approx \frac{[H^+]^2}{0.10} = 1.8 \times 10^{-5} \text{ mol dm}^{-3}$$

$$\Rightarrow [H^+] = \sqrt{1.8 \times 10^{-6}} = 1.34 \times 10^{-3} \text{ mol dm}^{-3}$$

pH = 2.87

Examples of other weak acid calculations

The pH of a 0.020 mol dm^{-3} solution of a weak acid is 3.9. Find the K_a of the acid.

$$K_a = \frac{[H^+]^2}{(0.020 - [H^+])} \approx \frac{10^{-3.9} \times 10^{-3.9}}{0.020} = 7.92 \times 10^{-7} \text{ mol dm}^{-3}$$

An acid whose K_a is 4.1×10^{-6} mol dm^{-3} has a pH of 4.5. Find the concentration of the acid.

$$[HA] = \frac{[H^+]^2}{K_a} = \frac{10^{-4.5} \times 10^{-4.5}}{4.1 \times 10^{-6}} = 2.44 \times 10^{-4} \text{ mol dm}^{-3}$$

WEAK BASES

The reaction of a weak base can be written:

$$B(aq) + H_2O(l) \rightleftharpoons BH^+(aq) + OH^-(aq)$$

Since the concentration of water is constant:

$$K_b = \frac{[BH^+] \times [OH^-]}{[B]}$$

where K_b is the base dissociation constant.

Example of calculation

The K_b value for ammonia is 1.8×10^{-5} mol dm^{-3}. Find the pH of a 0.010 mol dm^{-3} solution.

Since [NH$_4^+$] = [OH$^-$] then

$$K_b = \frac{[OH^-]^2}{[NH_3]} \approx \frac{[OH^-]^2}{0.010}$$

$$= 1.8 \times 10^{-5}$$

$$[OH^-] = \sqrt{1.8 \times 10^{-7}}$$

$$= 4.24 \times 10^{-4} \text{ mol dm}^{-3}$$

$$\Rightarrow pOH = -\log_{10} 4.24 \times 10^{-4}$$

$$= 3.37 \Rightarrow pH = 14 - 3.37 = 10.6$$

BUFFER CALCULATIONS

The equilibrium expression for weak acids also applies to acidic buffer solutions, e.g. ethanoic acid/sodium ethanoate solution.

$$K_a = \frac{[H^+] \times [CH_3COO^-]}{[CH_3COOH]}$$

The essential difference is that now the concentrations of the two ions from the acid will not be equal. Since the sodium ethanoate is completely dissociated the concentration of the ethanoate ions in solution will be almost the same as the concentration of the sodium ethanoate as very little will come from the acid.

If logarithms are taken and the equation is rearranged then:

$$pH = pK_a + \log_{10} \frac{[CH_3COO^-]}{[CH_3COOH]}$$

Two facts can be deduced from this expression. Firstly the pH of the buffer does not change on dilution as the concentration of the ethanoate ions and the acid will be affected equally. Secondly the buffer will be most efficient when [CH$_3$COO$^-$] = [CH$_3$COOH]. At this point, which equates to the half equivalence point when ethanoic acid is titrated with sodium hydroxide, the pH of the solution will equal the pK_a value of the acid.

Calculate the pH of a buffer containing 0.20 mol of sodium ethanoate in 500 cm^3 of 0.10 mol dm^{-3} ethanoic acid (given that K_a for ethanoic acid = 1.8×10^{-5} mol dm^{-3}).

[CH$_3$COO$^-$] = 0.40 mol dm^{-3};
[CH$_3$COOH] = 0.10 mol dm^{-3}

$$K_a \approx \frac{[H^+] \times 0.40}{0.100} = 1.8 \times 10^{-5} \text{ mol dm}^{-3}$$

[H$^+$] = 4.5×10^{-6} mol dm^{-3}

pH = 5.35

Calculate what mass of sodium propanoate must be dissolved in 1.0 dm^3 of 1.0 mol dm^3 propanoic acid (pK_a = 4.87) to give a buffer solution with a pH of 4.5.

$$[C_2H_5COO^-] = \frac{K_a \times [C_2H_5COOH]}{[H^+]} = \frac{10^{-4.87} \times 1.0}{10^{-4.5}}$$

$$= 0.427 \text{ mol dm}^{-3}$$

Mass of NaC$_2$H$_5$COO required = $0.427 \times 96.07 = 41.0$ g

IB QUESTIONS – OPTION A – HIGHER PHYSICAL ORGANIC CHEMISTRY

The questions below are intended both for those Standard Level students who are taking the Option and for all Higher Level students.

1. Two isomers **A** and **B** of empirical formula C_5H_{12} give mass spectra showing peaks corresponding to molecular ions and other major fragments with mass-to-charge values as follows:

A	72	57	42	27	12
B	72	57	43	29	15

(a) (i) Give the molecular formula of **A** and **B**. [1]

   (ii) Give the structural formulas of the **three** isomers of this hydrocarbon. [3]

   (iii) With respect to the mass spectrum of **A**, identify, with a brief explanation, **one** fragment which would cause the observed mass losses. [2]

   (iv) With respect to the mass spectrum of **B**, identify **one different** fragment which would cause the observed mass losses. [1]

   (v) Give the structures of **A** and **B** and name each of them. [4]

(b) Briefly describe and explain the ^1H NMR spectrum of isomer **A**. Indicate the number of different chemical environments of the hydrogen atoms in isomer **B**. [3]

(c) State why infrared spectroscopy is less useful than mass spectroscopy and ^1H NMR spectroscopy in distinguishing between **A** and **B**. [1]

2. The following equation represents the reaction between hydrogen and nitrogen(II) oxide. The reaction is first order with respect to hydrogen and second order with respect to nitrogen(II) oxide.

$$2H_2(g) + 2NO(g) \rightarrow 2H_2O(g) + N_2(g)$$

(a) Write down the rate equation and give the overall order of this reaction. [4]

(b) What would be the effect on the reaction rate of doubling the concentration of **both** reactants? [1]

(c) The depletion of ozone, O_3, in the upper atmosphere can be caused by the reaction of automobile exhaust gases, such as NO, with ozone. The reaction between $O_3(g)$ and $NO(g)$ has been studied and the following data were obtained at 25°C.

Experiment	[NO(g)]/ mol dm^{-3}	[O$_3$(g)]/ mol dm^{-3}	Rate/ mol dm^{-3} s^{-1}
1	1.00×10^{-6}	3.00×10^{-6}	0.660×10^{-4}
2	1.00×10^{-6}	6.00×10^{-6}	1.32×10^{-4}
3	3.00×10^{-6}	9.00×10^{-6}	5.94×10^{-4}
4	4.50×10^{-6}	7.20×10^{-6}	

(i) Give the rate equation for the reaction between NO(g) and O_3(g), showing your reasoning. [3]

(ii) Calculate the value of the rate constant, k, stating its units. [2]

(iii) Calculate the rate of the reaction for Experiment 4. [1]

3. (a) Write an equation to show the ionisation of propanoic acid in water. [1]

(b) Give the equilibrium expression for this reaction. [1]

(c) Using information from Table 16 in the Data Booklet, determine the pH of a 0.200 mol dm^{-3} aqueous solution of propanoic acid. State the approximation that you have made in arriving at your answer. [3]

(d) What mass of sodium propanoate, $Na^+CH_3CH_2COO^-$, is required in 500 cm^3 of a solution of 0.200 mol dm^{-3} propanoic acid to give a pH of 4.87? [2]

(e) Explain, with equations, why the pH of the above solution in (d) will remain relatively constant even if small amounts of acid or alkali are added. [2]

4. 2-chloro-2-methylpropane $C(CH_3)_3Cl$ is a tertiary halogenoalkane.

(a) Give the equation for the substitution reaction between 2-chloro-2-methylpropane and warm dilute aqueous sodium hydroxide. [1]

(b) Give the rate expression for the above reaction. [1]

(c) Name and give the stepwise mechanism for this reaction. [3]

(d) State and explain whether the rate of the reaction will increase or decrease if:

   (i) 1-chlorobutane is reacted with sodium hydroxide solution under the same conditions. [2]

   (ii) 2-iodo-2-methylpropane is reacted with sodium hydroxide solution under the same conditions. [2]

Pharmaceutical products

THE EFFECTS OF DRUGS AND MEDICINES

For centuries man has used natural materials to provide relief from pain, heal injuries, and cure disease. Many of these folk remedies have been shown to be very effective and the active ingredients isolated and identified.

Morphine was extracted from the poppy *Papaver somniferum* early in the nineteenth century and later salicylic acid, the precursor of aspirin, was isolated from willow bark. The words 'drug' or 'medicine' are commonly applied to these substances, but they have different connotations in different countries and are difficult to define precisely. For example, the pineal gland in humans, a small lump of tissue that resides at the base of the brain, produces a substance called melatonin. This substance is known to bring on the onset of sleep and is often taken by people suffering from 'jet-lag'. As it occurs naturally in very low amounts in many foods it is classed as a food in America and can readily be bought. However, it is unavailable in Europe since it is classed as a drug because of its potential to modify physiological functions in humans. Generally a drug or medicine is any chemical (natural or man-made), which does one or more of the following:

• alters incoming sensory sensations
• alters mood or emotions
• alters the physiological state (including consciousness, activity level, or co-ordination).

Drugs and medicines are commonly (but not always) taken to improve health. They accomplish this by assisting the body in its natural healing process. The mechanism of drug action is still not fully understood, and there is evidence that the body can be 'fooled' into healing itself through the 'placebo' effect.

RESEARCH, DEVELOPMENT, AND TESTING OF NEW PRODUCTS

The research and development of new drugs is a long and expensive process. Traditionally a new product is isolated from an existing species, or synthesized chemically and then subjected to thorough laboratory and clinical pharmacological studies to demonstrate its effectiveness. Before studies are allowed on humans it must be tested on animals to determine the **lethal dose** required to kill fifty percent of the animal population, known as the LD_{50}. The effective dose required to bring about a noticeable effect in 50% of the population is also obtained, so that the safe dose to administer can be determined. The drug can then be used in an initial clinical trial on humans. This is usually on volunteers as well as on patients, half of whom are given the drug and half of whom are given a placebo. This initial trial is closely monitored to establish the drug's safety and possible side effects.

Drugs usually have unwanted **side effects**, for example aspirin can cause bleeding of the stomach and morphine, which is normally used for pain relief, can cause constipation. Side effects can be relative depending on why the drug is taken. People with diarrhoea are sometimes given a kaolin and morphine mixture, and people who have suffered from a heart

METHODS OF ADMINISTERING DRUGS

In order to reach the site where their effects are needed the majority of drugs must be absorbed into the bloodstream. The method of administering the drug determines the route taken and the speed with which it is absorbed into the blood. The four main methods are: by mouth (oral); inhalation; through the anus (rectal), and by injection (parenteral).

Drugs may also be applied topically so that the effect is limited mainly to the site of the disorder, such as the surface of the skin. Such drugs may come in the form of creams, ointments, sprays, and drops.

The three different methods of injection

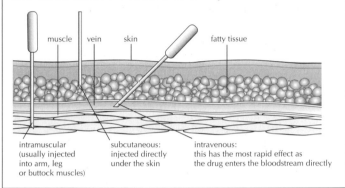

muscle   vein   skin   fatty tissue

intramuscular (usually injected into arm, leg or buttock muscles)

subcutaneous: injected directly under the skin

intravenous: this has the most rapid effect as the drug enters the bloodstream directly

THALIDOMIDE – HOW IT CAN ALL GO WRONG

In 1958 a German pharmaceutical company launched a massive publicity campaign for a new tranquilizer to combat 'morning sickness' in pregnant women. The drug was sold world-wide under brand names such as Thalidomide and Contergan. In many countries it was sold without prescription and marketed as completely innocuous. Reports of severe adverse side-effects began to appear in 1959, and it later transpired that as early as 1956 clinical trials by the company itself had revealed problems. Nevertheless because it was so profitable the company continued to market the drug heavily and sales kept increasing until it was withdrawn in 1961. By that time many children had been born with absent or severely malformed limbs.

attack are advised to take aspirin as it is effective as an anti-clotting agent. The severity of the complaint will determine an acceptable **risk-to-benefit ratio**. If an effective treatment is found for a life threatening disease then a high risk from side effects will be more acceptable.

The **tolerance** of the drug is also determined. Drug tolerance occurs as the body adapts to the action of the drug. A person taking the drug needs larger and larger doses to achieve the original effect. The danger with tolerance is that as the dose increases so do the risks of dependence and the possibility of reaching the lethal dose.

If the drug passes the initial clinical trial it will then go through a rigorous series of further phases, where its use is gradually widened in a variety of clinical situations. If it passes all these trials it will eventually be approved by the drug administration of a particular country, for use either as an OTC (over the counter) drug or for use only through prescription by a doctor.

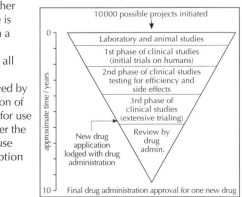

10 000 possible projects initiated

Laboratory and animal studies

1st phase of clinical studies (initial trials on humans)

2nd phase of clinical studies testing for efficiency and side effects

3rd phase of clinical studies (extensive trialing)

Review by drug admin.

New drug application lodged with drug administration

approximate time / years

Final drug administration approval for one new drug

Antacids

DIGESTION

The process of digestion involves the breakdown of food into molecules that can be utilized by individual cells in the body. The process starts in the mouth through the mechanical action of chewing and by the action of the digestive enzyme, amylase, present in saliva. Much of the digestive process, however, takes place in the stomach, a collapsible muscular bag which can hold between 2 and 4 litres of food.

The walls of the stomach are lined with a layer of cells which secrete mucus, pepsinogen (a precursor for the enzyme pepsin that breaks down proteins into peptides), and hydrochloric acid, collectively known as gastric juices. The hydrogen ion concentration of the hydrochloric acid normally lies between 3×10^{-2} mol dm^{-3} and 3×10^{-3} mol dm^{-3} giving a pH value between 1.5 and 2.5. The wall of the stomach is protected from the action of the acid by a lining of mucus. Problems can arise if the stomach lining is damaged or when too much acid is produced, which can eat away at the mucus lining.

TREATMENT OF INDIGESTION

The discomfort caused by excess acid is known as indigestion and may result from overeating, alcohol, smoking, anxiety, or in some people from eating certain types of food. Some drugs, such as aspirin, can also irritate the stomach lining and can result in ulceration of the stomach walls by the gastric acid. Antacids are used to combat excess stomach acid. They are most effective if taken between one and three hours after eating, as food typically remains in the stomach for up to four hours after a meal.

Antacids are essentially simple bases, such as metal oxides, hydroxides, carbonates, or hydrogencarbonates. They work by neutralizing the acid, preventing inflammation, relieving pain and discomfort, and allow the mucus layer and stomach lining to mend. When used in the treatment of ulcers they prevent acid from attacking the damaged stomach lining and so allow the ulcer to heal. Common examples include $Al(OH)_3$, $NaHCO_3$, $CaCO_3$, and 'milk of magnesia', which is a mixture of MgO and $Mg(OH)_2$. Typical neutralization reactions are:

$$NaHCO_3(s) + HCl(aq) \rightarrow NaCl(aq) + CO_2(g) + H_2O(l)$$

$$MgO(s) + 2HCl(aq) \rightarrow MgCl_2(aq) + H_2O(l)$$

$$Al(OH)_3(s) + 3HCl(aq) \rightarrow AlCl_3(aq) + 3H_2O(l)$$

SIDE EFFECTS

Although relatively harmless, antacids can have side effects. Magnesium compounds can cause diarrhoea, whereas aluminium compounds can cause constipation. Aluminium compounds can interfere with the absorption of phosphate from the diet causing possible bone damage if taken in high doses over a long period. Sodium hydrogen carbonate produces carbon dioxide gas, which may cause bloating and belching.

Antacids are commonly combined with alginates and anti-foaming agents. Alginates float on the contents of the stomach to produce a neutralizing layer. This prevents heartburn, which is caused when the stomach acid rises up the oesophagus. Anti-foaming agents are used to help prevent flatulence. The most usual anti-foaming agent is dimethicone.

WORKED EXAMPLE

Which would be the most effective in combating indigestion – a spoonful of liquid containing 1.00 g of magnesium hydroxide, or a spoonful of liquid containing 1.00 g of aluminium hydroxide?

M_r for $Mg(OH)_2 = 24.30 + (2 \times 17.01) = 58.33$

M_r for $Al(OH)_3 = 26.98 + (3 \times 17.01) = 78.01$

Amount of $Mg(OH)_2$ in 1.00 g $= \frac{1.00}{58.33} = 0.0171$ moles

Amount of $Al(OH)_3$ in 1.00 g $= \frac{1.00}{78.01} = 0.0128$ moles

$$Mg(OH)_2(s) + 2HCl(aq) \rightarrow MgCl_2(aq) + 2H_2O(l)$$

Therefore amount of HCl neutralized by 1.00 g of $Mg(OH)_2$
$= 2 \times 0.0171 = 0.0342$ moles

$$Al(OH)_3(s) + 3HCl(aq) \rightarrow AlCl_3(aq) + 3H_2O(l)$$

Therefore amount of HCl neutralized by 1.00 g of $Al(OH)_3$
$= 3 \times 0.0128 = 0.0384$ moles

Therefore the aluminium hydroxide would be slightly more effective.

Analgesics

MILD ANALGESICS

Aspirin

For a long time the bark of the willow tree (*Salix alba*) was used as a traditional medicine to relieve the fever symptoms of malaria. In the 1860s chemists showed that the active ingredient in willow bark is salicylic acid (2-hydroxybenzoic acid) and by 1870 salicylic acid was in wide use as a pain killer (analgesic) and fever depressant (antipyretic). However, salicylic acid has the undesirable side effect of irritating and damaging the mouth, oesophagus, and stomach membranes. In 1899 the Bayer Company of Germany introduced the ethanoate ester of salicylic acid, naming it 'Aspirin'.

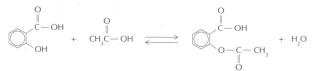

Aspirin is thought to work by preventing a particular enzyme, prostaglandin synthase, being formed at the site of the injury or pain. This enzyme is involved in the synthesis of prostaglandins, which produce fever and swelling, and the transmission of pain from the site of an injury to the brain. Because of its anti-inflammatory properties aspirin can also be taken for arthritis and rheumatism. Aspirin also has an ability to prevent blood clotting and is sometimes taken to prevent strokes or the recurrence of heart attacks. Recent research suggests that aspirin may also be effective in preventing prostate cancer.

Aspirin can, however, have side effects. The most common side effect is that it causes bleeding in the lining of the stomach. A few people are allergic to aspirin with just one or two tablets leading to bronchial asthma. The taking of aspirin by children under twelve has been linked to Reye's disease – a potentially fatal liver and brain disorder with the symptoms of vomiting, lethargy, irritability, and confusion. Exceeding the safe dose of aspirin can be fatal as the salicylic acid leads to acidosis due to a lowering of the pH of the blood.

OTHER MILD ANALGESICS

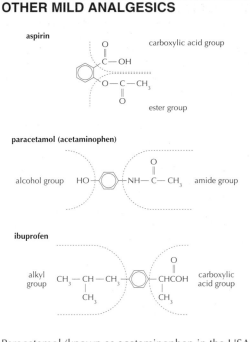

Paracetamol (known as acetaminophen in the USA) is often preferred to aspirin as a mild pain reliever, particularly for young children, as its side-effects are less problematical although in rare cases it can cause kidney damage and blood disorders. Serious problems can arise, however, if an overdose is taken. Even if the overdose does not result in death it can cause brain damage and permanent damage to the liver and kidneys.

STRONG ANALGESICS

Strong analgesics are only available on prescription and are given to relieve the severe pain caused by injury, surgery, heart attack, or chronic diseases, such as cancer. They work by interacting temporarily with receptor sites in the brain, with the result that pain signals within the brain and spinal cord are blocked. The most important naturally occurring strong analgesics are morphine and codeine found in the opium poppy. The active part of the morphine molecule has been identified. Codeine and semi-synthetic (obtained by simple structural modifications to morphine) opiates, such as heroin, and totally synthetic compounds (e.g. 'demerol') all possess this basic structure and function as strong analgesics.

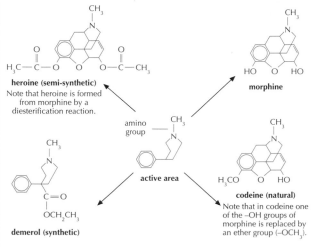

Effects of opiates

Diethanoylmorphine (heroin) is a more powerful painkiller than morphine but also more addictive. All opiates cause addiction and lead to tolerance, where there is a danger of reaching the lethal limit.

Short term effects
- Induce a feeling of euphoria (sense of well-being)
- Dulling of pain
- Depress nervous system
- Slow breathing and heart rate
- Cough reflex inhibited
- Nausea and vomiting (first time users)
- High doses lead to coma and/or death

Long term effects
- Constipation
- Loss of sex drive
- Disrupts menstrual cycle
- Poor eating habits
- Risk of AIDS, hepatitis, etc. through shared needles
- Social problems, e.g. theft, prostitution

Withdrawal symptoms occur within 6 to 24 hours for addicts if the supply of the drug is stopped. These include hot and cold sweats, diarrhoea, anxiety, and cramps. One treatment to wean addicts off their addiction is to use methadone as a replacement for heroin. Methadone is also an amine and functions as an analgesic but does not produce the euphoria craved by addicts.

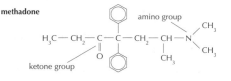

Depressants

EFFECTS OF DEPRESSANTS

Depressants are drugs which depress the central nervous system by interfering with the transmission of nerve impulses in the nerve cells (neurones). Depressants slow down the functions of the body including mental activity. In low doses they induce a feeling of calm and relieve anxiety and may induce sleep, but in larger doses they can cause loss of consciousness, coma, and death. The most commonly taken depressant is alcohol (ethanol). Rather confusingly depressants are sometimes described as anti-depressants, because they relieve the symptoms of mental depression.

USE AND ABUSE OF ALCOHOL (ETHANOL)

Medically alcohol is used as an antiseptic before injections and is also used to harden the skin. Drinking large and regular amounts of alcohol can cause psychological and physical dependence, known as alcoholism. The social costs of road accidents, violent behaviour, and family breakdowns due to alcohol consumption are huge.

Short term effects In moderate quantities it gives the drinker a feeling of relaxation and confidence and increases sociability. It dilates small blood vessels leading to flushing and a feeling of warmth. With increasing amounts judgement and concentration become progressively impaired. Violent behaviour is possible. Speech becomes slurred and loss of balance occurs. Loss of consciousness may follow at high concentrations and there is a risk of death from inhalation of vomit or stoppage of breathing.

Long term effects Long term heavy drinking can lead to severe liver disease including cirrhosis and liver cancer. It is linked with coronary heart disease, high blood pressure, strokes and an increasing risk of dementia. During pregnancy it can cause miscarriage and lead to fetal abnormalities. Sudden discontinuation of alcohol by heavy users can lead to delirium tremens (the 'DTs'), which includes severe shaking that can last up to four days.

SYNERGISTIC EFFECTS OF ALCOHOL

Ethanol can interact with, and considerably enhance the effect of other drugs because it depresses the central nervous system itself. This synergistic effect can be fatal, particularly when alcohol is taken together with benzodiazepines, narcotics, barbiturates, and solvents. With aspirin it increases the risk of stomach bleeding.

OTHER DEPRESSANTS

Other depressants commonly prescribed to reduce anxiety and relieve stress or to help insomnia include the benzodiazepines and prozac. They do not however remove the causes and are usually only prescribed for a limited period while counselling or psychotherapy are put in place, as they can induce dependence. They are also used as a premedication in hospitals before general surgery.

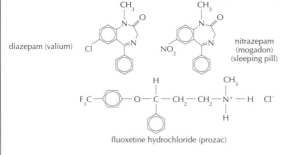

diazepam (valium)

nitrazepam (mogadon) (sleeping pill)

fluoxetine hydrochloride (prozac)

LEGAL LIMITS FOR DRIVING

In many countries the legal limit for driving is a blood alcohol concentration (BAC) of 80 mg of ethanol per 100 cm³ of blood. In some countries it is even lower. After drinking, the concentration of alcohol in the blood increases for some time as the ethanol is absorbed, then it slowly decreases as it is metabolized and excreted. A unit of alcohol is roughly equal to:

- 280 cm³ ($\frac{1}{2}$ pint) of beer or lager
- a glass of wine
- a measure of spirits.

Estimate of the alcohol consumption needed to exceed a legal limit of 80 mg of ethanol per 100 cm³ of blood for a man of average weight (for a woman the quantities need to be reduced by about 30%)

Number of units	Time taken drinking / hours
4.0	1
5.0	2
6.0	3
6.5	4
7.5	5

DETECTION OF ALCOHOL IN BREATH, BLOOD, AND URINE

At the roadside a motorist may be asked to blow into a breathalyser. This may involve acidified potassium (or sodium) dichromate(VI) crystals turning green as they are reduced by the alcohol to Cr^{3+}, or the use of a fuel cell where the ethanol is oxidized to produce electricity. None of these are accurate enough to be used in court. At the police station a blood or urine sample may be taken and sent to a forensic science laboratory for analysis using gas liquid chromatography.

infrared intoximeter

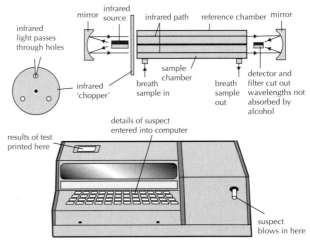

Modern intoximeters can now be used in a police station to accurately measure the amount of alcohol in the breath. They are based on the principle that the C–H bonds in ethanol absorb infrared radiation of a particular wavelength –3.39 micrometres (μm). The suspect blows a sample of breath into a chamber. Infrared radiation from a heated source passes through a 'chopper' (a rotating slotted disc), which makes the beam then pass alternately through the sample chamber and a chamber containing no breath sample. The intensities of the two emerging beams are compared and the amount of radiation absorbed by the sample is then converted into micrograms of ethanol per 100 cm³ of breath.

Stimulants

Stimulants are drugs that increase a person's state of mental alertness.

AMPHETAMINES

Amphetamine is chemically related to adrenaline, the 'flight or fight' hormone. It is a sympathomimetic drug, that is, one which mimics the effect of stimulation on the sympathetic nervous system. This is the part of the nervous system which deals with subconscious nerve responses, such as speeding up the heart and increasing sweat production.

Amphetamines were initially used to treat narcolepsy (an uncontrollable desire for sleep) and were issued to airmen in World War II to combat fatigue. In the 1950s and 1960s they were used as anti-depressants and slimming pills. Regular use can lead to both tolerance and dependence. Short-term effects include increase in heart rate and breathing, dilation of the pupils, decrease in appetite followed by fatigue and possible depression as the effects wear off. Long-term effects include weight loss, constipation, and emotional instability.

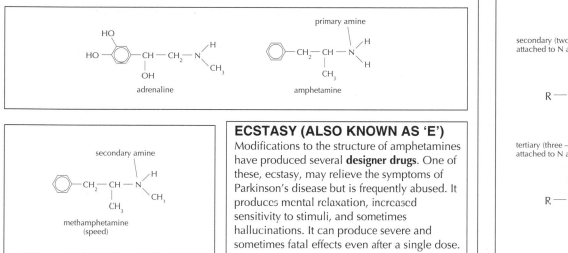

CLASSIFICATION OF AMINES

primary (one – R group attached to N atom)

secondary (two – R groups attached to N atom)

tertiary (three – R groups attached to N atom)

ECSTASY (ALSO KNOWN AS 'E')

Modifications to the structure of amphetamines have produced several **designer drugs**. One of these, ecstasy, may relieve the symptoms of Parkinson's disease but is frequently abused. It produces mental relaxation, increased sensitivity to stimuli, and sometimes hallucinations. It can produce severe and sometimes fatal effects even after a single dose.

CAFFEINE

Caffeine is the most widely used stimulant in the world. It is present in coffee, tea, chocolate, and cola drinks and is also found in some pain killers and other medicines. There is evidence that consuming 400 mg of caffeine a day, or more, can cause dependence and physical side effects.

Caffeine content of different products

cup of ground coffee	80–120 mg
cup of instant coffee	65 mg
cup of tea	40 mg
can of cola	40 mg
bar (100 g) of plain chocolate	80 mg

Caffeine is a diuretic (causes frequent urination) and increases alertness, concentration, and restlessness. It is included in

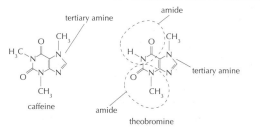

many common painkillers as it speeds up their effects. Like nicotine, morphine, codeine, and cocaine, caffeine is an alkaloid. Alkaloids are nitrogen-containing compounds of plant origin containing heterocyclic rings (rings containing other atoms as well as carbon) and a tertiary amine group. A compound with a similar structure to caffeine, which is also found in chocolate, is theobromine (although note that it contains no bromine!).

NICOTINE

It is the nicotine in tobacco that is largely responsible for causing approximately one third of the world's population to be addicted to smoking. Stopping smoking can produce temporary withdrawal symptoms that include a craving for tobacco, nausea, weight gain, insomnia, irritability, and depression. Like amphetamines nicotine is sympathomimetic. It increases concentration and relieves tension and the physical effects include increased heart rate and blood pressure, and reduction in urine output. The long term effects include increased risk of heart disease and coronary thrombosis. Its stimulatory effects may also lead to excess production of stomach acid leading to an increased risk of

peptic ulcers. The other well-known risks of smoking include chronic lung diseases, adverse effects on pregnancy, and cancers of the lung, mouth, and throat.

nicotine

Antibacterials

INFECTIOUS ORGANISMS

Many different types of micro-organism can cause disease, among them bacteria, viruses, fungi, yeasts, and protozoa. A typical bacterium consists of a single cell with a protective wall made up of a complex mixture of proteins, sugars, and lipids. Inside the cell wall is the cytoplasm, which may contain granules of glycogen, lipids, and other food reserves. Each bacterial cell contains a single chromosome consisting of a strand of deoxyribonucleic acid (DNA). Some bacteria are aerobic – that is they require oxygen and are more likely to infect surface areas, such as the skin or respiratory tract. Others are anaerobic and multiply in oxygen-free or low oxygen surroundings, such as the bowel. Not all bacteria cause disease and some are beneficial. Many exist on the skin or in the bowel without causing ill effects and some cannot live either in or on the body.

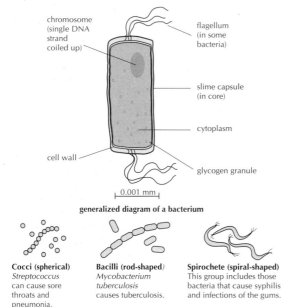

chromosome (single DNA strand coiled up)

flagellum (in some bacteria)

slime capsule (in core)

cytoplasm

cell wall

glycogen granule

|— 0.001 mm —|

generalized diagram of a bacterium

Cocci (spherical)
Streptococcus can cause sore throats and pneumonia.

Bacilli (rod-shaped)
Mycobacterium tuberculosis causes tuberculosis.

Spirochete (spiral-shaped)
This group includes those bacteria that cause syphilis and infections of the gums.

DISCOVERY OF PENICILLIN

Antibacterials are chemicals, which prevent the growth and multiplication of bacteria. The first effective antibacterial, the dye trypan red, was developed by Paul Ehrlich to cure sleeping sickness. In 1910 Ehrlich also developed an arsenic containing compound, salvarsan, which was effective against syphilis. In 1935 the first 'sulfa dug' prontosil, which was effective against streptococcal bacteria, was developed.

However, the real discovery in the fight against bacteria was made by Alexander Fleming in 1928. Fleming, a bacteriologist, was working with cultures of *Staphylococcus aureus*, a bacterium that causes boils and other types of infections. He left an open petri dish containing one of the cultures in the laboratory, while he went away on holiday. Upon his return he noticed that a mould had developed and had inhibited growth of the bacterium. He deduced that the mould (*Penicillium notatum*) produced a compound (which he called penicillin), which inhibited the growth of bacteria. Although he published his results Fleming did not pursue his discovery. It was Howard Florey and Ernest Chain who overcame the problems associated with isolating and purifying penicillin. In 1941 they used penicillin on a policeman who was dying of septicaemia. They recorded a dramatic improvement in his condition, but unfortunately their meagre supply ran out before the policeman was cured and he relapsed and died. The search was on to produce penicillin in bulk. It was solved in America by growing strains of the penicillin mould in large tanks containing corn-steep liquor. In the 1950s the structure of penicillin was determined and this enabled chemists to synthesize different types of penicillin and other antibiotics (antibacterials originating from moulds) in the laboratory without recourse to moulds.

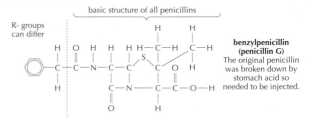

basic structure of all penicillins

R- groups can differ

benzylpenicillin (penicillin G)
The original penicillin was broken down by stomach acid so needed to be injected.

NARROW AND BROAD SPECTRUM ANTIBIOTICS

The penicillins are **narrow spectrum antibiotics**, effective against only certain types of bacteria. Other types of antibiotics, such as the tetracyclines (e.g. aureomycin and terramycin), are effective against a much broader range of bacteria (broad spectrum antibiotics). When a doctor is confronted with a patient suspected to be suffering from a bacterial infection the organism needs to be identified by taking blood, sputum, urine, pus, or stool samples. This takes time so initially a broad spectrum antibiotic may be prescribed. Once the bacterium is known the treatment may be switched to a narrow spectrum antibiotic, which is the recommended treatment for the identified organism.

MECHANISM OF ACTION OF ANTIBIOTICS

There are two main mechanisms by which antibiotics destroy bacteria. Penicillins and the cephalosporins prevent bacteria from making normal cell walls. Other antibiotics act inside the bacteria interfering with the chemical activities essential to their life function.

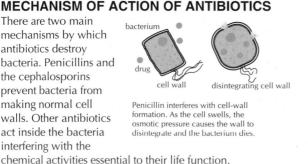

bacterium

drug

cell wall

disintegrating cell wall

Penicillin interferes with cell-wall formation. As the cell swells, the osmotic pressure causes the wall to disintegrate and the bacterium dies.

OVERUSE AND RESISTANCE TO PENICILLINS

When penicillin became readily available to doctors it was often given to cure minor illnesses, such as a sore throat. Certain bacteria were resistant to penicillin and were able to multiply. Their resistance was due to the presence of an enzyme called penicillinase, which could deactivate the original penicillin, penicillin G. To combat this chemists developed other penicillins whereby the active part of the molecule is retained but the side chain is modified. However, as bacteria multiply and mutate so fast it is a continual battle to find new antibiotics, which are effective against an ever more resistant breed of 'super bugs'.

The use of antibiotics in animal feedstocks has also contributed to this problem. Healthy animals are given antibiotics to prevent risk of disease, but the antibiotics are passed on through the meat and dairy products to humans, increasing the development of resistant bacteria.

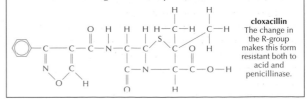

cloxacillin
The change in the R-group makes this form resistant both to acid and penicillinase.

Antivirals

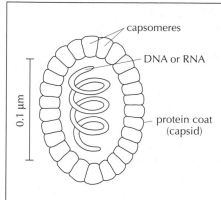

capsomeres

DNA or RNA

protein coat (capsid)

0.1 μm

STRUCTURE OF A VIRUS

There are many different types of virus with varying shape and structure. All viruses, however, have a central core of DNA or RNA (ribonucleic acid) surrounded by a coat (capsid) of regularly packed protein units (capsomeres), each containing many protein molecules. Unlike bacteria they have no nucleus or cytoplasm and are therefore not cells. They do not feed, excrete, or grow and they can only reproduce inside the cells of living organisms using materials provided by the host cell.

MULTIPLICATION OF VIRUSES

Athough viruses can survive outside the host they can only replicate by penetrating the living host cell and injecting their DNA or RNA into the cell's cytoplasm. The virus then 'takes over' the biochemical machinery inside the cell. This causes the cell to die or become seriously altered and causes the symptoms of the viral infection. The cell is made to produce new DNA or RNA and forms large numbers of new viruses. These are then released and move on to infect other healthy cells.

MODE OF ACTION OF ANTIVIRAL DRUGS

Common viral infections include the common cold, influenza, and childhood diseases, such as mumps and chicken pox. Fortunately the body's own defence mechanism is usually strong enough to overcome infections such as these and drugs are given more to remove the associated pain, fight the fever or to counteract secondary infections. One difficulty in treating viral infections is the speed with which the virus multiplies. By the time the symptoms have appeared the viruses are so numerous that antiviral drugs will have little effect. During the past few years some drugs have been developed to fight specific viral infections. They can work in different ways. Some work by altering the cell's genetic material so that the virus cannot use it to multiply. An example of this is acyclovir, which is applied topically to treat cold sores caused by the herpes virus. Its structure is similar to deoxyguanosine, one of the building blocks of DNA. It tricks the viral enzymes into using it as a building block for the viral DNA and thus prevents the virus from multiplying. However, it is difficult to eliminate the virus completely so the infection may flare up again at a later date.

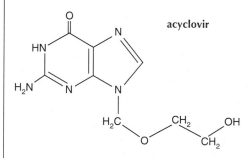

acyclovir

Others work by preventing the new viruses formed from leaving the cell. One such drug is amantadine, which is active against the influenza virus. One of the enzymes used by all influenza viruses to stick to the host cell wall as it leaves is called neuraminidase, and the drug works by inhibiting the active site on this enzyme. One of the problems with developing antiviral drugs is that the viruses themselves are regularly mutating – this is particularly true with the Human Immunodeficiency Virus (HIV).

AIDS

AIDS (Acquired Immune Deficiency Syndrome) is caused by a retrovirus – that is, it contains RNA rather than DNA. The virus invades certain types of cells, particularly the white blood cells, which normally activate other cells in the immune system, with the result that the body is unable to fight infection. Once it invades a healthy cell its first task is to make viral-DNA from the RNA template using an enzyme called reverse transcriptase. This is opposite to the process that takes place in normal cells in which RNA is made from a DNA template using transcriptase as the enzyme.

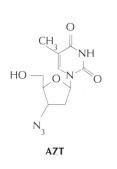

AZT

There are various ways in which a drug may be developed to eradicate the HIV virus. One drug that has met with some success is AZT (zidovudine). This combines with the enzyme that the HIV virus uses to build DNA from RNA and clogs up its active site. It is therefore a reverse-transcriptase inhibitor. Since it is only retroviruses that use this enzyme AZT does not affect normal cells.

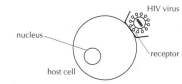

HIV virus

nucleus

receptor

host cell

1. Virus binds to a receptor on the host cell.
Possible action of antiviral
Binding site could be altered to prevent virus attaching to host cell.

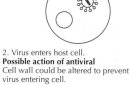

2. Virus enters host cell.
Possible action of antiviral
Cell wall could be altered to prevent virus entering cell.

viral RNA

reverse transcriptase

3. Virus loses its protective coat and releases RNA and reverse transcriptase.
Possible action of antiviral
Drugs might be developed which would prevent the virus from losing its protective coat.

altered form of RNA

4. The reverse transcriptase converts the viral RNA into a form which can enter the nucleus of the host cell so it can integrate with the cell's DNA.
Possible action of antiviral
AZT works by blocking the action of reverse transcriptase.

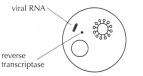

viral RNA

protein

5. The host cell produces new viral RNA and protein.
Possible action of antiviral
May be able to inhibit the production of new viral RNA and proteins by altering the genetic material of the virus.

new HIV viruses

dead host cell

6. The new RNA and proteins form new viruses which then leave the host cell. The host cell dies.
Possible action of antiviral
Develop a drug which prevents the new viruses from leaving the cell. This is how amantadine works against the influenza virus.

HL Importance of stereochemistry in drug design and action (1)

GEOMETRIC ISOMERISM

Geometric isomerism can exist in inorganic compounds as well as in organic compounds, e.g. the square planar compounds of diamminedichloroplatinum(II) can exist in both *cis-* and *trans-* forms.

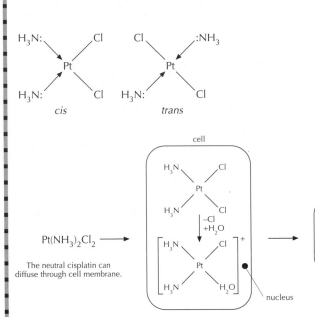

cis   *trans*

The *cis-* form (known as cisplatin) is highly effective in the treatment of testicular and ovarian cancers as well as other forms of cancer. Transplatin is not an effective anti-cancer drug. Cisplatin has no overall charge so can diffuse through the cancer cell membrane. Inside the cell it exchanges a chloride ion for a molecule of water to form $[Pt(NH_3)_2Cl(H_2O)]^+$. This complex ion enters the cell nucleus, where it binds to the DNA by exchanging another chloride ion to form $[Pt(NH_3)_2(DNA)(H_2O)]^{2+}$. This alters the cancer cell's DNA, so that when the cell tries to replicate it cannot be copied correctly and the cell dies.

$Pt(NH_3)_2Cl_2 \longrightarrow$

The neutral cisplatin can diffuse through cell membrane.

The complex ion inside the cell is charged so cannot diffuse out of the cell membrane.

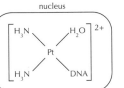

Complex ion enters the nucleus and starts to bind to DNA.

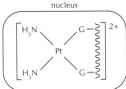

Binding to DNA complete as it links to two guanine bases on the DNA.

CHIRALITY

Asymmetric or chiral carbon atoms form two different optically active forms. Because of their different stereochemistry the two enantiomers can behave in totally different ways in the body. It is now realized that one of the enantiomers of thalidomide gives the benefits associated with the drug and it is the other enantiomer that is thought to be responsible for causing the fetal deformities. When new drugs are synthesized nowadays the pharmacological activity of both forms is studied separately.

It can be difficult to prepare just one of the isomers and often a racemic mixture (50:50 mixture of both isomers) is prepared first and then separated. A new technique using **chiral auxilliaries** now makes it possible to synthesize just the desired isomer. Attaching an auxilliary, which is itself optically active, to the starting material creates the stereochemical conditions necessary for the reaction to form only one enantiomer. After the desired product has been formed the auxilliary is removed and recycled.

One drug for which this technique has been used to great effect is the anti-cancer drug taxol. Although it does occur naturally in yew trees the commercial semi-synthesis of taxol provides the necessary quantity to meet the demand.

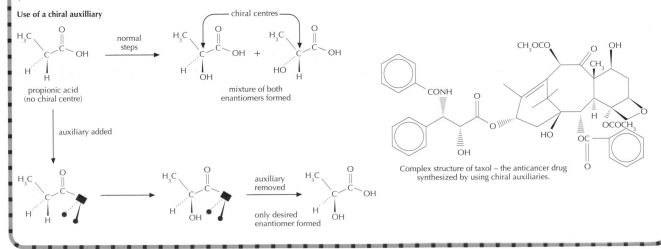

Use of a chiral auxilliary

propionic acid (no chiral centre)

mixture of both enantiomers formed

auxiliary added

auxiliary removed

only desired enantiomer formed

Complex structure of taxol – the anticancer drug synthesized by using chiral auxiliaries.

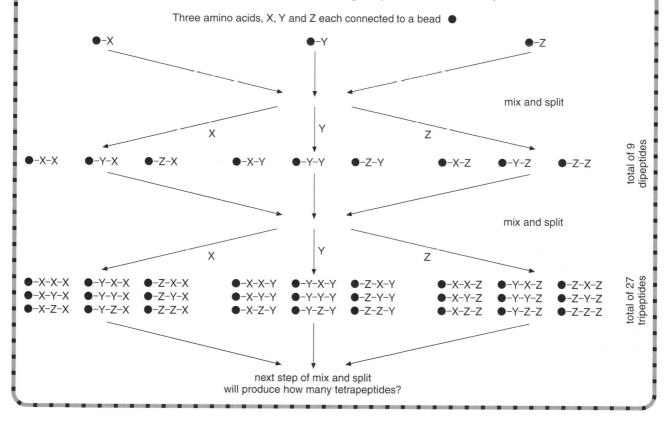 HL Importance of stereochemistry in drug design and action (2)

COMBINATORIAL CHEMISTRY

Drug companies possess 'libraries' of compounds which have been screened for drug activity. In the past it could take many years to build up such a library as the substances were synthesized, purified, and screened by traditional methods. The advent of very sensitive techniques, such as mass spectrometry, to identify extremely small amounts of substances has lead to the development of 'combinatorial libraries'. In combinatorial chemistry very large numbers of related compounds can be prepared quickly. One way is to use computer controlled syringes to carry out repetitive chemical techniques but much use is now made of solid phase organic chemistry.

The starting material for the reaction is covalently bonded to very small beads (100 micrometres diameter) of polystyrene-based resin. A process called 'mix and split' is then used. Imagine the process for just three amino acids. After the first coupling, all the resin beads are then split into individual portions for the next step so that when reacted again all the nine possible combinations of dipeptides are formed. After another step all 27 possible combinations of tripeptides have been formed. If the process is scaled up for all the 20 naturally occurring amino acids then each cycle produces 400 dipeptides, 8000 tripeptides, and 160 000 tetrapeptides, etc. By using a large excess of the second and subsequent amino acids the reaction can be made to give a high yield. The final products can be purified easily by filtering off the beads from the reaction mixture and washing. Preliminary screening for drug activity can then take place either *in vitro* or *in vivo* by measuring the ability of a compound to affect enzymes and bind to receptor cells.

This process was done first for amino acids but has been extended to cover many other types of active molecules, such as those containing the benzodiazepine group in depressants to form very large combinatorial libraries. Once a particular substance is identified as potentially useful it can be made on a larger scale. In the future it may become unnecessary to actually make the initial compounds. By knowing the exact shape both of the active site on the enzyme or receptor and the shapes of the active groups within a potential drug the use of virtual computer modelling can produce a virtual library.

LOCAL ANAESTHETICS

Local anaesthetics (sometimes called local analgesics) block pain in a specific area when injected under the skin or applied topically, but do not affect the overall level of consciousness.

COCAINE

The first anaesthetic to be used in medicine was cocaine. It comes from a South American plant called *Erythoxylum coco*, and its leaves have been chewed for centuries by Indians from Peru and Ecuador to deaden pain and increase endurance. It was first used clinically in 1884. However, as well as being very addictive it can have unpleasant side effects, such as anxiety, nausea, and headaches. In a few cases it can cause breathing difficulties, convulsions, coma, and even death.

Cocaine also acts as a stimulant. It used to be legal and was used socially in Victorian times and recommended by Sigmund Freud to his patients. Even though it is now illegal cocaine abuse has risen rapidly during the past few decades. Cocaine and its synthetic derivatives work by suppressing nerve transmissions. They do this by blocking the action of acetylcholine, a neurotransmitter which allows repetitive impulses to travel along nerves. They also decrease the blood supply to the area by constricting the blood vessels.

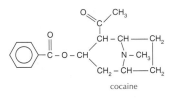

cocaine

PROCAINE AND LIDOCAINE

Due to its undesirable side effects, derivatives of cocaine were synthesized which retain the anaesthetic properties of cocaine but which do not also affect the brain or act as stimulants. Two of the most important compounds which fulfil these criteria are procaine and lidocaine. Both have similar structures to cocaine and both are used in dentistry and for minor surgery.

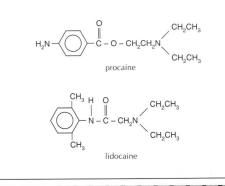

procaine

lidocaine

GENERAL ANAESTHETICS

A general anaesthetic renders the patient unconscious so that they can feel no pain. Before 1840 major surgery, which often meant amputation, was carried out without any anaesthetic except perhaps alcohol and a lead bullet or leather belt to bite on. The first anaesthetic to be used was ethoxyethane (ether) in 1846. Ether is highly inflammable and very strong smelling so it was rapidly superseded by trichloromethane (chloroform). For the next hundred years chloroform or nitrogen(I) oxide (N_2O) were widely used in surgery, and a nitrogen(I) oxide and air mixture is still used by some dentists today. During that period various other compounds, such as cyclopropane, were tried, however, all have disadvantages. Like ether, cyclopropane is prone to ignite and explode, nitrogen(I) oxide is not very efficient and trichloromethane can lead to liver damage.

In the 1950s a research programme was initiated to find a safer alternative. The objectives were that the anaesthetic should be stable, safe, and non-inflammable. In addition of course all anaesthetics must have a low boiling point and must be capable of being inhaled easily. Several fluorinated hydrocarbons fulfilled these criteria and the most successful one was 2-bromo-2-chloro-1,1,1-trifluoroethane (halothane), which is now widely used all over the world.

However, halothane is a CFC (chlorofluorocarbon) and can initiate damage to the ozone layer so research is still continuing to find more environmentally friendly alternatives. There are still risks associated with using general anaesthetics and advances in medicine, such as 'key hole surgery' and the use of lasers, are diminishing the need for them.

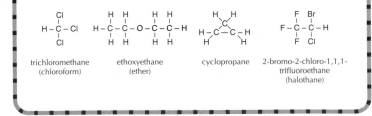

trichloromethane (chloroform)   ethoxyethane (ether)   cyclopropane   2-bromo-2-chloro-1,1,1-trifluoroethane (halothane)

PARTIAL PRESSURES OF COMPONENT GASES

It is important that anaesthetists administer the correct gas and air mixture. Too much anaesthetic and the patient may suffer oxygen starvation or even death, too little and the patient may recover consciousness prematurely and feel pain. The partial pressure of a component gas in a mixture is equal to the pressure it would exert if it occupied the total volume on its own. Dalton's law states that the total pressure exerted by a mixture of ideal gases is equal to the sum of the partial pressures of all the component gases.

e.g. consider a 5 dm³ flask containing N_2O at 3 atm pressure connected by a tap to a 2 dm³ flask containing O_2 at 4 atm pressure. When the tap is opened, the total pressure of the mixture will be $3\frac{2}{7}$ atm.

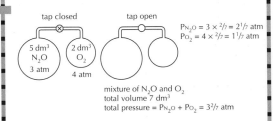

$P_{N_2O} = 3 \times \frac{2}{7} = 2\frac{1}{7}$ atm
$P_{O_2} = 4 \times \frac{2}{7} = 1\frac{1}{7}$ atm

mixture of N_2O and O_2
total volume 7 dm³
total pressure $= P_{N_2O} + P_{O_2} = 3\frac{2}{7}$ atm

Partial pressures can also be expressed in terms of mole fractions. In a mixture of two gases A and B where the number of moles of A is n_A and the number of moles of B is n_B then:

$$P_A = \frac{n_A}{n_A + n_B} \times P_{total}$$

$$P_B = \frac{n_B}{n_A + n_B} \times P_{total} \text{ and } P_{total} = P_A + P_B$$

THE INDOLE RING STRUCTURE

The hallucinogenic drugs LSD, psilocybin, and psilocin are all amines and contain the indole ring structure. Mescaline is also an amine and hallucinogenic and contains a structure similar to indole, except that the five-membered nitrogen ring is not closed.

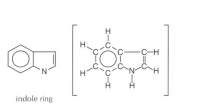

indole ring

LYSERGIC ACID DIETHYLAMIDE LSD

LSD does not occur naturally but is derived from ergot, a fungus that grows on wheat. Only 0.028 mg of LSD will produce a noticeable effect in most people. The short term effects, which last for about 12 hours, include restlessness, dizziness, the desire to laugh, and distortions in sound and visual perceptions. In some cases unpleasant hallucinations lead to feelings of despair and suicide may be attempted. LSD takers may believe they are able to fly and driving under the influence can be extremely dangerous.

The long term risks include severe depression and the possibility of recurrences of LSD effects ('flashbacks') months or even years afterwards. There appears to be no lasting physical ill effects but LSD may cause psychological dependence. LSD has been used medically in psychotherapy, but is no longer used legitimately as it may lead to psychosis in susceptible patients. It is thought that LSD works by blocking the action of serotonin, one of the compounds responsible for transmitting impulses across synapses in the brain.

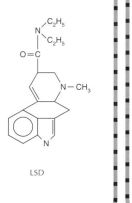

LSD

PSILOCYBIN

Psilocybin and the structurally related compound psilocin are found in the liberty cap mushroom *Psilocybe semilanceata* ('magic mushroom'). They are only mildly hallucinogenic causing slight alteration of perception and distortion of the senses together with feelings of exhilaration and insight. Although tolerance develops they are not addictive and it is not known if they can cause long term harm. The biggest dangers lie in the inability to recognize the mushrooms correctly, as similar-looking fungi are poisonous.

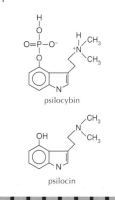

psilocybin

psilocin

MESCALINE

Mescaline is one of the oldest known hallucinogens. It comes from the peyote cactus found in South and Central America. Spanish explorers described the effects on the Mexican Indians in 1560 as 'terrible or ludicrous visions which lasted for two or three days'. Mescaline alters visual and auditory perceptions and appetite is reduced. It may cause unpleasant mental effects with people who are already anxious or depressed. Depending on how it is taken, other active substances in the plant may cause nausea and trembling. The effects of mescalin can be considerably enhanced when taken with alcohol. The long term effects are not clearly known but there is evidence that it could cause liver damage.

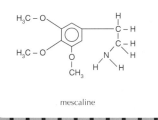

mescaline

MARIJUANA

Marijuana (also known as cannabis) is a mild hallucinogen made from the flowering tops, stems, leaves, and seeds of the hemp plant *Cannabis sativa*. Hashish, made from the resin of the plant is about five times stronger than marijuana. The active ingredient in marijuana is tetrahydrocannabinol (THC). The short term effects include a feeling of relaxation and enhanced auditory and visual perception. Loss of the sense of time, confusion, and emotional distress can also result. Hallucinations may occur in rare cases. It can have a synergistic effect and increase the risk of sedation with depressants. The long term effects include apathy and lethargy, and a lowering of fertility together with all the risks associated with tobacco smoking.

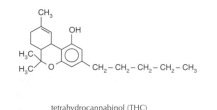

tetrahydrocannabinol (THC)

SHOULD MARIJUANA BE LEGALIZED?

Arguments for include:

- Medicinal use: research on the medical uses of cannabis is continuing but there is evidence that it is useful to prevent vomiting (anti-emetic), as a pain killer in cancer chemotherapy and AIDS, and for the treatment of glaucoma, epilepsy, Parkinson's disease, and Huntington's chorea.
- Lower crime rate – police freed to concentrate on other criminal activities.
- Freedom of the individual.

Arguments against include:

- Cost to society due to increased risk of heart disease and cancer through smoking.
- Claims by some that cannabis can lead on to harder drugs.
- Increased risk of dangerous driving while under the influence.

IB QUESTIONS – OPTION B – MEDICINES AND DRUGS

1. Viral infections are common but they are very difficult to treat by the use of drugs. One antiviral drug that has had some success, particularly in counteracting the effects of cold sores and shingles, is Acyclovir. Its structure is given below.

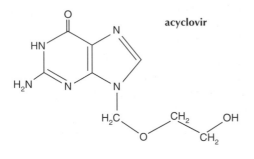

acyclovir

(a) Describe very briefly the basic structure of a virus. **[1]**

(b) Give **two** reasons why viruses are so difficult to deal with. **[2]**

(c) Identify the amide group in the structure of Acyclovir given above, by drawing a circle around it. **[1]**

(d) Explain why Acyclovir has a greater solubility in dilute acids than it has in water. **[2]**

(e) Many drugs, including Acyclovir, can be taken by mouth. However some drugs have to be administered by injection. Suggest **two** reasons why such drugs cannot be taken orally. **[2]**

2. The general structure of penicillin is given below.

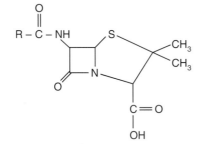

(a) The letter R in the above structure represents a side chain. State **one** reason why there are a number of different modifications of this side chain. **[1]**

(b) State why a prescribed course of penicillin should be completed. **[1]**

(c) Apart from the cost, what is an effect of the over prescription of penicillins? **[1]**

(d) What do you understand by the term 'broad based spectrum' antibiotics? **[1]**

(e) State which group of micro-organisms penicillins kill and briefly explain how they do this. **[3]**

3. The mild analgesic, aspirin, can be made from 2-hydroxybenzoic acid by reaction with ethanoyl chloride. The balanced equation for the reaction is,

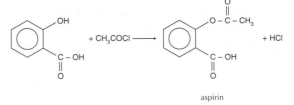

aspirin

(a) Write the molecular formula of 2-hydroxybenzoic acid. **[1]**

(b) Draw a circle around **one** of the functional groups in the structure of aspirin above. What is the name of the group and what is its effect on pH indicator solution? **[3]**

(c) Paracetamol can also be made from a simpler organic compound by reaction with ethanoyl chloride. Draw the structure of this compound. (The structure of paracetamol is shown in the IB Data Booklet). **[2]**

(d) Both aspirin and paracetamol are important mild analgesics, but like all drugs they have some disadvantages. Give **one** disadvantage in the use of each drug. **[2]**

4. (a) Cisplatin $Pt(NH_3)_2Cl_2$ is an effective anticancer drug. It bonds with the guanine in the DNA present in cancer cells and prevents it from replicating.

   (i) Draw the structure of *trans*-platin. **[1]**

   (ii) Describe the feature of guanine that enables it to bond with cisplatin and name the type of reaction that occurs when the bonds are formed. **[2]**

(b) Explain why it is important to carry out clinical trials on all the different enantiomers of a new drug. **[2]**

(c) Most reactions to form chiral compounds give a racemic mixture which then has to be separated into the two different enantiomers. Describe how a chiral auxilliary can be used to isolate the desired enantiomer of a particular drug. **[3]**

(d) The anticancer drug taxol can be synthesised using chiral auxilliaries. Part of its structure is shown below. Identify with an asterisk * **two** chiral centres. **[2]**

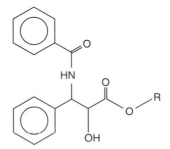

Diet

REQUIREMENTS FOR A HEALTHY DIET

The function of food is to keep the body functioning and healthy. It provides energy and replenishes chemicals. Good health requires a balanced diet that includes all the essential nutrients taken from as wide a variety of foods as possible. Nutrients can be divided into six main groups: proteins, carbohydrates, fats, vitamins, minerals, and water. The amount of each required depends on several factors, such as age, weight, gender, and occupation. A well-balanced diet consists of about 60% carbohydrate, 20–30% protein, and 10–20% fat. Foods containing these three components will also provide the essential vitamins and the fifteen essential minerals, which include calcium, magnesium, sodium, iron, and sulfur along with trace elements like iodine and chromium. In addition a daily intake of about 2 dm³ of water is required. Malnutrition can occur when either too little or too much of these essential components are taken.

Guidelines for healthy eating:
- eat a variety of foods
- maintain a healthy weight
- eat a diet low in fat, saturated fat, and cholesterol
- include plenty of fruit and vegetables
- use salt and sugar sparingly
- moderate the intake of alcohol.

Nutritional information for a 37.5 g serving of a typical wheat breakfast cereal.

Energy	540 kJ (128 kcal)	Sodium	0.1 g
Protein	4.2 g	Vitamins	
Carbohydrate	25.4 g	thiamin (B$_1$)	0.4 mg/32% RDA
(of which sugar)	(1.8 g)	riboflavin (B$_2$)	0.5 mg/32% RDA
Fat	1.0 g	niacin	5.7 mg/32% RDA
(of which saturates)	(0.2 g)	folic acid	63.8 µg/32% RDA
Fibre	3.9 g	Iron	4.5 mg/32% RDA
(soluble)	(1.2 g)		
(insoluble)	(2.7 g)	(RDA = recommended daily amount)	

ENERGY

The daily calorific intake for a moderately active woman is about 8400 kJ (2000 kcal) per day. For an adult male undertaking physical work this increases to about 14 700 kJ (3500 kcal). Energy is provided by fats, carbohydrates, and proteins. Carbohydrates provide the main source of energy but like proteins are already partially oxidized so do not provide as much energy weight-for-weight as fats which are used to store energy.

BENEFITS AND CONCERNS OF GM FOOD

Genetic engineering involves the process of selecting a single gene for a single characteristic and transferring that stretch of DNA from one organism to another. An example of genetically modified food is the FlavrSavr tomato. In normal tomatoes a gene is triggered when they ripen to produce a chemical that makes the fruit go soft and eventually rot. In the FlavrSavr tomato the gene has been modified to 'switch off' the chemical so that the fruit can mature longer on the vine for a fuller taste and have a longer shelf life.

Benefits:
- improve flavour, texture, nutritional value, and shelf life of food
- could incorporate anti-cancer substances and reduce exposure to less healthy fats
- make plants more resistant to disease, herbicides, and insect attack.

Concerns:
- outcome of alterations uncertain as not enough is known about how genes operate
- may cause disease as antibiotic-resistant genes could be passed to harmful micro-organisms
- genetically engineered genes may escape to contaminate normal crops with unknown effects
- may alter the balance of delicate ecosystems as food chains become altered.

FOOD CALORIMETRY

The energy content of a food can be found by burning the food in a food calorimeter. A known mass of the food is heated electrically and burned in a supply of oxygen. The heat produced is transferred through a copper spiral to water and the temperature increase of the water recorded. The 'water equivalent' of the whole system is calibrated and the calorific value of the food determined. Calorific values are typically recorded as kcal per 100 g or more commonly as kJ per 100 g. 1 kcal = 4.18 kJ.

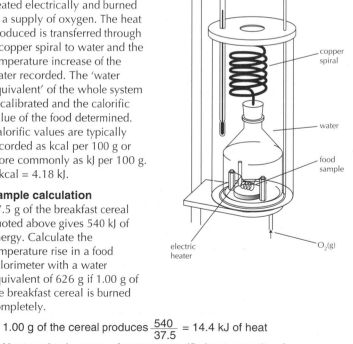

thermometer · stirrer · copper spiral · water · food sample · electric heater · O$_2$(g)

Sample calculation
37.5 g of the breakfast cereal quoted above gives 540 kJ of energy. Calculate the temperature rise in a food calorimeter with a water equivalent of 626 g if 1.00 g of the breakfast cereal is burned completely.

1.00 g of the cereal produces $\frac{540}{37.5}$ = 14.4 kJ of heat

Heat evolved = mass of water × specific heat capacity of water × temperature rise (ΔT)

$$\Delta T = \frac{14.4 \text{ kJ}}{0.626 \text{ kg} \times 4.18 \text{ kJ kg}^{-1} \text{ }^{\circ}\text{C}^{-1}} = 5.50 \text{ }^{\circ}\text{C}$$

Structure and uses of proteins

STRUCTURE OF PROTEINS

Proteins are large macromolecules made up of chains of 2-amino acids. About twenty 2-amino acids occur naturally. The amino acids bond to each other through condensation reactions resulting in the formation of a polypeptide, in which the amino acid residues are joined to each other by an amide link (peptide bond).

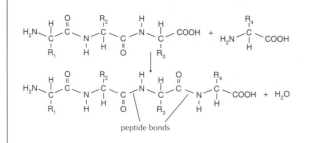

peptide bonds

Each protein contains a fixed number of amino acid residues connected to each other in strict sequence. This sequence, e.g. gly-his-ala-ala-leu- ... is known as the primary structure of proteins. The secondary structure describes the way in which the chain of amino acids folds itself due to intramolecular hydrogen bonding. The folding can either be α-helix in which the protein twists in a spiralling manner rather like a coiled spring, or β-pleated to give a sheet-like structure.

The tertiary structure describes the overall folding of the chains by interactions between distant amino acids to give the protein its three-dimensional shape. These interactions may be due to hydrogen bonds, van der Waals' attraction between non-polar side groups, and ionic attractions between polar groups. In addition two cysteine residues can form **disulfide bridges** when their sulfur atoms undergo oxidation.

Examples of interactions between side groups on polypeptide chains:

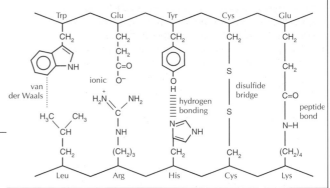

Separate polypeptide chains can interact together to give a more complex structure – this is known as the quaternary structure. Haemoglobin has a quaternary structure that includes four protein chains (two α-chains and two β-chains) grouped together around four haem groups.

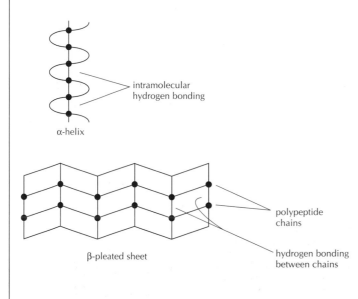

intramolecular hydrogen bonding

α-helix

polypeptide chains

hydrogen bonding between chains

β-pleated sheet

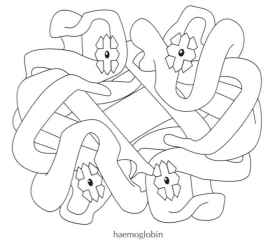

haemoglobin

USES OF PROTEINS

Proteins have many different functions in the body. They can act as biological catalysts for specific reactions (enzymes). They can give structure (e.g hair and nails consist almost entirely of polypeptides coiled into α-helices), and provide a source of energy. Some hormones are proteins, e.g. FSH (follicle stimulating hormone), responsible for triggering the monthly cycle in females.

Analysis of proteins

The primary structure of proteins can be determined either by paper chromatography or by electrophoresis. In both cases the protein must first be hydrolysed by hydrochloric acid to successively release the amino acids. The three-dimensional structure of the complete protein can be confirmed by X-ray crystallography.

PAPER CHROMATOGRAPHY

A small spot of the unknown amino acid sample is placed near the bottom of a piece of chromatographic paper. Separate spots of known amino acids can be placed alongside. The paper is placed in a solvent (eluent), which then rises up the paper due to capillary action. As it meets the sample spots the different amino acids partition themselves between the eluent and the paper to different extents, and so move up the paper at different rates. When the eluent has nearly reached the top, the paper is removed from the tank, dried, and then sprayed with an organic dye (ninhydrin) to develop the chromatogram by colouring the acids. The positions of all the spots can then be compared.

If samples of known amino acids are not available the R_f value (retention factor) can be measured and compared with known values as each amino acid has a different R_f value. It is possible that two acids will have the same R_f value using the same solvent, but different values using a different solvent. If this is the case the chromatogram can be turned through 90° and run again using a second solvent.

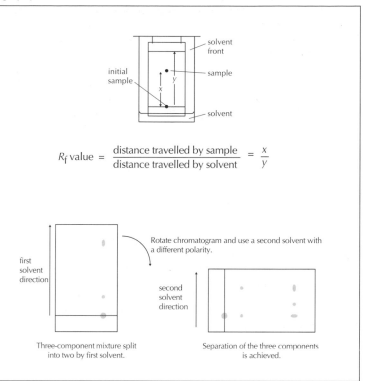

$$R_f \text{ value} = \frac{\text{distance travelled by sample}}{\text{distance travelled by solvent}} = \frac{x}{y}$$

Three-component mixture split into two by first solvent.

Separation of the three components is achieved.

ELECTROPHORESIS

The structure of amino acids alters at different pH values. At low pH (acid medium) the amine group will be protonated. At high pH (alkaline medium) the carboxylic acid group will lose a proton. For each amino acid there is a unique pH value (known as the **isoelectric point**) where the acid will exist as the zwitterion.

The medium on which electrophoresis is carried out is usually a polyacrylamide gel. So the process is known as PAGE (polyacrylamide gel electrophoresis). The sample is placed in the centre of the gel and a potential difference applied across it. Depending on the pH of the buffer the different amino acids will move at different rates towards the positive and negative electrodes. At its isoelectric point a particular amino acid will not move as its charges are balanced. When separation is complete the acids can be sprayed with ninhydrin and identified by comparing the distance they have travelled with standard samples, or from a comparison of their isolelectric points.

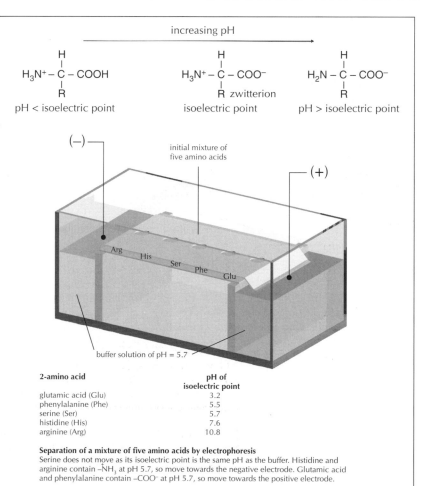

2-amino acid	pH of isoelectric point
glutamic acid (Glu)	3.2
phenylalanine (Phe)	5.5
serine (Ser)	5.7
histidine (His)	7.6
arginine (Arg)	10.8

Separation of a mixture of five amino acids by electrophoresis
Serine does not move as its isoelectric point is the same pH as the buffer. Histidine and arginine contain $-NH_3$ at pH 5.7, so move towards the negative electrode. Glutamic acid and phenylalanine contain $-COO^-$ at pH 5.7, so move towards the positive electrode.

Carbohydrates

Carbohydrates are produced by photosynthesis in plants. The overall stoichiometric process can be represented as:

$$6CO_2(g) + 6H_2O(l) \xrightarrow[\text{chlorophyll}]{\text{light}} \underset{\text{glucose}}{C_6H_{12}O_6(s)} + 6O_2(g)$$

Glucose and other simple carbohydrates are then converted into more complicated carbohydrates, such as starch and cellulose.

MONOSACCHARIDES

All monosaccharides have the empirical formula CH_2O. In addition they contain a carbonyl group ($\mathord{>}C{=}O$) and at least two –OH groups. They have between three and six carbon atoms. Monosaccharides with the general formula $C_5H_{10}O_5$ are known as pentoses (e.g. ribose) and monosaccharides with the general formula $C_6H_{12}O_6$ are known as hexoses (e.g. glucose).

Many structural isomers of monosaccharides are possible. In addition several carbon atoms are chiral (asymmetric) and give rise to optical isomerism. To make matters worse open chain structures and ring structures are possible. The form of glucose that is found in nature is known as D-glucose.

```
      ¹CHO
       |
  H—²C—OH
       |
 HO—³C—H
       |
  H—⁴C—OH
       |
  H—⁵C—OH
       |
      ⁶CH₂OH

   D-glucose
```

Straight chain formula of D-glucose

The ring structure of D-glucose can exist in two separate crystalline forms known as α-D-glucose and β-D-glucose. Note: the only difference is that the –OH group on the first carbon atom is inverted.

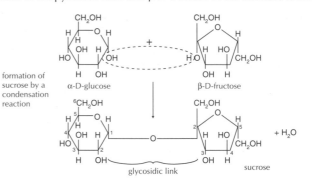

α-D-glucose   β-D-glucose

Six-membered ring monosaccharides are known as pyranoses. Hexoses can also have a furanose structure where they have a five-membered ring containing an oxygen atom.

MAJOR FUNCTIONS OF POLYSACCHARIDES IN THE BODY

Carbohydrates are used by humans:
- **to provide energy:** foods such as bread, biscuits, cakes, potatoes, and cereals are all high in carbohydrates
- **to store energy:** starch is stored in the livers of animals in the form of glycogen – also known as animal starch. Glycogen has almost the same chemical structure as amylopectin
- **as precursors** for other important biological molecules, e.g. they are components of nucleic acids and thus play an important role in the biosynthesis of proteins.

(Although humans cannot digest cellulose it plays an important role in the structure of plants. A high fibre diet (roughage) has been shown to prevent bowel cancer.)

POLYSACCHARIDES

Monosaccharides can undergo condensation reactions to form disaccharides and eventually polysaccharides. For example, sucrose, a disaccharide formed from the condensation of α-D-glucose in the pyranose form and β-D-fructose in the furanose form.

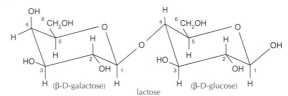

formation of sucrose by a condensation reaction

α-D-glucose   β-D-fructose

glycosidic link   sucrose $+ H_2O$

The link between the two sugars is known as a glycosidic link. In the case of sucrose the link is between the C-1 atom of glucose in the α-configuration and the C-2 atom of fructose. The link is known as an β-1,2 bond.

Lactose is a disaccharide in which the β-D-galactose is linked at the C-1 atom to the C-4 atom of β-D-glucose. This is called a β-1,4 bond.

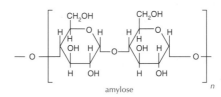

(β-D-galactose)   (β-D-glucose)

lactose

One of the most important polysaccharides is starch. Starch exists in two forms: amylose, which is water soluble, and amylopectin, which is insoluble in water.

Amylose is a straight chain polymer of α-D-glucose units with α-1,4 bonds:

amylose

Amylopectin also consists of α-D-glucose units but it has a branched structure with both α-1,4 and α-1,6 bonds:

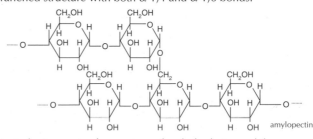

amylopectin

Most plants use starch as a store of carbohydrates and thus energy.

Fats

STRUCTURE AND MELTING POINTS OF FATS

Fats and oils are **triesters (triglycerides)** formed from the condensation reaction of propane-1,2,3-triol (glycerol) with long chain carboxylic acids (fatty acids).

Fats are solid triglycerides; examples include butter, lard, and tallow. Oils are liquid at room temperature and include castor oil, olive oil, and linseed oil. The essential chemical difference between them is that fats contain saturated carboxylic acid groups (i.e. they do not contain C=C double bonds). Oils contain at least one double bond and are said to be unsaturated. Most oils contain several double bonds and are said to be polyunsaturated.

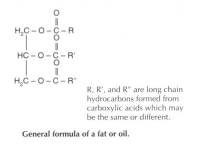

R, R', and R'' are long chain hydrocarbons formed from carboxylic acids which may be the same or different.

General formula of a fat or oil.

Generally polyunsaturated oils are thought to be better for health than fats as they reduce the risk of heart disease. A diet high in saturated fats can produce high levels of cholesterol in the body which can lead to the blocking of arteries.

Stearic acid (m.pt 69.6 °C) and linoleic acid (m.pt –5 °C) both contain the same number of carbon atoms and have similar molar masses. However, linoleic acid contains two double bonds. Generally the more unsaturated the fatty acid the lower its melting point. The regular tetrahedral arrangement of saturated acids means that they can pack together closely, so the van der Waals' forces holding molecules together are stronger as the surface area between them is greater. As the bond angle at the C=C double bonds changes from 109.5° to 120° in unsaturated acids it produces a 'kink' in the chain. They are unable to pack so closely and the van der Waals' forces between the molecules become weaker, which results in lower melting points. This packing arrangement is similar in fats and explains why unsaturated fats (oils) have lower melting points.

stearic acid – a saturated acid

$CH_3-(CH_2)_4-CH=CH-CH_2-CH=CH-(CH_2)_7COOH$

linoleic acid – an unsaturated acid

Name		Number of C atoms per molecule	Number of C=C bonds	Melting point / °C
saturated fatty acids				
lauric acid	$CH_3(CH_2)_{10}COOH$	12	0	44.2
myristic acid	$CH_3(CH_2)_{12}COOH$	14	0	54.1
palmitic acid	$CH_3(CH_2)_{14}COOH$	16	0	62.7
stearic acid	$CH_3(CH_2)_{16}COOH$	18	0	69.6
unsaturated fatty acids				
oleic acid	$CH_3(CH_2)_7CH=CH(CH_2)_7COOH$	18	1	10.5
linoleic acid	$CH_3(CH_2)_4CH=CHCH_2CH=CH(CH_2)_7COOH$	18	2	–5

DETERMINING THE NUMBER OF C=C BONDS IN AN UNSATURATED FAT

Unsaturated fats can undergo addition reactions. Margarine is made by the hydrogenation of vegetable oils so that it is a solid at room temperature. The addition of iodine to unsaturated fats can be used to determine the number of C=C double bonds since one mole of iodine will react quantitatively with one mole of double bonds. Iodine is coloured. As the iodine is added to the unsaturated fat the purple colour of the iodine will disappear as the addition reaction takes place. Often fats are described by their iodine number, which is the number of grams of iodine that add to 100 g of the fat.

unsaturated fat   diiodo-addition product

MAJOR FUNCTIONS OF FATS IN THE BODY

Fats provide a very efficient way for the body to store energy. Fats contain proportionately less oxygen than carbohydrates, so when they are oxidized in the body they release more energy. Fats are stored in adipose (fatty) tissue, which provides both thermal insulation and protection to parts of the body. Fats also form part of cell membranes.

SOAP

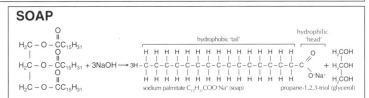

sodium palmitate $C_{15}H_{31}COO^-Na^+$ (soap)   propane-1,2,3-triol (glycerol)

In the presence of alkali fats can be hydrolysed to form soap. This process (known as **saponification**) is the reverse of esterification and soap is the sodium (or potassium) salt of the fatty acids produced. Soaps function because their long non-polar hydrophobic 'tail' dissolves in oil or grease to form a **micelle**. The micelle is surrounded by the polar hydrophilic 'heads' of the soap molecules, which make it soluble in water. Soaps do not work well in hard water as the Ca^{2+}

grease or oil on fibre — hydrophobic 'tails' mix with grease — hydrophilic 'heads' — grease-containing micelle lifted away from fibre — hydrophilic 'heads' dissolve micelle in water

and Mg^{2+} ions present in the water cause the precipitation of the insoluble calcium or magnesium salts of the fatty acids known as 'scum'. Synthetic detergents, such as alkyl benzene sulfonates, do not suffer from this problem as their calcium or magnesium salts are soluble. Soaps however do have the advantage that they are readily biodegradeable, whereas synthetic detergents (particularly those containing branched alkyl chains) cause more pollution.

$CH_3(CH_2)_{10}\bigcirc SO_3^-Na^+$

example of a synthetic detergent

Vitamins

GENERAL PROPERTIES OF VITAMINS

Vitamins are substances with very different structures. Apart from vitamin D the body is not capable of synthesizing vitamins and cannot function correctly without them, so they must be obtained from food. A poor diet may lead to deficiency diseases. Vitamins can be classified as fat soluble or water soluble.

The structure of fat soluble vitamins is characterized by long non-polar hydrocarbon chains or rings. These include vitamins A, D, E, F, and K. They can accumulate in the fatty tissues of the body. In some cases an excess of fat soluble vitamins can be as serious as a deficiency.

The molecules of water soluble vitamins, such as vitamin C and the eight B-group vitamins, contain hydrogen attached directly to electronegative oxygen or nitrogen atoms that can hydrogen bond with water molecules. They do not accumulate in the body so a regular intake is required.

Prolonged cooking destroys most vitamins. When boiling vegetables it is better to use small amounts of water and use the stock in gravy or soups to avoid loss of water soluble vitamins. Vitamins containing C=C double bonds and –OH groups are readily oxidized and keeping food refrigerated slows down this process.

VITAMIN A (RETINOL)

Vitamin A is found in cod liver oil, green vegetables, and fruit. Carrots also indirectly provide a good source as they contain β-carotene, which the body converts to vitamin A. Although it does contain one –OH group, vitamin A is fat soluble due to the long non-polar hydrocarbon chain. Unlike most other vitamins, it is not broken down readily by cooking. Vitamin A is an aid to night vision. In the body vitamin A (retinol) is oxidized to retinal.

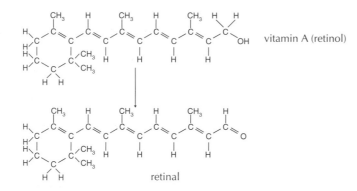

vitamin A (retinol)

retinal

The retinal combines with the protein opsin to form rhodopsin, the active agent for converting light signals into electrical signals that can travel along the optic nerve to the brain. A deficiency in vitamin A leads to night blindness. A serious deficiency can lead to xeropthalmia, a chronic form of conjunctivitis, which is the commonest form of blindness in the Third World.

VITAMIN D (CALCIFEROL)

Vitamin D is also fat soluble. It is found in fish liver oils and in egg yolk. It can be formed on the surface of the skin through the action of ultraviolet light in sunlight on 7-dehydrocholesterol.

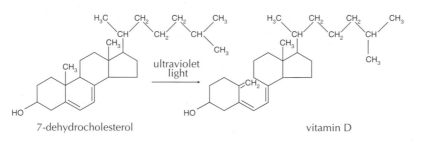

7-dehydrocholesterol   ultraviolet light   vitamin D

Vitamin D is involved in the uptake of calcium and phosphate ions from food into the body and especially in the formation of bone structure. A deficiency of vitamin D leads to bone softening and malformation – a condition known as rickets. Vitamin D can be destroyed through oxidation by some of the bleaching agents used in the manufacture of purified flour.

VITAMIN C (ASCORBIC ACID)

Vitamin C is found in fresh fruit and vegetables. Due to the large number of polar –OH groups it is soluble in water so is not retained for long by the body.

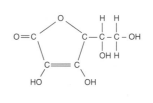

vitamin C (ascorbic acid)

It is broken down by cooking, so raw vegetables provide a better source than boiled vegetables. The various roles played by vitamin C are not fully understood. Claims that it is effective in preventing both the common cold and cancer appear to be unfounded. It does aid the healing of wounds and helps to prevent bacterial infections. It is involved in the biosynthesis of the protein collagen found in connective tissue, such as cartilage, ligaments, and tendons.

The most famous disease associated with a lack of vitamin C is scorbutus ('scurvy'). The symptoms are swollen legs, rotten gums, and bloody lesions. It was a common disease in sailors, who spent long periods without fresh food, until the cause was recognized. Vitamin C is easily oxidized. It can help to preserve food by being more readily oxidized than the food it is preserving.

Hormones

Hormones are chemicals produced in glands and transported to the site of action by the blood stream. The glands themselves are controlled by the pituitary gland, which in turn is controlled by the hypothalamus. Hormones act as chemical messengers and perform a variety of different functions. Examples of specific hormones include adrenaline, thyroxine, insulin, and the sex hormones.

ADRENALINE

Adrenaline (epinephrine) is produced in the adrenal glands – two small organs located above the kidneys. It is a stimulant closely related to the amphetamine drugs. It is released in times of excitement and causes a rapid dilation of the pupils and airways and increases heartbeat and the rate of release of sugar into the bloodstream. It is sometimes known as the 'fight or flight' hormone.

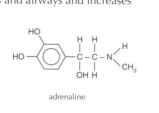

adrenaline

SEX HORMONES

The sex hormones are all steroids. Steroids contain a characteristic four-ring structure. The basic building block for the other steroids is cholesterol formed in the liver and found in all tissues, the blood, brain, and

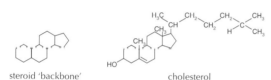

steroid 'backbone'   cholesterol

spinal cord. The male sex hormones are produced in the testes and comprise mainly testosterone and androsterone. They are anabolic – encouraging tissue, muscle, and bone growth – and androgenic – conferring

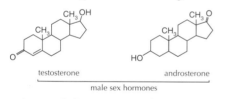

testosterone   androsterone

male sex hormones

the male sexual characteristics. The female sex hormones are structurally very similar with just small changes in the functional groups attached to the steroid framework. They are produced in the ovaries from puberty until menopause. The two main female sex hormones are oestradiol and progesterone. They are responsible for sexual development and for the menstrual and reproductive cycles in women.

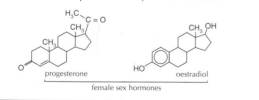

progesterone   oestradiol

female sex hormones

THYROXINE

Thyroxine is produced in the thyroid gland located in the neck. It is unusual in that it contains iodine. A lack of iodine in the diet can cause the thyroid gland to swell to produce the condition known as goitre. Thyroxine regulates the body's metabolism. Low levels of thyroxine cause hypothyroidism, characterized by lethargy as well as sensitivity to cold and a dry skin. An overactive thyroid gland can cause the opposite effect. This is known as hyperthyroidism with the symptoms of anxiety, weight loss, intolerance to heat, and protruding eyes.

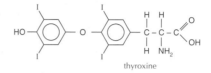

thyroxine

INSULIN

Insulin is a protein containing fifty-one amino acid residues. It is formed in the pancreas – an organ located at the back of the abdomen – and regulates blood sugar levels. In diabetics the levels of insulin are low or absent and glucose is not transferred sufficiently from the bloodstream to the tissues. This is known as hyperglycaemia and results in thirst, weight loss, lethargy, coma, and circulation problems. Long term sufferers of diabetes can suffer blindness, kidney failure, and need limbs amputated due to poor circulation. It is treated by reducing sugar intake and taking daily insulin injections. Too much insulin can cause hypoglycaemia, where the blood sugar level falls resulting in dizziness and fainting.

ORAL CONTRACEPTIVES

At the beginning of the menstrual cycle the pituitary releases follicle stimulating hormone (FSH). FSH travels to the ovaries causing the release of oestradiol, which prepares for the release of the ovum or egg and the build up of the uterine wall. After about two weeks a feedback system stops the release of FSH and triggers the release of the luteinizing hormone (LH). This travels to the ovaries and releases progesterone. The progesterone causes the egg to be transported to the uterus as well as continuing to build up the uterine wall. If the egg is fertilized it embeds itself into the uterine wall and hormone levels rise dramatically, otherwise hormone levels fall and menstruation begins.

The most common 'pill' contains a mixture of oestradiol and progesterone and mimics pregnancy by intentionally keeping the hormones at a high levels so that no more eggs are released. It is usual to take the pill for 21 days and then a placebo for 7 days so that a mild period will result, but without the risk that the hormone levels will fall and allow the unexpected release of an egg. Oestradiol and progesterone may also be given to post menopausal women as hormone replacement therapy (HRT) partly to prevent brittle bone disease (osteoporosis).

ANABOLIC STEROIDS

The anabolic steroids have similar structures to testosterone and build up muscle. They may be given to someone recuperating from a serious illness to build up muscles weakened by inactivity. Some athletes have abused these drugs as they can enhance athletic performance. Competitors are given random urine tests to detect these and other banned substances.

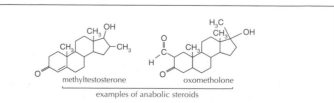

methyltestosterone   oxometholone

examples of anabolic steroids

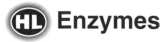

CATALYTIC ACTIVITY AND ACTIVE SITE

Enzymes are protein molecules that catalyse biological reactions. Each enzyme is highly specific for a particular reaction, and extremely efficient, often being able to increase the rate of reaction by a factor greater than 10^8. Enzymes work by providing an alternative pathway for the reaction with a lower activation energy, so that more of the reactant particles (substrate) will possess the necessary minimum activation energy.

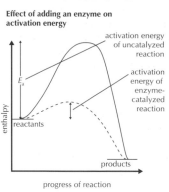

Effect of adding an enzyme on activation energy

activation energy of uncatalyzed reaction

activation energy of enzyme-catalyzed reaction

E_a

enthalpy

reactants

products

progress of reaction

The specificity of enzymes depends on their tertiary and quaternary structure. The part of an enzyme that reacts with the substrate is known as the active site. This is a groove or pocket in the enzyme where the substrate will bind. The site is not necessarily rigid but can alter its shape to allow for a better fit – known as the induced fit theory.

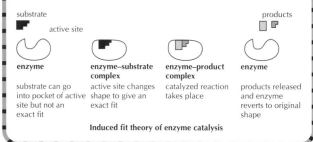

substrate

active site

products

enzyme
substrate can go into pocket of active site but not an exact fit

enzyme–substrate complex
active site changes shape to give an exact fit

enzyme–product complex
catalyzed reaction takes place

enzyme
products released and enzyme reverts to original shape

Induced fit theory of enzyme catalysis

COMPETITIVE AND NON-COMPETITIVE INHIBITION

Inhibitors are substances that slow down the rate of enzyme-catalyzed reactions. Competitive inhibitors resemble the substrate in shape, but cannot react. They slow down the reaction because they can occupy the active site on the enzyme thus making it less accessible to the substrate. Non-competitive inhibitors also bind to the enzyme, but not on the active site. This causes the enzyme to change its shape so that the substrate cannot bind. As the substrate concentration is increased the effect of competitive inhibitors lessens, as there is increased competition for the active sites by the substrate. With non-competitive inhibitors increasing the substrate concentration has no effect, as the enzyme's shape still remains altered.

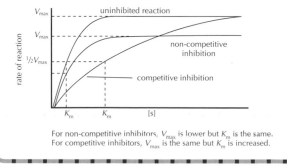

Effect of substrate concentration on inhibitors

V_{max} — uninhibited reaction

V_{max}

non-competitive inhibition

$1/2 V_{max}$

competitive inhibition

rate of reaction

K_m K_m [s]

For non-competitive inhibitors, V_{max} is lower but K_m is the same.
For competitive inhibitors, V_{max} is the same but K_m is increased.

ENZYME KINETICS

At low substrate concentrations the rate of reaction is proportional to the concentration of the substrate. However, at higher concentrations the rate reaches a maximum known as V_{max}. This can be

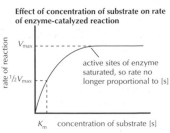

Effect of concentration of substrate on rate of enzyme-catalyzed reaction

V_{max}

rate of reaction

$1/2 V_{max}$

active sites of enzyme saturated, so rate no longer proportional to [s]

K_m concentration of substrate [s]

explained in terms of enzyme saturation. At low substrate concentrations there are enough active sites present for the substrate to bind to and react. Once all the sites are used up the enzyme cannot work any faster.

The Michaelis–Menten constant K_m is the substrate concentration when the rate of the reaction is $\frac{1}{2} V_{max}$. K_m for a particular enzyme with a particular substrate will always be the same. It indicates whether the enzyme functions efficiently at low substrate concentrations, or whether high substrate concentrations are necessary for efficient catalysis.

EFFECT OF TEMPERATURE, pH, AND HEAVY METAL IONS ON ENZYME ACTIVITY

The action of an enzyme depends on its specific shape. Increasing the temperature will initially increase the rate of enzyme-catalyzed reactions, as more of the reactants will possess the minimum activation energy. The optimum temperature for most enzymes is about 40 °C. Above this temperature enzymes rapidly become denatured as the weak bonds holding the tertiary structure together break.

At different pH values the charges on the amino acid residues change affecting the bonds between them, and so altering the tertiary structure and making the enzyme ineffective. Heavy metals can poison enzymes by reacting with –SH groups replacing the hydrogen atom with a heavy metal atom, or ion so that the tertiary structure is altered.

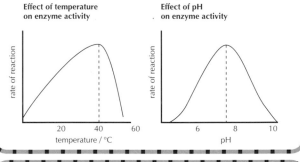

Effect of temperature on enzyme activity

rate of reaction

20   40   60

temperature / °C

Effect of pH on enzyme activity

rate of reaction

6   8   10

pH

USE OF ENZYMES IN BIOTECHNOLOGY

Biotechnology is the use of enzymes to catalyze reactions in industrial processes. Enzymes in yeast have been used for millenia to make alcohol and bread. Modern techniques use natural or genetically engineered enzymes immobilized on a solid carrier so that they can be reused many times. Uses include proteases in biological detergents to digest protein stains and fructose syrup production in which starch is broken down enzymatically into glucose and is then partially isomerized by another enzyme into fructose. Enzymes are also being developed to act as biosensors. Biosensors can detect the presence of substrate molecules, and send signals via an electronic transducer to give information about the amount of substrate present.

Nucleic acids

STRUCTURE OF NUCLEOTIDES AND NUCLEIC ACIDS

Almost all cells in the human body contain DNA (deoxyribonucleic acid). DNA and a related material RNA (ribonucleic acid) are macromolecules with relative molar masses of up to several million. Both nucleic acids are made up of repeating base-sugar-phosphate units called **nucleotides**. A nucleotide of DNA contains deoxyribose (a pentose sugar), a phosphate group, and one of four nitrogen-containing bases, adenine, guanine, cytosine, or thymine. RNA contains a different sugar, ribose, but also contains a phosphate group and four nitrogen-containing bases. Three of the bases are the same as those in DNA but the fourth, uracil, replaces thymine.

In DNA the polynucleotide units are wound into a helical shape with about 10 nucleotide units per complete turn. Two helices are then held together by hydrogen bonds between the bases to give the characteristic double helix structure. The hydrogen bonds are very specific. Cytosine can only hydrogen bond with guanine and adenine can only hydrogen bond with thymine (uracil in RNA).

sugars

bases (showing complementary hydrogen bonding)

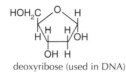

deoxyribose (used in DNA)

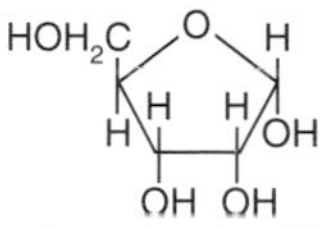

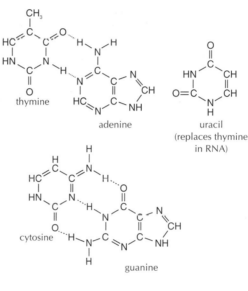

thymine

adenine

uracil (replaces thymine in RNA)

ribose (used in RNA)

cytosine

guanine

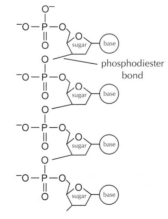

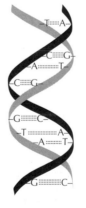

Nucleotides condense to form a polynucleotide. Each nucleotide is joined by a phosphodiester bond between C_3 of the sugar and the neighbouring phosphate group.

The double helix structure of DNA is shown here. Note the hydrogen bonds between the two different strands of polynucleotides.

THE GENETIC CODE

When cells divide the genetic information has to be replicated intact. The genetic information is stored in chromosomes found inside the nucleus. In humans there are 23 pairs of chromosomes. Chromosomes are effectively a very long DNA sequence. The DNA in the cell starts to partly unzip as hydrogen bonds between the bases break. Sugar base units will be picked up from the aqueous solution to form a complementary new strand. Because adenosine can only hydrogen bond with thymine (A–T) and cytosine can only hydrogen bond with guanine (C–G) the new strand formed will be identical to the original.

DNA resides wholly in the nucleus, whilst protein synthesis takes places in the cytoplasm. The information required to make complex proteins is passed from the DNA to RNA by a similar unzipping process, except that the new strand of RNA contains a different sugar and uracil in place of thymine.

The coded information held by the DNA lies in the sequence of bases. Each sequence of three bases represents one amino acid and is known as the **triplet code**. The triplet code allows for up to 64 permutations known as **codons**. This is more than sufficient to represent the 20 amino acids and several different codons may represent the same amino acid. Consecutive DNA codons of AAA, TAA, AGA, GTG, and CTT will transcribe to RNA codons of UUU, AUU, UCU, CAC, and GAA which will cause part of a strand of a protein to be formed that contains the amino acid residues -Phe-Ile-Ser-His-Glu-. In 2000 the human genome – the complete sequence of bases in human DNA – was finally determined and published on the internet.

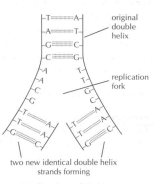

Replication of DNA

UUU	Phe	UCU	Ser	UAU	Tyr	UGU	Cys
UUC	Phe	UCC	Ser	UAC	Tyr	UGC	Cys
UUA	Leu	UCA	Ser	UAA	Terminator	UGA	Terminator
UUG	Leu	UCG	Ser	UAA	Terminator	UGG	Trp
CUU	Leu	CCU	Pro	CAU	His	CGU	Arg
CUC	Leu	CCC	Pro	CAC	His	CGC	Arg
CUA	Leu	CCA	Pro	CAA	Gln	CGA	Arg
CUG	Leu	CCG	Pro	CAG	Gln	CGG	Arg
AUU	Ile	ACU	Thr	AAU	Asn	AGU	Ser
AUC	Ile	ACC	Thr	AAC	Asn	AGC	Ser
AUA	Ile	ACA	Thr	AAA	Lys	AGA	Arg
AUG	Met	ACG	Thr	AAG	Lys	AGG	Arg
GUU	Val	GCU	Ala	GAU	Asp	GGU	Gly
GUC	Val	GCC	Ala	GAC	Asp	GGC	Gly
GUA	Val	GCA	Ala	GAA	Glu	GGA	Gly
GUG	Val	GCG	Ala	GAG	Glu	GGG	Gly

The genetic code carried by RNA

DNA PROFILING

DNA profiling (also known as DNA fingerprinting) was developed in 1985. In this process a small amount of cellular material is required. This may be obtained from blood, semen, hair, or saliva. The DNA is extracted and broken down into smaller fragments known as minisatellites using restriction enzymes. DNA not only contains coded triplets, but also has regions where there is no coded message in the base sequence. It is these regions where the splits occur to form the minisatellites which are unique to the person giving the sample (except in the case of identical twins). The fragments are separated into bands by gel electrophoresis. The resulting pattern is transferred to a nylon membrane and then labelled with radioactive ^{32}P which binds to particular bands of the DNA. X-ray film is then exposed to the radiation produced by the ^{32}P. The film is developed to give the characteristic 'fingerprint' of all the fragments. Use is made of this in court cases to positively identify murderers and rapists and to prove paternity. It also allows palaeontologists to map the evolutionary tree of extinct species.

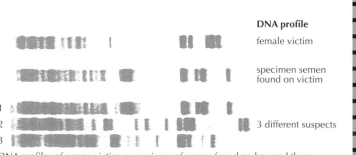

DNA profiles of a rape victim, a specimen of semen found on her, and three different suspects. Which one was the rapist?

Metal ions in biological systems

FUNCTIONS OF METAL IONS

There are a large number of metal ions that fulfil important biological roles in the body. The functions they fulfil depend on the different chemical and physical properties associated with the metal ions. Some depend on the different charge densities on two similar ions, some on the ability of the ion to show variable oxidation states, and some on the ability of the ion to form complexes with ligands.

Some important ions are:
- Na^+ and K^+ for nerve impulses and water balance
- Ca^{2+} for strong bones and teeth
- Cu^{2+} in enzymes, especially cytochrome oxidase, which helps convert food into energy
- Co^{2+} in vitamin B_{12}
- Fe^{2+} in haemoglobin.

SODIUM–POTASSIUM PUMP

Humans (and other mammals) have a much higher concentration of potassium ions inside the cell than outside and a much higher concentration of sodium ions outside the cell than inside. Some of the reasons for this are that cell protein synthesis requires high potassium ion concentrations, and nerve impulses require a concentration gradient between the inside and outside of a neurone (nerve cell). It would be expected that osmosis would occur through the partially permeable cell wall and the concentrations would equalize. To maintain the difference in concentrations the body uses a sodium–potassium pump, which requires energy obtained from the conversion of ATP (adenosine triphosphate) into ADP (adenosine diphosphate).

Protein structures in the cell wall act as one way valves continuously pumping sodium ions out of the cell and pumping potassium ions into the cell. The reason is mainly due to charge density. Both have a charge of +1 but the larger potassium ions can pass more easily through a medium of lower polarity, such as that provided by the protein pump. The pump itself is a protein molecule, which has three sites where Na^+ can bind and two sites where K^+ can bind. As it pumps it changes shape, due to phosphorylation from the ATP, so that Na^+ is removed from the cell and K^+ is drawn in.

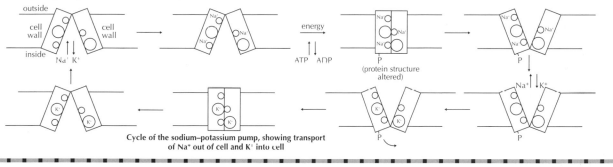

Cycle of the sodium–potassium pump, showing transport of Na^+ out of cell and K^+ into cell

ELECTRON TRANSPORT

Food does not burn in the body to produce energy, instead it is oxidized by a series of redox reactions involving the transport of electrons. The reactions take place in the mitochondria found inside cells. The oxidizing enzymes are called cytochromes. Cytochromes contain copper or iron. The metal ion in cytochrome oxidase is surrounded by a porphyrin ligand. This contains four nitrogen atoms each of which donates a pair of electrons, so that it occupies four of the sites around the metal ion and is said to be a tetradentate ligand. During each step of the the the overall oxidation of glucose the Fe^{3+} ion is reduced to Fe^{2+} (or the Cu^{2+} ion is reduced to Cu^+). In the reduction stage when oxygen is reduced to water the Fe^{2+} ion is oxidized back to Fe^{3+} (or Cu^+ is oxidized to Cu^{2+}).

Oxidation step:
$$C_6H_{12}O_6 + 6H_2O \longrightarrow 6CO_2 + 24H^+ + 24e$$
$$Fe^{3+} + e \longrightarrow Fe^{2+} \text{ (or } Cu^{2+} + e \rightarrow Cu^+)$$
metal ion reduced

Reduction step:
$$O_2 + 4H^+ + 4e \longrightarrow 2H_2O$$
$$Fe^{2+} \longrightarrow Fe^{3+} + e \text{ (or } Cu^+ \rightarrow Cu^{2+} + e)$$
metal ion oxidized

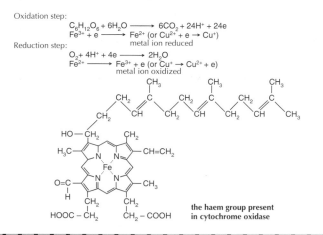

the haem group present in cytochrome oxidase

HAEMOGLOBIN AS AN OXYGEN CARRIER

The ability of iron to form complexes is also important in haemoglobin. Haemoglobin in the blood is responsible for carrying oxygen around the body during respiration. Haemoglobin contains four large polypeptide groups, and four iron ions each surrounded by flat porphyrin ligands known as haem groups. Haem is a prosthetic group, a group essential for the protein to be able to carry out its function.

At high oxygen concentrations haemoglobin binds to oxygen molecules. In the process the oxygen bonds onto the iron in the haem group as an extra ligand.

At low concentrations the reverse process occurs. Carbon monoxide and cyanide ions are poisonous as they form irreversible complex ions with the iron preventing it from carrying oxygen.

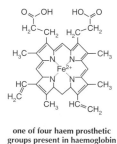

one of four haem prosthetic groups present in haemoglobin

1 (a) The structures of two important sex hormones, testosterone and oestradiol are given in the IB Data Booklet. Apart from the extra methyl group, state **two** different functional groups which are present in testosterone but are absent in oestradiol. Suggest **one** simple chemical test which you could use to distinguish between oestradiol and testosterone in a school laboratory. [3]

(b) Two synthetic steroids that are used in contraception are norethynodrel, the first commercially available oral contraceptive, and RU-486, the active ingredient of the 'morning after pill'.

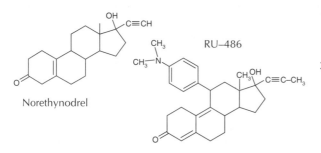

Norethynodrel   RU–486

Name **two** different functional groups that are present in both norethynodrel and RU-486 but are absent in progesterone. Name **two** other functional groups that are present in RU-486 but absent in both progesterone and norethynodrel. Describe briefly how norethynodrel functions as an oral contraceptive. [7]

2. The diagram below shows a triacyclglycerol (triglyceride) molecule.

$$CH_3(CH_2)_4CH=CHCH_2CH=CH(CH_2)_7CO-O-CH_2$$
$$CH_3(CH_2)_4CH=CHCH_2CH=CH(CH_2)_7CO-O-CH$$
$$CH_3(CH_2)_7CH=CH(CH_2)_7CO-O-CH_2$$

(a) To which of the main dietary groups does this molecule belong? [1]

(b) Is this molecule likely to have come from an animal or vegetable source? Explain your answer. [2]

(c) Is this molecule likely to be a solid or an oil? Explain your answer. [2]

(d) What are the major functions of this class of molecules in the body? [2]

3. The structures of vitamin A (retinol) and vitamin C (ascorbic acid) are given in the IB Data Booklet.

(a) (i) Name **two** functional groups which are present in retinol. [2]

(ii) By referring to the structures, classify vitamin A and vitamin C as water or fat soluble and account for the difference on the molecular level. [3]

(iii) State **one** physical symptom of each of vitamin A and vitamin C deficiency. State the common name given to vitamin C deficiency. [3]

(b) 0.014 moles of a particular oil was found to react exactly with 14.2 of iodine. Calculate the number of moles of iodine that reacted and state what can be deduced about the structure of the oil from this information. [3]

HL ───

4. (a) Iron combines reversibly with oxygen in haemoglobin. Apart from the ability to use different oxidation states, give **two** typical characteristics of transition metals that are shown by iron in haemoglobin. [2]

(b) The ability of haemoglobin to carry oxygen at body temperature depends on the concentration of oxygen, the concentration of carbon dioxide and on the pH. The graph shows how the percentage saturation of haemoglobin with oxygen changes with pH at different partial pressures of oxygen.

(i) The partial pressure of oxygen in active muscle is shown by the dotted line at 2.8 kPa. What is the difference in the percentage saturation of haemoglobin with oxygen in active muscle when the pH changes from 7.6 to 7.2? [1]

(ii) When the cells in muscles respire they excrete carbon dioxide and sometimes lactic acid as waste products. Explain how this affects the ability of haemoglobin to carry oxygen. [2]

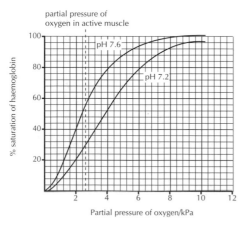

Primary air pollution

MAIN POLLUTANTS

Unpolluted air contains about 78% nitrogen, 21% oxygen, 1.0% argon, 0.03% carbon dioxide, and trace amounts of other gases together with up to 4% water vapour. An air pollutant can be described as a substance that is not normally present in air or a substance that is normally present but in excess amounts. The main *primary* air pollutants are carbon monoxide, oxides of nitrogen, sulfur dioxide, particulates, and hydrocarbons. *Secondary* pollutants are compounds formed when primary pollutants react in the air.

Pollutant	Natural source	Man-made source	Effect on health	Methods of reduction
Carbon monoxide (CO)	Incomplete oxidation of methane $CH_4 + 1\frac{1}{2}O_2 \rightarrow CO + 2H_2O$	Incomplete combustion of fossil fuels, e.g. $C_8H_{18} + 8\frac{1}{2}O_2 \rightarrow 8CO + 9H_2O$	Prevents haemoglobin from carrying oxygen by forming carboxyhaemoglobin	Use of lean burn engine, thermal exhaust reactor, or catalytic converter
Oxides of nitrogen (NO_x), e.g. N_2O, NO and NO_2	Electrical storms and biological processes	At high temperatures inside internal combustion engines $N_2 + O_2 \rightarrow 2NO$	Respiratory irritant leading to respiratory tract infections	Use of lean burn engine, recirculation of exhaust gases or catalytic converter
Sulfur dioxide (SO_2) (can be oxidized in the air to also form SO_3)	Oxidation of H_2S produced by volcanoes, and decay of organic matter	Combustion of sulfur-containing coal and smelting of sulfide ores $S + O_2 \rightarrow SO_2$	Respiratory irritant leading to respiratory tract infections	Removal of sulfur from fossil fuels before combustion. Alkaline scrubbing. Fluidized bed combustion
Particulates	Soot, ash, dust asbestos, sand, smoke, pollen, bacterial and fungal spores	Burning of fossil fuels, particularly coal and diesel	Can affect the respiratory system and cause lung diseases, such as emphysema, bronchitis, and cancer	Sedimentation chambers. Electrostatic precipitation
Hydrocarbons (C_xH_y or R–H)	Plants, e.g. rice. Many plants emit unsaturated hydrocarbons called terpenes	Unburned or partially burned gasoline and other fuels; solvents	Some (e.g. benzene) are carcinogenic. Can form toxic secondary pollutants, e.g. PANs (peroxyacylnitrates)	Catalytic converter

METHODS OF REMOVAL

Thermal exhaust reactor
Exhaust from the car engine is combined with more air and reacts due to the heat of the exhaust gases. Carbon monoxide is converted into carbon dioxide and unburned hydrocarbons are also combusted.

$$2CO(g) + O_2(g) \rightarrow 2CO_2(g)$$

Lean burn engines
By adjusting the carburettor the ratio air:fuel can be altered. The higher the ratio the less carbon monoxide emitted as more complete combustion occurs. Unfortunately this produces higher temperatures so more NO_x is produced. At lower ratios less NO_x but more carbon monoxide will be emitted.

Catalytic converter
The hot exhaust gases are passed over a catalyst of platinum, rhodium, or palladium. These fully oxidize carbon monoxide and unburned hydrocarbons and also catalyse the reaction between carbon monoxide and nitrogen oxide.

$$2CO(g) + 2NO(g) \rightarrow 2CO_2(g) + N_2(g)$$

Sulfur dioxide
Some sulfur is present in coal as metal sulfides (e.g. FeS) and can be removed physically by crushing the coal and mixing with water. The more dense sulfides sink to the bottom and the cleaned coal can be skimmed off. Sulfur is also removed from oil before it is refined by converting it into hydrogen sulfide.

Sulfur dioxide can be removed from the exhaust of coal burning plants by 'scrubbing' with an alkaline solution of limestone ($CaCO_3$) and lime (CaO). The resulting sludge is used for landfill or as gypsum ($CaSO_4.2H_2O$) to make plasterboard.

$$CaCO_3(s) + SO_2(g) \rightarrow CaSO_3(s) + CO_2(g)$$

$$CaO(s) + SO_2(g) \rightarrow CaSO_3(s)$$

$$2CaSO_3(s) + O_2(g) + 4H_2O(g) \rightarrow 2CaSO_4.2H_2O(s)$$

A more modern method known as fluidized bed combustion involves burning the coal on a bed of limestone which removes the sulfur as $CaSO_3$ or $CaSO_4$ as the coal burns.

wet alkaline scrubber

slurry of $CaCO_3$/CaO

input of gas containing CO_2

output of clean gas

sludge removed

Electrostatic precipitation
Particulates are solid or liquid particles suspended in the air. Larger particles can be allowed to settle under the influence of gravity in sedimentation chambers. For smaller particles an electrostatic precipitation chamber can be used. The charged particulates are attracted to the oppositely charged electrodes, which are shaken periodically so that the aggregated particulates fall to the bottom of the precipitator where they can be removed.

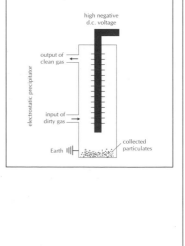

high negative d.c. voltage

electrostatic precipitator

output of clean gas

input of dirty gas

Earth

collected particulates

Ozone depletion (1)

FORMATION AND DEPLETION OF OZONE IN THE STRATOSPHERE

The ozone layer occurs in the stratosphere between about 12 km and 50 km above the surface of the Earth. Stratospheric ozone is in dynamic equilibrium with oxygen and is continually being formed and decomposed. The strong double bond in oxygen is broken by high energy ultraviolet light from the Sun to form atoms. These oxygen atoms are called radicals as they possess an unpaired electron and are very reactive. One oxygen radical can then react with an oxygen molecule to form ozone.

$$O{=}O(g) \xrightarrow{\text{UV (high energy)}} 2O^{\bullet}(g)$$
$$O^{\bullet}(g) + O_2(g) \rightarrow O_3(g)$$

The bonds in ozone are weaker so ultraviolet light of less energy will break them. When they are broken the reverse process happens and the ozone breaks down back to an oxygen molecule and an oxygen radical. The radical can then react with another ozone molecule to form two oxygen molecules.

$$O_3(g) \xrightarrow{\text{UV (lower energy)}} O_2(g) + O^{\bullet}(g)$$
$$O_3(g) + O^{\bullet}(g) \rightarrow 2O_2(g)$$

Overall the rate of production of ozone is equal to the rate of ozone destruction – this is known as a steady state. Because the formation and depletion of ozone absorbs a wide range of ultraviolet radiation the ozone layer serves to protect the surface of the Earth from this damaging radiation.

height (km)

90 THERMOSPHERE (MESOPAUSE)

MESOSPHERE

50 (STRATOPAUSE)

STRATOSPHERE

20 (TROPOPAUSE)

12

TROPOSHERE ⎫ contains 95% of mass of air
0 SURFACE ⎭

The Earth's atmosphere

Lewis structures of oxygen and ozone

⟨O=O⟩

Double bond in oxygen is stronger so requires more energy to break.

Two Lewis structures can be draw for ozone – known as resonance hybrids. The average oxygen-to-oxygen bond is between a single and a double bond so requires less energy to break.

LOWERING OF OZONE CONCENTRATION IN THE STRATOSPHERE

Measurements of the concentration of ozone in the stratosphere have shown that the amount of ozone in the 'ozone layer' has been decreasing. This is particularly true over both the south and north poles. The concentration above both the Antarctic and Arctic is seasonal. The biggest 'holes' occur during the winter and early spring. This decrease is due to particular chemicals produced and released by man. The main culprits are the CFCs – chlorofluorocarbons, the most common being dichlorodifluoromethane CCl_2F_2 also known as freon or CFC–12. CFCs were developed and used for refrigerants, propellants for aerosols, foaming agents for expanding plastics, and cleaning solvents. Other substances that also damage the ozone layer are oxides of nitrogen NO_x formed from internal combustion engines, power stations, and jet aeroplanes.

dichlorodifluoromethane, a chlorofluorocarbon (CFC)

ENVIRONMENTAL EFFECTS OF OZONE DEPLETION

Ultraviolet light has sufficient energy to damage biological molecules, such as amino acids, proteins, and nucleic acids. Because of ozone depletion more ultraviolet light has been reaching the Earth's surface.

Effect on humans
- Increase in sunburn.
- Increase in melanoma and non-melanoma skin cancer.
- Increase in eye cataracts and blindness.

Effect on marine ecosystems
- Marine phytoplankton loss. Since they produce much biomass this provides less food for other marine organisms and loss of the CO_2 'sink'.

Effect on plants
- More susceptible to disease.
- Growth inhibited.
- Photosynthesis inhibited.

Effect on weather
- Stratospheric convection currents altered. This affects wind and oceanic circulation.

ALTERNATIVES TO CFCS

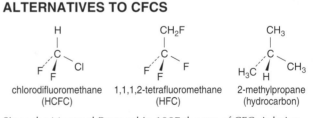

chlorodifluoromethane (HCFC)

1,1,1-tetrafluoromethane (HFC)

2-methylpropane (hydrocarbon)

Since the Montreal Protocol in 1987 the use of CFCs is being phased out. Even so, because of their low reactivity they are expected to remain in the atmosphere for at least the next eighty years. They destroy the ozone layer because the ultraviolet light breaks the relatively weak C–Cl bond. Substitutes for CFCs must have similar properties but not contain a bond that can be broken in ultraviolet light to form radicals. The properties required are: low reactivity, low toxicity, and low inflammability, as well as no weak C–Cl bonds. They also should not absorb infrared radiation otherwise they will act as greenhouse gases.

The most immediate replacements are HCFCs – hydrochlorofluorocarbons, such as CHF_2Cl, as they decompose more readily and do not build up in the stratosphere. Other alternatives actively being investigated are HFCs – hydrofluorocarbons, such as CF_3CH_2F, and hydrocarbons such as 2-methylpropane C_4H_{10}, for refrigerants, but they suffer from being flammable and also contribute to global warming

Greenhouse effect and global warming

THE GREENHOUSE EFFECT

A steady state equilibrium exists between the energy reaching the Earth from the Sun and the energy reflected by the Earth back into space. This regulates the mean average temperature of the Earth's surface. The incoming radiation is short wave ultraviolet and visible radiation. Some is reflected back into space and some is absorbed by the atmosphere before it reaches the surface. The energy reflected back from the Earth's surface is longer wavelength infrared radiation. Not all of the radiation escapes. Greenhouse gases in the atmosphere allow the passage of the incoming short wave radiation but absorb some of the reflected infrared radiation and re-radiate it back to the Earth's surface. Because of its abundance water is the main greenhouse gas.

The bonds in the carbon dioxide molecule absorb radiation of a different wavelength than the bonds in water molecules. Although it only constitutes 0.03% of the atmosphere, carbon dioxide therefore plays a key role in keeping the average global temperature at about 15 °C.

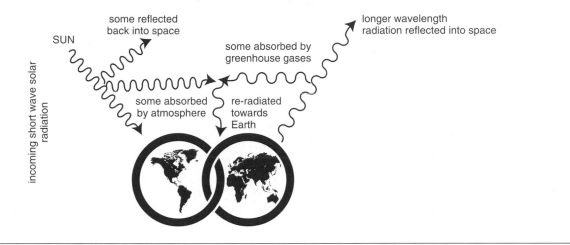

GREENHOUSE GASES

The contribution to global warming made by different greenhouse gases will depend both on their concentration in the atmosphere (abundance) and on their ability to absorb infrared radiation. Apart from water, carbon dioxide contributes about 50% to global warming. CFCs are thousands of times better at absorbing heat than CO_2 but because their concentrations are so low their effect is limited to about 14%.

Gas	Main source	Heat trapping effectiveness compared with CO_2	Overall contribution to increased global warming
H_2O	Evaporation of oceans and lakes	0.1	–
CO_2	Combustion of fossil fuels and biomass	1	50%
CH_4	Anaerobic decay of organic matter caused by intensive farming	30	18%
N_2O	Artificial fertilizers and combustion of biomass	150	6%
O_3	Secondary pollutant in photochemical smogs	2000	12%
CFCs	Refrigerants, propellants, foaming agents, solvents	10 000–25 000	14%

INFLUENCE OF GREENHOUSE GASES ON GLOBAL WARMING

Most of the greenhouse gases have natural as well as man-made sources. As man has burned more fossil fuels the concentration of carbon dioxide in the air has risen steadily. Readings taken from Mauna Loa in Hawaii show a steady increase in the concentration of carbon dioxide by about 1 ppm (0.0001%) each year for the past forty years. During the same period measurements of the Earth's mean temperature also show a general increase. During the past 100 years the mean temperature of the Earth has increased by about 1 °C although there were some years where the mean temperature fell rather than rose. Evidence from ice core samples in Greenland shows that there have also been large fluctuations in global temperature in the past; however, most scientists now accept that the current global warming is a direct consequence of the increased emission of greenhouse gases.

The predicted consequences of global warming are complex and there is not always agreement. The two most likely effects are:

1. Changes in agriculture and biodistribution as the climate changes.

2. Rising sea-levels due to thermal expansion and the melting of polar ice caps and glaciers.

PARTICULATES

Particulates, such as soot and volcanic dust, can have the opposite effect to greenhouse gases. They cool the Earth by scattering the short wave radiation from the Sun and reflecting it back into space. The lowering of mean global temperatures during the 1940s and 1960s has been attributed to the increased volcanic activity during these periods.

Acid rain

OXIDES OF SULFUR SO$_x$

Sulfur dioxide occurs naturally from volcanoes. It is produced industrially from the combustion of sulfur-containing fossil fuels and the smelting of sulfide ores.

$$S(s) + O_2(g) \rightarrow SO_2(g)$$

In the presence of sunlight sulfur dioxide is oxidized to sulfur trioxide.

$$SO_2(g) + \tfrac{1}{2}O_2(g) \rightarrow SO_3(g)$$

The oxides can react with water in the air to form sulfurous acid and sulfuric acid:

$$SO_2(g) + H_2O(l) \rightarrow$$
$$H_2SO_3(aq)$$
and
$$SO_3(g) + H_2O(l) \rightarrow$$
$$H_2SO_4(aq)$$

OXIDES OF NITROGEN NO$_x$

Nitrogen oxides occur naturally from electrical storms and bacterial action. Nitrogen monoxide is produced in the internal combustion engine and in jet engines.

$$N_2(g) + O_2(g) \rightarrow 2NO(g)$$

Oxidation to nitrogen dioxide occurs in the air.

$$2NO(g) + O_2(g) \rightarrow 2NO_2(g)$$

The nitrogen dioxide then reacts with water to form nitric acid and nitrous acid:

$$2NO_2(g) + H_2O(l) \rightarrow$$
$$HNO_3(aq) + HNO_2(aq)$$

or is oxidized directly to nitric acid by oxygen in the presence of water:

$$4NO_2(g) + O_2(g) + 2H_2O(l) \rightarrow 4HNO_3(aq)$$

ACID RAIN

Pure rainwater is naturally acidic with a pH of 5.65 due to the presence of dissolved carbon dioxide. Carbon dioxide itself is *not* responsible for acid rain since acid rain is defined as rain with a pH less than 5.6. It is the oxides of sulfur and nitrogen dissolved in the water which cause acid rain. The term acid rain also covers precipitation as snow, hail, fog, and dew.

VEGETATION

Increased acidity in the soil leaches important nutrients, such as Ca^{2+}, Mg^{2+}, and K^+. Reduction in Mg^{2+} can cause reduction in chlorophyll and consequently lowers the ability of plants to photosynthesize. Many trees have been seriously affected by acid rain. Symptoms include stunted growth, thinning of tree tops, and yellowing and loss of leaves. The main cause is the aluminium leached from rocks into the soil water. The Al^{3+} ion damages the roots and prevents the tree from taking up enough water and nutrients to survive.

LAKES AND RIVERS

Increased levels of aluminium ions in water can kill fish. Aquatic life is also highly sensitive to pH. Below pH 6 the number of sensitive fish, such as salmon and minnow, decline as do insect larvae and algae. Snails cannot survive a pH less than 5.2 and below pH 5.0 many microscopic animal species disappear. Below pH 4.0 lakes are effectively dead. The nitrates present in acid rain can also lead to eutrophication.

BUILDINGS

Stone, such as marble, that contains calcium carbonate is eroded by acid rain. With sulphuric acid the calcium carbonate reacts to form calcium sulfate, which can be washed away by rainwater thus exposing more stone to corrosion. Salts can also form within the stone that can cause the stone to crack and disintegrate.

$$CaCO_3(s) + H_2SO_4(aq) \rightarrow$$
$$CaSO_4(aq) + CO_2(g) + H_2O(l)$$

HUMAN HEALTH

The acids formed when NO$_x$ and SO$_x$ dissolve in water irritate the mucus membranes and increase the risk of respiratory illnesses, such as asthma, bronchitis, and emphysema. In acidic water there is more probability of poisonous ions, such as Cu^{2+} and Pb^{2+}, leaching from pipes and high levels of aluminium in water may be linked to Alzheimer's disease.

METHODS TO LOWER OR COUNTERACT THE EFFECTS OF ACID RAIN

1. Lower the amounts of NO$_x$ and SO$_x$ formed, e.g. by improved engine design, the use of catalytic converters, and removing sulfur before, during, and after combustion of sulfur-containing fuels.

2. Switch to alternative methods of energy (e.g. wind and solar power) and reducing the amount of fuel burned, e.g. by reducing private transport and increasing public transport and designing more efficient power stations.

3. Liming of lakes – adding calcium oxide or calcium hydroxide (lime) neutralizes the acidity, increases the amount of calcium ions and precipitates aluminium from solution. This has been shown to be effective in many, but not all, lakes where it has been tried.

Water suitable for drinking

SUPPLY OF FRESH WATER

Water is a very abundant substance. About 72% of the Earth's surface is covered by oceans at an average depth of 4 km. However, about 97% of all water is salt water. The rest consists of 2.1% locked up in glaciers, 0.6% fresh water in rivers and lakes, and 0.6% groundwater. Hence only about 1% of all the water on the planet is available as a ready source of fresh water. Since 1950 the amount of annually renewable fresh water has fallen from over $16\,000\ m^3$ per human to about $7000\ m^3$. This would be still more than adequate if the fresh water were distributed evenly over the planet. Unfortunately it is not and lack of sufficient fresh water in many countries is a serious problem.

The main use of water depends on the country. In industrial countries, such as the United States, the main use is for industry (about 59%) and agriculture (33%) with public supplies taking about 8%. In less industrialized countries about two thirds of the fresh water consumption is used for agriculture.

TREATMENT OF FRESH WATER FOR DRINKING

Water fit for drinking must be free of disease-causing agents (pathogens), have no unpleasant tastes, odours, colours or turbidity and contain no harmful dissolved substances. Harmful bacteria are killed by the addition of an oxidizing agent. This is usually chlorine or ozone.

	Chlorine, Cl_2	Ozone, O_3
Action	$Cl_2(aq) + H_2O(l) \rightarrow HCl(aq) + HClO(aq)$ Hypochlorous acid oxidizes the bacteria $HClO(aq) + 2H^+(aq) + 2e \rightarrow HCl(aq) + H_2O(l)$	O_3 itself oxidizes the bacteria $O_3(aq) + 2H^+(aq) + 2e \rightarrow O_2(g) + H_2O(l)$
Advantages	Cheap, longer retention time.	More effective, tasteless.
Disadvantages	Can form chlorinated organic compounds in the water (e.g. $CHCl_3$) which are carcinogenic.	More expensive, shorter retention time.

METHODS OF OBTAINING FRESH WATER FROM SEA WATER

Distillation

Many hot countries distil sea water to obtain fresh water. One of the largest plants in Israel produces over 3 million litres a day. The sea water is heated in a series of coiled pipes and then introduced into a partially evacuated chamber. Under the reduced pressure some of the sea water boils instantly. The water vapour produced is condensed by contact with cold water pipes carrying sea water. In this way the heat released when the water condenses is used to preheat more sea water.

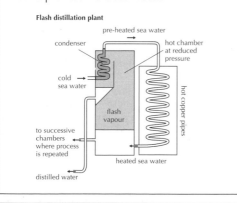

Flash distillation plant

Reverse osmosis

A high pressure is applied to the solution side of a partially permeable membrane made of cellulose ethanoate. Water is forced out of the salt solution through the membrane leaving the salts behind. The challenge is to make a membrane that will withstand the high pressure and permit a reasonable flow of water. Commercial plants use pressures of up to 70 atm and one cubic metre of membrane can produce about 250 000 litres of fresh water each day.

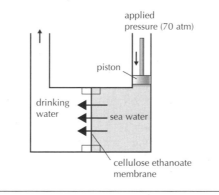

Ion exchange

Sea water is passed through columns containing natural materials with silicate type structures (zeolites) or synthetic ion exchange resins. H^+ ions are exchanged for Na^+ ions and OH^- ions are exchanged for the Cl^- ions in the sea water. The columns can be regenerated using sulfuric acid for cationic exchange resins and sodium hydroxide for anionic exchange resins. The process produces good quality drinking water but is relatively expensive.

SAVING AND RECYCLING WATER

Since the main use for water is for agriculture and industrial purposes it is not necessary that all the water used should be fit for drinking. Water used to cool industrial plants can be recycled. Instead of allowing water to flow into the sea, dams and reservoirs can be built to store water for agricultural use although this can cause other environmental problems. Individuals can save water by flushing the toilet less often, showering instead of bathing, and using waste water to water plants and lawns. Water companies can conserve water by replacing or repairing leaky pipes.

Average daily domestic water consumption (in litres per person per day)	
USA	397
Japan	379
Switzerland	270
UK	150
Germany	145
India	25

Note that to sustain a reasonable quality of life about 80 litres of water per person per day are required.

Dissolved oxygen in water

IMPORTANCE OF DISSOLVED OXYGEN IN WATER

At a pressure of one atm and at a temperature of 20 °C the maximum solubility of oxygen in water is only about 9 ppm (i.e 0.009 g dm^{-3}). Although this value is small it is crucial since most aquatic plants and animals require oxygen for aerobic respiration. Fish require the highest levels and bacteria require the lowest levels. Fish need at least 3 ppm in order to be able to survive but to maintain a balanced and diversified aquatic community the oxygen content should not be less than 6 ppm.

BIOLOGICAL OXYGEN DEMAND (BOD)

The biological (or biochemical) oxygen demand is a measure of the dissolved oxygen (in ppm) required to decompose the organic matter in water biologically. It is often measured over a set time period of five days. Water that has a high BOD without the means of replenishing oxygen will rapidly not sustain aquatic life. A fast flowing river can recover its purity as the water becomes oxygenated through the mechanical action of its flow. Lakes have relatively little flow and reoxygenation is much slower or will not happen at all. Pure water has a BOD of less than 1 ppm, water with a BOD above about 5 ppm is regarded as polluted.

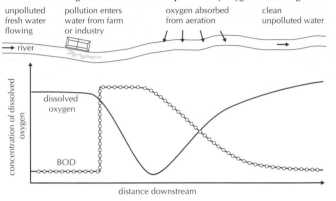

How a fast flowing river can recover from pollution by oxygen-demanding waste

MEASUREMENT OF BOD

The BOD of a sample of water can be determined by the Winkler method. The sample of the water is saturated with oxygen so the initial concentration of dissolved oxygen is known. A measured volume of the sample is then incubated at a fixed temperature for five days while micro-organisms in the water oxidize the organic material. An excess of a manganese(II) salt is then added to the sample. Under alkaline conditions manganese(II) ions are oxidized to manganese(IV) oxide by the remaining oxygen.

$$2Mn^{2+}(aq) + 4OH^-(aq) + O_2(aq) \rightarrow 2MnO_2(s) + 2H_2O(l)$$

Potassium iodide is then added which is oxidized by the manganese(IV) oxide in acidic solution to form iodine.

$$MnO_2(s) + 2I^-(aq) + 4H^+(aq) \rightarrow Mn^{2+}(aq) + I_2(aq) + 2H_2O(l)$$

The iodine released is then titrated with standard sodium thiosulfate solution.

$$I_2(aq) + 2S_2O_3^{2-}(aq) \rightarrow S_4O_6^{2-}(aq) + 2I^-(aq)$$

By knowing the number of moles of iodine produced the amount of oxygen present in the sample of water can be calculated and hence its concentration.

EUTROPHICATION

Nitrates from intensive animal farming and excess use of artificial fertilizers together with phosphates from artificial fertilizers and detergents accumulate in lakes. They act as nutrients and increase the growth of plants and algae. This can also happen in slow moving areas of sea water.

Normally when plants and algae die they decompose aerobically and form carbon dioxide and water. However, if the growth is excessive and the dissolved oxygen is not sufficient to cope anaerobic decomposition will occur. The hydrides formed, such as ammonia, hydrogen sulfide, and phosphine, not only smell foul but they poison the water. More species will die resulting in more anaerobic decay and the lake itself becomes devoid of life – a process known as **eutrophication**.

AEROBIC AND ANAEROBIC DECOMPOSITION

If sufficient oxygen is present organic material will decay aerobically and oxides or oxyanions are produced. Anaerobic decay involves organisms which do not require oxygen. The products are in the reduced form and are often foul smelling and toxic.

Element	Aerobic decay product	Anaerobic decay product
Carbon	CO_2	CH_4 (marsh gas)
Nitrogen	NO_3^-	NH_3 and amines
Hydrogen	H_2O	CH_4, NH_3, H_2S, and H_2O
Sulfur	SO_4^{2-}	H_2S ('rotten eggs' gas)
Phosphorus	PO_4^{3-}	PH_3 (phosphine)

THERMAL POLLUTION

The solubility of oxygen in water is temperature dependent. As the temperature is increased the solubility drops. At the same time the metabolic rate of fish and other organisms increases so the demand for oxygen increases. Many industries use water as a coolant and the careless discharge of heated water into rivers can cause considerable thermal pollution.

Solubility of oxygen in water at different temperatures

Temperature / °C	Solubility of O_2 in fresh water / ppm	Solubility of O_2 in sea water / ppm
0	14.71	11.71
10	11.42	9.28
20	9.14	7.50
30	7.50	6.21

Waste water treatment

RAW SEWAGE

In some countries raw sewage is still discharged untreated into rivers and the sea. It is eventually degraded by micro-organisms but the process takes time and can cause pollution of beaches and swimming areas. In remote areas sewage is often discharged into cesspits or septic tanks where it is broken down before leaching into the ground. Waste water treated at sewage works contains floating matter, suspended matter, colloidal matter, and a range of micro-organisms. The aim of the treatment is to remove this additional material and recycle the fresh water.

PRIMARY TREATMENT

Primary treatment effectively removes about 60% of the solid material and about a third of the oxygen-demanding wastes. The incoming sewage is passed through coarse mechanical filters to remove large objects like sticks, paper, condoms, and rags. It is then passed into a grit chamber where sand and small objects settle. From there it passes into a sedimentation tank where suspended solids settle out as sludge. A mixture of calcium hydroxide and aluminium sulfate is added to aid this process.

$$Al_2(SO_4)_3(aq) + 3Ca(OH)_2(aq) \rightarrow 2Al(OH)_3(s) + 3CaSO_4(aq)$$

The two chemicals combine to form aluminium hydroxide which precipitates carrying with it suspended dirt particles – a process known as **flocculation**. Grease is removed from the surface tank by skimming. The effluent is then discharged into a waterway or passed on for secondary treatment.

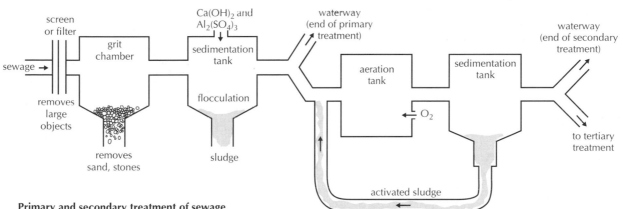

Primary and secondary treatment of sewage

SECONDARY TREATMENT

Secondary treatment removes up to 90% of the oxygen-demanding wastes (organic material and dangerous bacteria). The principle is to degrade the waste aerobically using oxygen and bacteria. One method uses trickle filters. The organic material is degraded by bacteria as the waste water trickles through a bed of stones. A more effective method is the activated sludge process. The sewage is aerated with pure oxygen in a sedimentation tank. The sludge that settles out contains active micro-organisms that digest organic waste and some of it is recycled. The water that emerges is then discharged into a waterway where it will be disinfected with chlorine or ozone before being fit for drinking.

TERTIARY TREATMENT

Although it is expensive there is an increasing need for tertiary treatment. Primary and secondary treatments do not remove heavy metal ions, nitrates and phosphates, or residual amounts of organic compounds.

Precipitation

Heavy metal ions and phosphates can be removed by precipitation.

Aluminium sulfate or calcium oxide can be used to precipitate phosphates.

$$Al^{3+}(aq) + PO_4^{3-}(aq) \rightarrow AlPO_4(s)$$

$$3Ca^{2+}(aq) + 2PO_4^{3-}(aq) \rightarrow Ca_3(PO_4)_2(s)$$

Heavy metal ions can be precipitated as insoluble hydroxides or basic salts by the addition of calcium hydroxide or sodium carbonate,

e.g. $Cr^{3+}(aq) + 3OH^-(aq) \rightarrow Cr(OH)_3(s)$

However, some metal hydroxides redissolve due to the formation of soluble complexes so some heavy metal ions, such as Zn^{2+}, Hg^{2+}, and Cd^{2+}, are precipitated as insoluble sulfides by bubbling hydrogen sulfide into the water,

e.g. $Cd^{2+}(aq) + H_2S(g) \rightarrow CdS(s) + 2H^+(aq)$

Activated carbon beds

The remaining dissolved organic material can be removed by allowing the water to flow down a bed of activated carbon. To reactivate the carbon it is periodically heated to high temperatures to oxidize the adsorbed organic compounds to carbon dioxide and water and regenerate the carbon surface.

Removal of nitrates

Nitrates are difficult to remove by chemical means as all nitrates are soluble so precipitation cannot be used. Ion exchange columns made of zeolites can be used to exchange hydroxide ions for nitrate ions but this is very expensive for large volumes of water. They are more readily removed by biological methods. Anaerobic denitrifying bacteria can reduce them to nitrogen or the water can be passed through algal ponds where the algae utilize the nitrate as a nutrient.

 # Smog

CONDITIONS FOR THE FORMATION OF SMOGS

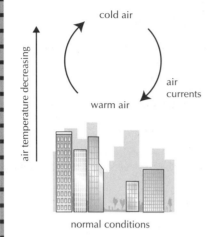

cold air

air temperature decreasing

air currents

warm air

normal conditions

Smog is a poisonous mixture of smoke, fog, air, and other chemicals. It tends to form in large cities and is favoured by lack of wind. It also tends to occur more in cities which are 'bowl shaped', i.e. surrounded by higher ground in all directions. Both of these factors tend to prevent the movement of air. Smogs are most likely to occur when there is a **temperature inversion**. Normally the temperature decreases with altitude. Warm air rises taking the pollutants with it and is replaced by cleaner cooler air. However, sometimes atmospheric conditions cause a layer of still warm air to blanket a layer of cooler air. The trapped pollutants cannot rise and if the conditions persists the amount of pollutants in the warm air near the ground can increase to dangerous levels.

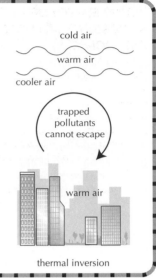

cold air

warm air

cooler air

trapped pollutants cannot escape

warm air

thermal inversion

REDUCING AND PHOTOCHEMICAL SMOGS

The classic London 'pea soup' smog that occurred before the introduction of clean air controls in the mid 1950s was a reducing smog, principally due to the combustion of coal and oil. This produced sulfur dioxide mixed with soot, fly ash, and partially oxidized organic material. This type of smog is now less common. Photochemical smogs occur in cities, such as Los Angeles, where exhaust from internal combustion engines concentrates in the atmosphere. Oxides of nitrogen and hydrocarbons give the air a characteristic yellow/brown colour. In sunlight these are converted into secondary pollutants.

OXIDATION OF SULFUR DIOXIDE IN SMOGS

Sulfur dioxide formed in reducing smogs is oxidized catalytically. A wide variety of catalysts can bring about this conversion. In the presence of particulates, such as metal ions or droplets of hydrocarbons, heterolytic catalysis can occur using ozone as the oxidizing agent. It then forms sulfuric acid.

e.g. $SO_2(g) + O_3(g) \rightarrow SO_3(g) + O_2(g)$
$SO_3(g) + H_2O(l) \rightarrow H_2SO_4(aq)$

Free radicals and oxides of nitrogen can also catalyse the oxidation of SO_2. This is an example of homogeneous catalysis (catalyst and reactants in the same phase). For example, the hydroxyl radical can be formed from the breakdown of ozone by UV light.

$O_3(g) \rightarrow O_2(g) + O^•(g)$ then
$O(g) + H_2O(l) \rightarrow 2OH^•(g)$

Note that this is a very reactive free radical containing an unpaired electron (it is not the hydroxide ion OH^-). This radical can then undergo further radical propagation reactions with sulfur dioxide.

$OH^•(g) + SO_2(g) \rightarrow HSO_3^•(g)$
$HSO_3^•(g) + O_2(g) \rightarrow HSO_5^•(g)$
$HSO_5^•(g) + NO(g) \rightarrow HSO_4^•(g) + NO_2(g)$

Termination occurs with the formation of a mixture of sulfuric and nitric acid (acid rain).

$HSO_4^•(g) + NO_2(g) + H_2O(l) \rightarrow$
$H_2SO_4(aq) + HNO_3(aq)$

FORMATION OF SECONDARY POLLUTANTS IN PHOTOCHEMICAL SMOGS

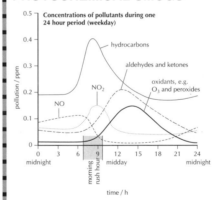

Concentrations of pollutants during one 24 hour period (weekday)

pollution / ppm

hydrocarbons

aldehydes and ketones

NO_2

oxidants, e.g. O_3 and peroxides

NO

0   3   6   9   12   15   18   21   24
midnight   midday   midnight

morning rush hour

time / h

During the early morning rush hour there is a build up of hydrocarbons and nitrogen oxides from car exhausts. As the Sun comes out the nitrogen dioxide absorbs sunlight and breaks down to form radicals.

$NO_2(g) \rightarrow NO(g) + O^•(g)$

These oxygen radicals can react with oxygen to form ozone and with water to form hydroxyl radicals. These secondary photochemical oxidants can react with a variety of molecules including nitrogen oxides to form nitric acid and hydrocarbons to form peroxides ROOR, aldehydes RCHO, and ketones RCOR. Peroxides are extremely reactive and aldehydes and ketones reduce visibility by condensing to form aerosols.

$O^•(g) + H_2O(l) \rightarrow 2OH^•(g)$ (formation of hydroxyl radicals)
$OH^•(g) + NO_2(g) \rightarrow HNO_3(aq)$ (nitric acid formation)
$OH^•(g) + RH(g) \rightarrow R^•(g) + H_2O(l)$ (radical propagation)
$R^•(g) + O_2(g) \rightarrow ROO^•$ (peroxide radical formation)

Chain termination can occur when peroxide radicals react with nitrogen dioxide to form **peroxyacylnitrates**, PANs. These compounds are eye irritants and are toxic to plants.

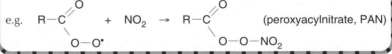

e.g.   R—C

O

O—O^•

+ NO_2 →   R—C

O

O—O—NO_2

(peroxyacylnitrate, PAN)

WAVELENGTH OF ULTRAVIOLET LIGHT NECESSARY FOR OZONE AND OXYGEN DISSOCIATION

In the earlier section on ozone depletion it was stated that the ozone layer absorbs a wide range of ultraviolet light. The average bond enthalpy at 298 K for the $O{=}O$ bond is given in the IB Data Book as 496 kJ mol^{-1}. For just one double bond this equates to 8.235×10^{-19} J. The wavelength of light that corresponds to this enthalpy value (E) can be calculated by combining the expressions $E = hf$ and $c = \lambda f$ to give

$$\lambda = \frac{hc}{E} \quad \text{where } h \text{ is Planck's constant, and } c \text{ is the velocity of light.}$$

$$\lambda = \frac{6.626 \times 10^{-34}\,(\text{J s}) \quad 2.998 \times 10^8\,(\text{m s}^{-1})}{8.235 \times 10^{-19}\,(\text{J})} = 241 \text{ nm}$$

This is in the high energy region of the ultraviolet spectrum. Ozone can be described as two resonance hybrids. An alternative bonding model is to consider the π electrons to be delocalized over all three oxygen atoms. In both models the bond order is 1.5, i.e. between an O—O single bond ($\Delta H = 146$ kJ mol^{-1}) and an $O{=}O$ double bond so ultraviolet light with a longer wavelength (lower energy) is absorbed in breaking the ozone bond. The actual wavelength required is 330 nm. Working backwards this gives the strength of the O—O bond in ozone as 362 kJ mol^{-1}.

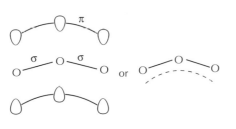

delocalized π bond in ozone

CATALYSIS OF OZONE DESTRUCTION BY CFCS AND NO$_x$

CFCs catalyze the destruction of ozone because the high energy ultraviolet light in the stratosphere causes the homolytic fission of the C—Cl bond to produce chlorine radicals. Note that it is the C—Cl bond that breaks, not the C—F bond, as the C—Cl bond strength is weaker. These radicals then break down ozone molecules and regenerate more radicals so that the process continues until the radicals eventually escape or terminate. It has been estimated that one molecule of a CFC can catalyze the breakdown of up to 100 000 molecules of ozone.

$$CCl_2F_2(g) \longrightarrow CClF_2^{\bullet}(g) + Cl^{\bullet}(g) \quad \text{(radical initiation)}$$

$$Cl^{\bullet}(g) + O_3(g) \longrightarrow ClO^{\bullet}(g) + O_2(g) \quad \text{(propagation}$$
$$ClO^{\bullet}(g) + O^{\bullet}(g) \longrightarrow Cl^{\bullet}(g) + O_2(g) \quad \text{of radicals)}$$

Evidence to support this mechanism is that the increase in the concentration of chlorine monoxide in the stratosphere over the Antarctic has been shown to mirror the decrease in the ozone concentration.

Nitrogen oxides also catalytically decompose ozone by a radical mechanism. The overall mechanism is complex. Essentially oxygen radicals are generated by the breakdown of NO_2 in ultraviolet light. They then react with ozone.

$$NO_2(g) \longrightarrow NO(g) + O^{\bullet}(g)$$
$$O^{\bullet}(g) + O_3(g) \longrightarrow 2O_2(g)$$

The overall reaction can be simplified as:

$$NO_2(g) + O_3(g) \longrightarrow NO(g) + 2O_2(g)$$

REASONS FOR GREATER OZONE DEPLETION IN THE POLAR REGIONS

During the winter the temperatures get very low in the stratosphere above the poles. At these low temperatures the small amount of water vapour in the air freezes to form crystals of ice. The crystals also contain small amounts of molecules, such as HCl and $ClONO_2$. It is believed that catalytic reactions occur on the surface of the ice crystals to produce species such as hypochlorous acid (HClO) and chlorine (Cl_2). With the advent of spring the Sun causes these molecules to break down to produce Cl$^{\bullet}$ radicals which catalyze the destruction of ozone. The largest holes occur during early spring. As the Sun warms the air the ice crystals disperse, warmer winds blow into the region, and the ozone concentration gradually increases again.

SUN-SCREENING COMPOUNDS

Sunlight can be harmful to the skin. At longer wavelengths it causes temporary reddening but at shorter wavelengths it can cause damage to proteins making the skin look leathery and wrinkled. It can also break bonds in DNA resulting in damage to genes and lead to skin cancer. Sunscreens block this damaging radiation. The atmosphere itself is an effective sunscreen but as the ozone concentration decreases more harmful ultraviolet light is reaching the Earth's surface. Glass is also an effective sunscreen so sunburn does not occur indoors behind a glass window. The body produces its own sunscreen – melanin. This is a natural dark brown pigment in the skin and its concentration is increased by the action of sunlight on the skin. Commercial sunscreens, such as 4-aminobenzoic acid (maximum absorption at 265 nm), are substances that are able to absorb light of a particular frequency. These contain conjugated double bonds (alternate $C{=}C$ and C—C bonds) with delocalized π electrons. The ultraviolet light is absorbed by exciting the π electrons in these bonds to higher energy levels. Some people now prefer to use sunblocks. These contain white pigments, such as zinc oxide or titanium(IV) oxide, which form a barrier to sunlight by reflecting and scattering the light.

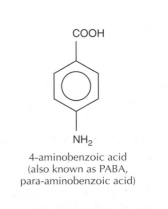

4-aminobenzoic acid
(also known as PABA, para-aminobenzoic acid)

TOXICITY

Measuring toxicity can be complex as different individuals can have very different reactions to toxins. Acute toxicity refers to a large dose of toxin over a short period of time, whereas chronic toxicity refers to low doses of toxin absorbed over a long period of time. One convenient way to measure toxicity is the **lethal dose 50%**, LD_{50}. This is the dose required to kill fifty percent of the population upon which the toxin is tested on. It is usually measured in micrograms per kilogram of body mass. The main disadvantage of this method is that the tests are done on animals and different species react differently to different poisons. Another disadvantage is that very large doses of relatively non-toxic substances are required to kill 50% of the population.

The **maximum daily tolerance** or **threshold daily dose** is a measure of the level of toxin that a person can be exposed to without being harmed. This is rather vague as people differ and a **safe limit** or **acceptable daily intake** is sometimes quoted. This is obtained by dividing the maximum daily tolerance by a **safety factor** which may be as much as one hundred although even this is subjective.

Lethal doses (LD_{50}) of dioxin for different animal species (from BMA *Hazardous Waste and Human Health* (1991), by permission of Oxford University Press.

Animal	LD_{50} (µg/kg/body wt)
Guinea pig	1.0
Rat (male)	22
Rat (female)	45
Monkey	> 77
Rabbit	115
Mouse	114
Dog	> 300
Bullfrog	> 500
Hamster	5000

TOXINS IN POLLUTED WATER

CADMIUM, MERCURY, AND LEAD
(other heavy metals in polluted water include Cr, Ni, Cu, and Zn)

	Cadmium	Mercury	Lead
Source	Water effluent in zinc mining areas, rechargeable batteries, metal plating, orange pigment in enamels and paints.	Fungicides for seed dressings, batteries, industrial electrolysis of brine using mercury electrode.	Use of lead tetraethyl in leaded gasoline, old paintwork, lead water pipes, car batteries.
Health and environmental effects	Highly toxic as can replace zinc in enzymes. Acute effects may cause vomiting, stomach pains and diarrhoea. Chronic effects include kidney damage and destruction of red blood cells.	Mercury is a neurotoxin and tends to accumulate in the tissues of the liver and kidneys and cause damage to these organs. Organomercury compounds concentrate in fatty tissues and affect the blood and brain.	Acute lead poisoning includes severe abdominal pains, kidney failure, and death. Chronic lead poisoning in children can affect balance and lead to mental retardation. In adults it can cause hypertension leading to heart disease.

Pesticides
These include insecticides, such as DDT, herbicides, such as paraquat, and fungicides.

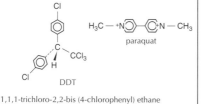

paraquat

DDT

1,1,1-trichloro-2,2-bis (4-chlorophenyl) ethane
(former name **d**ichloro**d**iphenyl**t**richloroethane)

Dioxin
This is formed when waste materials containing organochlorine compounds are not incinerated at high enough temperatures. It is very persistent in the environment and extremely toxic as it accumulates in fat and liver cells. It can also cause malformations in fetuses. It was one of the herbicides present in Agent Orange used during the Vietnam war.

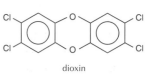

dioxin

Polychlorinated biphenyls, PCBs
These contain from one to ten chlorine atoms attached to a biphenyl molecule. They are chemically stable and have high electrical resistance so are used in transformers and capacitors. Like dioxin they persist in the environment and accumulate in fatty tissue. They affect reproductive efficiency, impair learning ability in children and are thought to be carcinogenic.

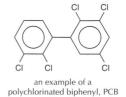

an example of a polychlorinated biphenyl, PCB

NITRATES
Nitrates enter drinking water from intensive animal farming and from the use of artificial fertilizers. They also enter the water from acid rain. Because all nitrates are soluble they are not easily removed during treatment of sewage and so tend to accumulate. The maximum limit of nitrates in drinking water has been set at 50 mg dm^{-3} by the World Health Organisation. At higher levels there is a risk to very small babies of infantile methaemoglobinaemia (blue baby syndrome). This is because there is less acid in babies' stomachs so more of the nitrate is converted to nitrite by bacteria. The nitrite ion will oxidize the iron in haemoglobin so that it can no longer bind with oxygen and the baby suffers from oxygen starvation. Even in adults some of the nitrates are converted into nitrites which in turn are converted into nitrosamines. Nitrosamines are known to be carcinogenic.

$$R_2N-N=O \text{ (nitrosamine)}$$

IB QUESTIONS – OPTION D – ENVIRONMENTAL CHEMISTRY

1. When the pH of rain water falls below about 5.6 it is known as *acid rain*.

(a) What is the ratio of the hydrogen ion concentration in acid rain with a pH of 4 compared to water with a pH of 7? **[1]**

(b) One of the two major acids present in acid rain originates mainly from the burning of coal. **Name** this acid and give equations to show how it is formed. **[3]**

(c) The second major acid responsible for acid rain originates mainly from internal combustion engines. **Name** this acid and state **two** different ways in which its production can be reduced. **[3]**

(d) Acid rain has caused considerable damage to buildings and statues made of marble ($CaCO_3$). Write an equation to represent the reaction of acid rain with marble. **[1]**

2. (a) In order to survive, fish require water containing dissolved oxygen. Discuss briefly how an increase in each of the following factors affects the amount of dissolved oxygen in a lake. **[3]**

(i) Temperature　(ii) Organic pollutants
(iii) Nitrates and phosphates

(b) Define *Biological Oxygen Demand (BOD)*. **[1]**

(c) In a method to find the concentration of dissolved oxygen, manganese(IV) oxide is formed. This is then used to release iodine which is titrated with standard thiosulfate solution. The equations for these three steps are:

$2Mn^{2+}(aq) + 4OH^-(aq) + O_2(g) \rightarrow 2MnO_2(s) + 2H_2O(l)$

$MnO_2(s) + 2I(aq) + 4H^+(aq) \rightarrow$
$\quad Mn^{2+}(aq) + I_2(aq) + 2H_2O(l)$

$I_2(aq) + 2S_2O_3^{2-}(aq) \rightarrow S_4O_6^{2-}(aq) + 2I^-(aq)$

1000 cm^3 of a sample of water was processed by this method. It was found that 10.0 cm^3 of 0.100 mol dm^{-3} $Na_2S_2O_3$ solution were required to react with the iodine produced. Calculate the concentration of dissolved oxygen in **g dm^{-3}** in the water sample. **[3]**

3. Global warming is one of the major dilemmas confronting humans today.

(a) Identify one substance produced by human activity that has been suggested to contribute to global warming. Give its source and explain briefly why the human contribution is so significant. **[3]**

(b) Describe on the molecular level how the substance in (a) exerts its effect. **[2]**

(c) Name one air pollutant that can counteract the effect of the substances in **(a)** and describe how it does so. **[2]**

4. (a) Use the value for the average bond enthalpy for oxygen (in kJ mol^{-1}) given in Table 10 and information from Tables 1 and 2 of the Data Booklet to calculate the maximum wavelength of light that is able to decompose oxygen. Give your answer to three significant figures. **[3]**

(b) Ozone absorbs radiation with a wavelength shorter than about 320 nm. Use Lewis structures to explain why ozone can be decomposed by light with a longer wavelength than that required to decompose oxygen. **[2]**

(c) (i) Explain, with equations, how a CFC, such as dichlorodifluoromethane, is able to deplete ozone in the ozone layer. **[3]**

(ii) Suggest an explanation to account for the fact that the depletion of ozone in the ozone layer is greatest as the winter months end over polar regions. **[2]**

5. (a) (i) Explain the meaning of the term LD_{50}. **[2]**

(ii) State **one** advantage and **one** disadvantage of the use of LD_{50}. **[2]**

(b) Lead and nitrates represent a health hazard in polluted water. Identify a source of **each** of these pollutants, state a health hazard caused by **each** pollutant, and indicate a way by which the concentration of **each** can be reduced. **[6]**

Initial overview

RAW MATERIALS

silicon 27.7%

oxygen 46.6%

aluminium 8.13%

iron 5.00%

calcium 3.63%

sodium 2.83%

titanium 0.44%

magnesium 2.09%

potassium 2.59%

(the remaining elements contribute <1.0%)

Elemental composition of the Earth's crust by mass

The major raw materials are:

Fossil fuels petroleum oil, natural gas, and coal. These provide the carbon source for the organic chemical industry. In many countries hydrocarbons from petroleum and natural gas are the major feedstock but coal is still abundant although at the current time more expensive.

Metallic ores and minerals some minerals (e.g. silica) are spread almost uniformly around the world but others occur more sparsely. The location of particular minerals can have a very large effect on the wealth or otherwise of a particular region or country. Metals are normally found as compounds, such as oxides or sulfides, in the Earth's crust although less reactive metals (e.g. silver and gold) are found uncombined. When the composition of each mineral is analysed by mass it can be seen that the Earth's crust is made up almost entirely of just nine elements. Almost half of the total mass of the Earth's crust is made up of just one element – oxygen.

Energy an abundant supply of energy is a necessity for the production of bulk chemicals. Energy is obtained directly from oil, gas, or coal, and increasingly from other sources, such as wind power and hydroelectric power. Many industries require electrical energy for electrolytic processes. Modern industries have become energy efficient both to reduce costs and to minimize the effect on the environment.

Air air is used both as a coolant and as a raw material in oxidation reactions (manufacture of sulfuric acid and nitric acid) and in fertilizer production (manufacture of ammonia).

Water water is used a heat exchange material, a solvent, and as a raw material. It can be used as liquid water or as steam.

LOCATION OF CHEMICAL INDUSTRIES

Heavy chemical industries were traditionally located around ports as this allowed for the easy import of raw materials and the export of products. Flowing fresh water for cooling and a ready source of energy (e.g. coal) were also important factors. With the advent of good railway systems and fast roads for transporting materials other factors have become more important. These include access to a good workforce and access to development grants. Grants are often given as inducements to entice new industries to areas of high unemployment.

Many industries and governments now have green policies so consideration needs to be given to selecting sites where minimum damage will be done to the local environment and to the local population. Many modern industries manufacture small volumes of high value speciality chemicals so are less affected by many of these considerations. All industries can only survive if there is a market for their products.

INTERMEDIATE AND CONSUMER PRODUCTS

It has been said that the economy of a country can be judged by the amount of sulfuric acid it produces. Sulfuric acid is still the most produced chemical in the world with annual production in the region of 150 million tonnes. However, much of it is used as an intermediate for other materials, such as fertilizers and explosives. This is true of many other bulk chemicals. Speciality chemicals for direct consumer consumption are produced in much smaller quantities – and sold at much higher prices per tonne or gram. These include such products as pharmaceuticals, silicon chips, dyes, food additives, and agrochemicals.

INCREASING IMPORTANCE OF BIOTECHNOLOGY

Biotechnology is beginning to play an increasingly important part in the manufacture of new substances. Enzymes have been used for centuries to manufacture ethanol by the fermentation of sugars and starches. Modern biotechniques use natural or genetically engineered enzymes that are immobilized on a solid carrier so that they can be reused many times. Uses include proteases in biological detergents to digest protein strains and fructose syrup production.

The first successful gene-splicing and gene-cloning experiments to produce recombinant DNA led to the synthesis of human insulin in the late 1970s. The production of genetically modified food by transferring stretches of DNA from one organism to another and the use of biosensors are discussed in 14. Option C – Human Biochemistry. Since the publication in 2000 of the human genome – the complete sequence of bases in human DNA – many companies have applied for patents to use this knowledge in a variety of ways.

Principles of extraction and production of metals

> **PURIFICATION OF METAL ORES**
>
> Very few metal ores are found in the pure state. They are usually contaminated with earthy materials, such as sand, rocks, and clay. The different methods of purifying the ore rely on the fact that the ore has a different density to the impurities. The crude ore is usually first crushed into small pieces. For dense uncombined metals, such as gold, the traditional method is 'panning'. The mixture containing tiny fragments of gold is mixed with water. The heavier gold particles sink to the bottom while the less dense earthy particles are washed away. One important method is 'froth flotation'. The crushed ore is added to water containing frothing agents. These are oils, such as pine oil or creosote. When air is blown through the mixture particles of the ore become attached to air bubbles. These float to the surface and collect as a froth which is removed periodically whereas the more dense materials sink to the bottom.

EXTRACTION OF METALS FROM THEIR ORES

Methods of extracting some important metals

Reactivity series	Main ore	Method of extraction
Sodium	NaCl	Electrolysis of molten NaCl with $CaCl_2$ added to lower melting point. $Na^+(l) + e \rightarrow Na(l)$
Aluminium	Al_2O_3	Electrolysis of Al_2O_3 in molten cryolite. $Al^{3+}(l) + 3e \rightarrow Al(l)$
Zinc	ZnS	Roast to form the oxide $2ZnS(s) + 3O_2(g) \rightarrow 2ZnO(s) + 2SO_2(g)$ then reduce oxide chemically with carbon monoxide $ZnO(s) + CO(g) \rightarrow Zn(s) + CO_2(g)$ or electrolytically $Zn^{2+}(l) + 2e \rightarrow Zn(l)$
Iron	Fe_2O_3	Heat with carbon monoxide $Fe_2O_3(s) + 3CO(g) \rightarrow 2Fe(l) + 3CO_2(g)$
Lead	PbS	Heat to form the oxide then reduce with carbon $PbO(s) + C(s) \rightarrow Pb(l) + CO(g)$
Copper	Cu or $CuFeS_2$	Heat in air to give copper and sulfur dioxide
Silver	Ag or Ag_2S	Metal deposited during the electrolytic refining of copper
Gold	Au	Metal found uncombined

The essential process involved in the extraction of metals from their ores is reduction. The ease with which this process occurs depends on the position of the metal in the reactivity series. Metals very low in the reactivity series, such as gold, silver, and copper, may be found uncombined. Metals slightly higher in the series can be obtained either by simply heating the ore strongly or by using chemical reducing agents, such as carbon or carbon monoxide. Titanium is required in a pure state and sodium metal is used as the reducing agent. Generally metals that can be obtained by relatively simple chemical reduction have been known since ancient times.

It is much harder and much more expensive to reduce chemically the more reactive metals and it took the discovery of electricity in the nineteenth century before these could be produced commercially. It is still an expensive process as molten ores must be used. Aqueous solutions of metals ions will produce hydrogen at the negative electrode (cathode) rather than the metal as hydrogen is lower in the reactivity series and hydrogen ions will be discharged preferentially. To reduce the energy required, and hence the cost of production, impurities are often added to the molten electrolyte to lower the melting point.

Apart from the reactivity series other factors that will influence the method of extraction include the accessibility of the ore, the ease of purification of the ore, the availability of energy resources, and the demand for the metal itself. Mines which at one time were not viable economically are sometimes reopened when the demand (and therefore the price) of the metal increases. Another factor to be considered is the use of any by-products. Many ores are sulfides and the roasting produces the sulfur dioxide required as feedstock in sulfuric acid production.

Iron

THE BLAST FURNACE

Iron is produced by reducing iron ores in a blast furnace. Traditionally the reducing agents are carbon monoxide and carbon. A modern blast furnace, which uses carbon monoxide, carbon, and hydrogen as the reducing agents, is capable of producing 10 000 tonnes of molten iron per day. Most of the molten 'pig' iron produced is converted directly into steel but some is cooled to make cast iron goods, such as engine cylinder blocks.

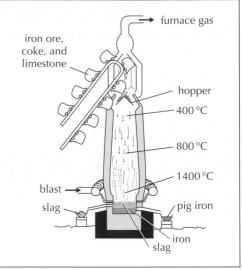

Raw materials

The solid charge is fed through the hopper via a conveyer belt. It consists of:

1. Iron ore (mainly haematite Fe_2O_3, magnetite Fe_3O_4, or hydrated oxides, e.g. geothite FeOH.OH and iron oxide obtained by roasting iron sulfides, e.g. iron pyrites, FeS_2) or scrap (recycled) iron.
2. Coke made by heating coal in the absence of air.
3. Limestone $CaCO_3$ to remove high melting point impurities by forming 'slag'.

In addition preheated air is blown in through nozzles, known as tuyeres, at the bottom of the furnace. This air is enriched with oxygen and may also include hydrocarbons, such as oil or natural gas, to replace up to 40% of the coke.

ESSENTIAL PROCESSES

Coke burns to form carbon monoxide:

$$2C(s) + O_2(g) \rightarrow 2CO(g)$$

In the reducing conditions incomplete combustion of the added hydrocarbons occurs:

e.g. $CH_4(g) + \frac{1}{2}O_2 \rightarrow CO(g) + 2H_2(g)$

The reducing gases pass up the furnace where they reduce the iron oxides in a series of stages depending on the temperature and composition of the gas. Examples of overall reactions taking place include:

$$Fe_2O_3(s) + 3CO(g) \rightarrow 2Fe(l) + 3CO_2(g)$$
$$Fe_3O_4(s) + 4H_2(g) \rightarrow 3Fe(l) + 4H_2O(g)$$
$$FeO(s) + CO(g) \rightarrow Fe(l) + CO_2(g)$$

In addition coke itself can reduce iron ore,

e.g. $Fe_2O_3(s) + 3C(s) \rightarrow 2Fe(l) + 3CO(g)$

The partially oxidized gases (furnace gas) which emerge from the top of the furnace are used as a fuel to preheat the air blasted in through the tuyeres.

At high temperatures the limestone decomposes:

$$CaCO_3(s) \rightarrow CaO(s) + CO_2(g)$$

The carbon dioxide reacts with coke to produce carbon monoxide

$$CO_2(g) + C(s) \rightarrow 2CO(g)$$

and the coke can also react with water from hydrocarbons to produce more carbon monoxide and hydrogen:

$$H_2O(g) + CO_2(g) \rightarrow H_2(g) + CO(g)$$

The calcium oxide reacts with high melting point impurities to form a complex aluminosilicate 'slag' that also contains most of the sulfur impurities,

e.g. $CaO(s) + SiO_2(s) \rightarrow CaSiO_3(l)$

At the very high temperatures at the bottom of the furnace the molten iron and liquid slag separate into two layers with the less dense slag on top. Both are tapped off as more charge is added to the furnace in a continuous process. The molten iron (known as pig iron) contains phosphorus and sulfur together with small amounts of other elements, such as manganese and silicon and about 4 to 5% carbon. The slag is used for road making or is treated to make by-products, such as cement and thermal insulation.

Steel

CONVERSION OF IRON INTO STEEL

Most steel is made by the basic oxygen process. Molten iron is added to a vessel known as an oxygen converter. Preheated oxygen is injected at high pressure into the vessel and the impurities are oxidized,

e.g. $$2C + O_2 \rightarrow 2CO_2$$
$$P_4 + 5O_2 \rightarrow P_4O_{10}$$
$$Si + O_2 \rightarrow SiO_2$$

Apart from carbon monoxide the products react with added lime to form a slag. As the reactions are highly exothermic the temperature is controlled by adding scrap steel. The dissolved oxygen in the steel must be removed by adding controlled amounts of aluminium or silicon before the steel is suitable for casting or rolling. During this process other elements, such as chromium and nickel, are also added to form the precise alloy required.

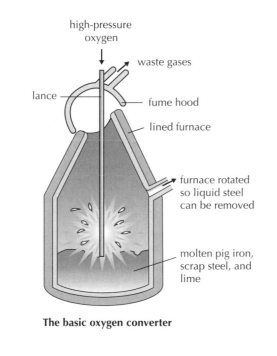

The basic oxygen converter

PROPERTIES AND USES OF STEEL ALLOYS

Steel is an alloy of iron, carbon, and other metallic and non-metallic elements. It has a wide range of uses and by adjusting its composition it can be tailor-made with specific properties. For example, chromium increases the resistance of steel to corrosion. Stainless steel, used for kitchen knives and sinks, etc., contains about 18% chromium and 8% nickel. Toughened steel for use in drill bits, which need to retain a sharp cutting edge at high temperatures, contains up to 20% molybdenum.

RECYCLING OF IRON AND STEEL

The worldwide annual production of steel is about 750 million tonnes. The process of iron and steel making is very energy intensive. This is expensive and uses up precious energy resources. Damage to the environment is caused by extraction of iron ores and by the iron and steel making processes.

To conserve both energy and the environment efficient schemes are now in operation to recycle used or unwanted scrap iron or steel. Currently about 310 million tonnes of the steel produced worldwide comes from recycled steel. This represents an energy saving equivalent to about 160 million tonnes of coal and the conservation of about 200 million tonnes of iron ore.

Aluminium

PRODUCTION OF ALUMINIUM

The worldwide production of aluminium amounts to 20 million tonnes. It is primarily made by the electrolytic reduction of aluminium oxide.

The main ore of aluminium is bauxite. The aluminium is mainly in the form of the hydroxide $Al(OH)_3$ and the principal impurites are iron(III) oxide and titanium hydroxide. The impurities are removed by heating powdered bauxite with sodium hydroxide solution. The aluminium hydroxide dissolves because it is amphoteric.

$$Al(OH)_3(s) + NaOH(aq) \rightarrow NaAlO_2(aq) + 2H_2O(l)$$

The aluminate solution is filtered leaving the impurities behind. Seeding with aluminium hydroxide then reverses the reaction. The pure recrystallized aluminium hydroxide is then heated to produce aluminium oxide (alumina).

$$2Al(OH)_3(s) \rightarrow Al_2O_3(s) + 3H_2O(l)$$

In a separate process hydrogen fluoride is added to the aluminate solution followed by sodium carbonate to precipitate cryolite (sodium hexafluoroaluminate(III), Na_3AlF_6).

$$NaAlO_2(aq) + 6HF(g) + Na_2CO_3(aq) \rightarrow Na_3AlF_6(s) + 3H_2O(l) + CO_2(g)$$

The electrolysis of molten alumina takes places in an open-topped steel container lined with graphite. Alumina has a melting point of 2045 °C so it is mixed with cryolite. This lowers the melting point to about 950 °C so that much less energy is required. The aluminium is produced on the graphite lining which acts as the negative electrode (cathode). Molten aluminium is more dense than cryolite so it collects at the bottom of the cell where it can be syphoned off periodically.

$$Al^{3+}(l) + 3e \rightarrow Al(l)$$

The positive electrode is made of blocks of graphite. As the oxide ions are oxidized some of the oxygen formed reacts with the graphite blocks so that they have to be renewed regularly.

$$2O^{2-}(l) \rightarrow O_2(g) + 4e$$
$$C(s) + O_2(g) \rightarrow CO_2(g)$$

A modern cell can produce up to two tonnes of aluminium per day.

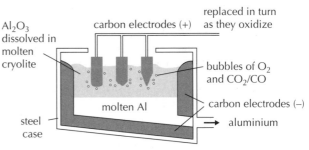

Production of aluminium by electrolysis

PROPERTIES AND USES

The aluminium produced from the electrolysis is more than 99% pure with the main impurities being small amounts of iron and silicon. Aluminium is malleable and can be shaped easily. It is an excellent conductor of heat and electricity. It has a much lower density than iron and yet can form alloys that are stronger than steel. Unlike iron it is resistant to corrosion as it forms a protective layer of aluminium oxide. The thickness of the oxide layer can be further increased by a process known as anodizing.

Some common uses of aluminium

Use	Examples
Transport	Superstructures of trains, ships, aeroplanes, and container vehicles
	Alloy engines for cars and motorcycles, car wheels
Construction	Window frames, doors, roofing
Power transmission	Overhead electricity cables, capacitor foil
Kitchen utensils	Kettles, saucepans
Packaging	Drink cans, foil wrapping
Chemical industry	$Al(OH)_3$ – flame retarder, paper making
	$Al_2(SO_4)_3$ – flocculant in sewage treatment and to precipitate PO_4^{3-}
	Al_2O_3 – catalyst and catalytic support material, abrasive

ENVIRONMENTAL IMPACT OF ALUMINIUM PRODUCTION

The production of aluminium requires very large quantities of electricity. A cheap source of electricity is therefore essential and aluminium plants are usually located either where hydroelectric power is readily available or near sources of coal or natural gas that can be utilized for electricity generation. As with iron, the recycling of aluminium is an important method of saving energy and minimizing the damage to the environment. Iron and steel can easily be separated from aluminium by using a magnet. The cost of recycling aluminium is low as it requires only about 5% of the energy that would be needed to produce the same amount of aluminium from bauxite. Currently about 60% of the aluminium produced in Europe comes from recycled aluminium.

The oil industry

IMPORTANCE OF OIL AS A CHEMICAL FEEDSTOCK

Crude oil is a mixture of many different hydrocarbons, most of which are alkanes. It was formed millions of years ago from the decay of marine organisms. It is often found together with natural gas. The economy of the modern world depends to a large extent on oil. Its main use is as a fuel to supply the world's energy demands but about 10% of it is used, after refining, as chemical feedstock. This makes it the most significant source of organic chemicals including plastics, pesticides, food additives, pharmaceuticals, detergents, cosmetics, dyes, and solvents. It is a limited resource and future generations may question why most of it is being burned with all the attendant problems of pollution rather than being used to make useful products.

REMOVAL OF SULFUR

Crude oil contains appreciable amounts of sulfur, mainly in the form of hydrogen sulfide, caused by the anaerobic decay of sulfur-containing amino acids in the original marine organisms. This must be removed before the oil can be refined as it can poison catalysts by blocking their active sites. It is also removed to prevent acid rain pollution caused by the production of sulfur dioxide when the oil is burned. It can be removed by a reversible acid–base reaction with potassium carbonate.

$$H_2S(g) + CO_3^{2-}(aq) \rightleftharpoons HS^-(aq) + HCO_3^-(aq)$$

The hydrogen sulphide recovered from the solution can then be burned under controlled conditions to give sulfur dioxide, which can be used to make sulfuric acid. It is also possible to react hydrogen sulfide directly with sulfur dioxide by a redox reaction to make elemental sulfur.

$$2H_2S(g) + SO_2(g) \rightarrow 3S(l) + 2H_2O(g)$$

CRACKING

Cracking is the process conducted at high temperatures whereby large hydrocarbons are broken down into smaller, more useful molecules. The products are usually alkanes and alkenes. For example, decane can be broken down to form octane and ethene.

$$C_{10}H_{22}(g) \rightarrow C_8H_{18}(g) + C_2H_4(g)$$

The alkanes are usually branched isomers (e.g. 2,2,4-trimethylpentane) and are added to gasoline (petrol) to improve the octane rating. The alkenes are used to make other chemicals, particularly addition polymers.

Steam cracking

The feedstock is preheated, vaporized, and mixed with steam and then converted to low molecular mass alkenes at 1000–1150 K.

Catalytic cracking

The use of a silica/alumina catalyst enables the cracking to take place at the relatively lower temperature of about 750 K.

Hydrocracking

The feedstock is mixed with hydrogen at a pressure of about 80 atm and cracked over a platinum or silica/alumina catalyst. This gives a high yield of branched alkanes, cycloalkanes, and aromatic compounds for use in 'green' unleaded gasoline (petrol).

FRACTIONAL DISTILLATION OF OIL

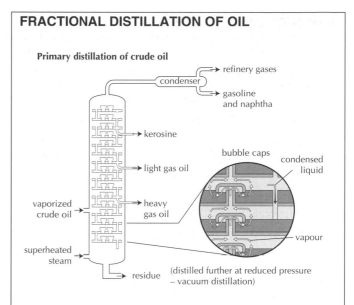

Primary distillation of crude oil

(distilled further at reduced pressure – vacuum distillation)

A single substance with a low boiling point can be separated from involatile impurities by simple distillation. When a mixture of volatile liquids with different boiling points is heated the vapour will be richer in the more volatile substances. By condensing the vapour then revaporizing the condensed liquid the new vapour will be even richer in the more volatile substances. By repeating the process many times the mixture can be separated into its different components according to their boiling point ranges – a process known as **fractional distillation**.

Crude oil is heated to about 450 °C and the vaporized hydrocarbons move up a fractionating column. Superheated steam is added to maintain the temperature and ensure the vapour moves up through the column. Bubble caps in the column allow the hot rising vapour to pass through the liquid that has condensed at each level. At each level the condensed higher boiling point components are tapped off or re-routed to lower levels. The composition and major uses of each fraction is given in 17. Option F – Fuels and Energy.

REFORMING

More useful branched alkanes can be obtained from straight-chain hydrocarbons by mixing them with hydrogen and heating them at 770 K over a platinum/alumina catalyst at high pressure. This process, known as **isomerization**, is a particular form of reforming. Other reforming processes in which straight-chain alkanes are reformed into molecules with the same number of carbon atoms include **cyclization** to make ring molecules and **aromatization** to make benzene. The other important product in both cyclization and aromatization is hydrogen, which can be used in the Haber process to make ammonia.

Examples of reforming

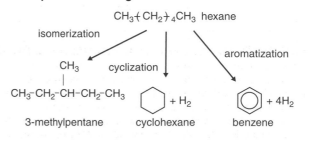

Polymers

EXAMPLES OF POLYMERS

Polymers can be formed either by addition reactions of alkenes or by condensation reactions between two compounds each containing two reactive functional groups.

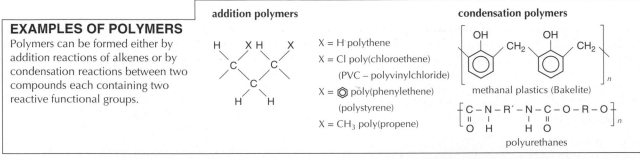

addition polymers

X = H polythene

X = Cl poly(chloroethene)
(PVC – polyvinylchloride)

X = ⬡ poly(phenylethene)
(polystyrene)

X = CH₃ poly(propene)

condensation polymers

methanal plastics (Bakelite)

polyurethanes

RELATIONSHIP BETWEEN STRUCTURAL FEATURES AND PROPERTIES OF POLYMERS

Polymers can also be divided into thermoplastics and thermosets. Thermoplastics can be remoulded each time they are heated. Thermosets cannot be softened or remoulded once they are formed and will be permanently destroyed if heated to a high temperature. Generally the longer the chain length of a polymer the higher the melting point and strength. Apart from the length of the chain there are three other main structural features of polymers that affect their particular properties.

Branching

Depending on the reaction conditions ethene can form high density or low density polythene. In high density polythene there is little branching. This gives long chains that can fit together closely making the polymer stronger, denser, and more rigid than low density polythene. The presence of side chains in low density polythene results in a more resilient and flexible structure making it ideal for the production of film products, such as food wrappings.

Orientation of alkyl groups

In poly(propene) the methyl groups can all have the same orientation along the polymer chain – **isotactic**. Due to the regular structure isotactic polymers are more crystalline and tough. Isotactic poly(propene) is a thermoplastic and can be moulded into objects, such as car bumpers, and drawn into fibres for clothes and carpets. In **atactic** poly(propene) the chains are more loosely held so the polymer is soft and flexible, making it suitable for sealants and roofing materials.

isotactic poly(propene) – all methyl groups orientated in same direction

atactic poly(propene) – methyl groups arranged randomly

Cross-linking

In Bakelite the benzene ring bonds in more than one position to form a three-dimensional cross-linked structure. This gives it considerable strength and such a high melting point that it decomposes before melting (a property of a thermoset). It also makes it very unreactive. Due to its high electrical resistance it was used as the casing for early radios and is now used as one of the ingredients in worktops and printed circuit-board insulation.

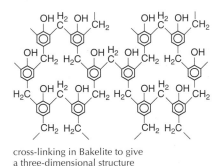

cross-linking in Bakelite to give a three-dimensional structure

---

MODIFICATIONS

Plasticizers

Plasticizers are small molecules that can fit between the long polymer chains. They act as lubricants and weaken the attraction between the chains, making the plastic more flexible. By varying the amount of plasticizer added PVC can form a complete range of polymers from rigid to fully pliable.

Volatile hydrocarbons

If pentane is added during the formation of polystyrene and the product heated in steam the pentane vaporizes producing expanded polystyrene. This light material is a good thermal insulator and is also used as packaging as it has good shock-absorbing properties.

Blowing agents

Air can be blown in the mixture of condensing monomers during the production of polyurethane to produce polyurethane foam. This is used for cushioning material, shoe soles, and thermal insulation.

ADVANTAGES AND DISADVANTAGES OF POLYMER USE

The examples above illustrate how polymers can be tailor-made to perform a variety of functions based on properties such as strength, density, thermal and electrical insulation, flexibility, and lack of reactivity. There are, however, some disadvantages.

1. **Depletion of natural resources** The majority of polymers are carbon based. Currently oil is the major source of carbon although in the past it was coal. Both are fossil fuels and are in limited supply.
2. **Disposal** Because of their lack of reactivity plastics are not easily disposed of. Some, particularly PVC and poly(propene), can be recycled and others (e.g. nylon) are weakened and eventually decomposed by ultraviolet light. Plastics can be burned but if the temperature is not high enough poisonous dioxins can be produced along with toxic gases, such as hydrogen cyanide and hydrogen chloride.
3. **Biodegradability** Most plastics do not occur naturally and are not degraded by micro-organisms. By incorporating natural polymers, such as starch, into plastics they can be made more biodegradeable. However, in the anaerobic conditions present in landfills biodegradation is very slow or will not occur at all.

 **Silicon**

EXTRACTION AND PURIFICATION OF SILICON

Silicon and its compounds are important as they have many uses including microprocessors, zeolites in ion exchange resins, glass and silicones. The element is obtained by reduction of its ore silica SiO_2 using carbon.

$$SiO_2(s) + C(s) \rightarrow Si(s) + CO_2(g)$$

This produces impure silicon. Silicon used for microprocessors needs to be extremely pure. To achieve this the impure silicon is reacted with chlorine to give silicon tetrachloride. This is a liquid as the non-polar molecules are only held together by weak van der Waals' forces. The liquid silicon tetrachloride is distilled to remove involatile impurities then reduced back to silicon with hydrogen. A process known as **zone refining** is then used to make the silicon extremely pure. A heater is passed along a rod of impure silicon. This melts the silicon. After the heater has passed the silicon resolidifies. The impurities dissolve better in molten silicon so are transferred along to the end of the rod.

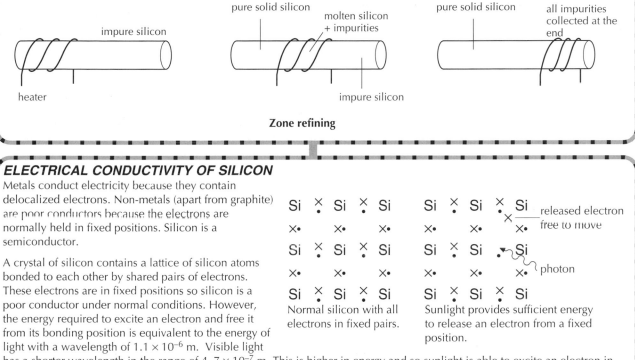

Zone refining

ELECTRICAL CONDUCTIVITY OF SILICON

Metals conduct electricity because they contain delocalized electrons. Non-metals (apart from graphite) are poor conductors because the electrons are normally held in fixed positions. Silicon is a semiconductor.

A crystal of silicon contains a lattice of silicon atoms bonded to each other by shared pairs of electrons. These electrons are in fixed positions so silicon is a poor conductor under normal conditions. However, the energy required to excite an electron and free it from its bonding position is equivalent to the energy of light with a wavelength of 1.1×10^{-6} m. Visible light has a shorter wavelength in the range of $4-7 \times 10^{-7}$ m. This is higher in energy and so sunlight is able to excite an electron in silicon. The electron is then free to move through the crystal lattice making it an electrical conductor. This is the basis of the photoelectric effect and is the theory behind solar powered batteries. In practice the process is not very efficient and the cost of purifying the silicon is high. However solar cells are not polluting and do not use up valuable fossil fuel reserves.

Normal silicon with all electrons in fixed pairs.

Sunlight provides sufficient energy to release an electron from a fixed position.

DOPING OF SILICON TO OBTAIN N-TYPE AND P-TYPE SEMICONDUCTORS

One method of improving the efficiency of the photoelectric effect is by doping. This process involves adding very small amount of atoms of other elements usually from group 3 (Al, Ga, or In) or from group 5 (P or As). When a group 5 element is added the extra electron can move easily throughout the crystal lattice making it a better conductor compared with pure silicon. Such doping produces an n-type semiconductor because the conductivity is due to negative electrons. When a group 3 element is added the element now has one fewer electron than silicon. This produces a 'hole' in the lattice. When a free electron moves into this hole it produces a new hole where the electron was formerly located. The hole can be regarded as a positive carrier so the semiconductor is known as a p-type.

n-type semiconductor

p-type semiconductor

An n, p junction which acts as a rectifier

At a junction between the two different types of semiconductors electrons can flow from an n-type to a p-type as they are moving from a negatively charged area to a positively charged region. However, they cannot flow the other way. The junction thus allows the current to flow in one direction only. Such a junction is known as a rectifier and can be used to convert alternating current into direct current. This property is used in transistors and silicon chips.

EXPLANATION OF ELLINGHAM DIAGRAMS

A reaction will be spontaneous if the value for the standard Gibbs free energy change ($\Delta G^{\ominus}$) is negative.

Consider the formation of a metal oxide from one mole of oxygen:

$$2M(s) + O_2(g) \rightarrow 2MO(s)$$

The oxidation of metals by oxygen is an exothermic process so the standard enthalpy change for the reaction ($\Delta H^{\ominus}$) is negative. The entropy change ($\Delta S^{\ominus}$) will also be negative as the system is becoming more ordered. From the expression $\Delta G^{\ominus} = \Delta H^{\ominus} - T\Delta S^{\ominus}$ it can be seen that at low temperatures the value for $\Delta G^{\ominus}$ will be negative but at higher temperatures it becomes less negative as the factor ($-T\Delta S^{\ominus}$) becomes more positive.

Ellingham diagrams show how the value of $\Delta G_f^{\ominus}$ for the formation of metal oxides changes with temperature. Normally they are plotted per mole of oxygen gas. All metals show similar slopes. When the metal reaches its boiling point (e.g. at 1181 K for zinc) there is a noticeable change in the slope of the line as there is a significant change in the disorder of the system. Note that the line for the formation of silver oxide becomes positive at about 400 K

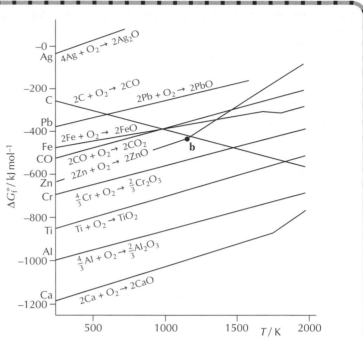

An Ellingham diagram for the formation of metal oxides from one mole of oxygen and for the formation of carbon monoxide from oxygen and carbon dioxide from carbon monoxide.
Point **b** corresponds to the boiling point of zinc.

(127 °C). At this point the reverse reaction becomes spontaneous and silver oxide will decompose spontaneously to metallic silver when heated above this temperature. However, for most metals the line does not become positive until very high temperatures are reached so most metals cannot be obtained from their ores by simply heating.

Metal oxides can be reduced by carbon or carbon monoxide. The plots for the formation of carbon monoxide from carbon and carbon dioxide from carbon monoxide are also included on the diagram.

$$2C(s) + O_2(g) \rightarrow 2CO(g)$$
$$2CO(g) + O_2(g) \rightarrow 2CO_2(g)$$

The line for the formation of carbon monoxide slopes *downwards* because one mole of gas is being converted into two moles of gas so the system is becoming more disordered (i.e. the factor $-T\Delta S^{\ominus}$ becomes more negative with increasing temperature).

Consider the reduction of iron(II) oxide by carbon monoxide at 500 K:

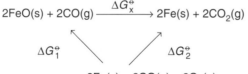

Using the Ellingham diagrams at 500 K $\Delta G_1^{\ominus} = -450$ kJ mol^{-1} and $\Delta G_2^{\ominus} = -480$ kJ mol^{-1}.

By Hess' law: $\Delta G_x^{\ominus} + \Delta G_1^{\ominus} = \Delta G_2^{\ominus}$
$$\Rightarrow \Delta G_x^{\ominus} = \Delta G_2^{\ominus} - \Delta G_1^{\ominus} = -480 - (-450) = -30 \text{ kJ mol}^{-1}$$

Because the value for $\Delta G_x^{\ominus}$ is negative the reduction of iron(II) oxide by carbon monoxide is spontaneous at 500 K and the reaction will occur. As the temperature increases the two lines get closer together until they converge at 1000 K. Above this temperature $\Delta G_x^{\ominus}$ will be positive and the reaction will not proceed. However, at 1000 K the line for the formation of carbon monoxide from carbon also crosses the iron(II) oxide line. At temperatures above 1000 K this line lies below the iron(II) oxide line. In the blast furnace iron(II) oxide will be reduced by **carbon**, not carbon monoxide, at temperatures above 1000 K.

$$\text{FeO(s) + C(s)} \rightarrow \text{Fe(s) + CO(g)} \quad \Delta G^{\ominus} \text{ negative above 1000 K}$$

Generally any metal oxide will be reduced by a carbon–oxygen system when the $\Delta G_f^{\ominus}$ value of the carbon–oxygen system is more negative than the $\Delta G_f^{\ominus}$ value for the metal–oxygen system.

Thus zinc oxide can be reduced by carbon at temperatures above 1200 K and by carbon monoxide at temperatures above 1450 K. The temperature needs to be raised to 1500 K before chromium can be formed from the reduction of chromium(III) oxide by carbon. For aluminium and calcium the temperature is much higher and so electrolytic methods must be used.

Mechanisms in the organic chemicals industry

MECHANISMS OF THERMAL AND CATALYTIC CRACKING

Cracking is used to convert large hydrocarbon molecules into smaller more useful molecules. Cracking is usually carried out at high temperatures either in the presence of steam to give high yields of alkenes or in the presence of a catalyst to give higher yields of cyclic and branched alkanes.

Steam cracking

At the higher temperatures involved in steam cracking (1000–1150 K) the hydrocarbon molecules gain sufficient energy to bring about homolytic fission of the carbon–carbon bonds. The cracking mechanism therefore involves free radicals.

$$-CH_2-CH_2- \rightarrow -CH_2^{\bullet} + -CH_2^{\bullet}$$

When the mixture is cooled the electrons and hydrogen atoms rearrange to give alkenes as the most abundant product.

e.g. the radical can split:
$$RCH_2-CH_2-CH_2^{\bullet} \rightarrow RCH_2^{\bullet} + CH_2=CH_2$$

or it can lose hydrogen:
$$RCH_2-CH_2-CH_2^{\bullet} \rightarrow RCH_2-CH=CH_2 + H^{\bullet}$$

The hydrogen radical can continue to propagate the chain.

Catalytic cracking

Catalytic cracking involves the use of aluminosilicate catalysts on a fluidized bed. The temperature is lower (about 700–800 K). The mechanism is ionic as carbocations are produced. These primary or secondary carbocations then rearrange to give branched alkanes by forming more stable tertiary carbocations, e.g.

$$CH_3-CH_2-\overset{+}{C}H-CH_2-CH_2CH_3 \rightarrow$$

$$CH_3-CH_2-\overset{+}{\underset{\underset{CH_3}{|}}{C}}-CH_2-CH_3$$

(The reason for the extra stability of the tertiary carbocation is explained in 19. option H – Further Organic Chemistry.)

MANUFACTURE OF LOW DENSITY POLYTHENE (LDPE)

The manufacture of low density polythene is carried out at very high pressures (1000–3000 atm) at a temperature of about 500 K. An initiator, such as an organic peroxide or a trace of oxygen, is added. Under these conditions free radicals are formed.

$$R-O-O-R \rightarrow 2RO^{\bullet} \qquad \text{free radical formation}$$

$$RO^{\bullet} + H_2C=CH_2 \rightarrow R-O-CH_2-CH_2^{\bullet} \qquad \text{propagation}$$

$$R-O-CH_2-CH_2^{\bullet} + H_2C=CH_2 \rightarrow R-O-CH_2-CH_2-CH_2-CH_2^{\bullet} \qquad \text{etc.}$$

Termination takes place when two radicals combine. The average polymer molecule contains between about 4×10^3 and 4×10^4 carbon atoms with many short branches. The branches affect both the degree of crystallinity and the density of the material. LDPE generally has a density of about 0.92 g cm^{-3} and is used mostly for packaging.

MANUFACTURE OF HIGH DENSITY POLYTHENE (HDPE)

High density polythene is manufactured by polymerising ethene at a low temperature (about 350 K) and pressure (1–50 atm) using a Ziegler–Natta catalyst. The catalyst is a suspension of titanium(III) or titanium(IV) chloride together with an alkyl–aluminium compound (e.g. triethylaluminium Al(C$_2$H$_5$)$_3$). The mechanism is complex and still not thoroughly understood. Essentially it involves the insertion of the monomer between the catalyst and the growing polymer chain. This is known as co-ordination polymerization (sometimes described as anionic polymerization). The titanium atom is attached to one end of the growing hydrocarbon chain and uses its empty d orbitals to form a co-ordination complex with the π electrons of the new incoming ethene molecule.

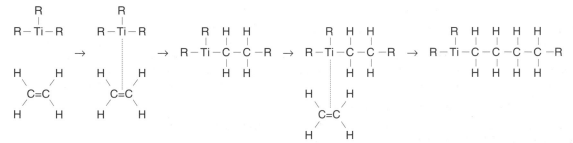

A simplified reaction sequence of the Ziegler–Natta catalysed polymerization of ethene is shown above.

The resulting polymer consists mainly of linear chains with very little branching. This gives it a higher density (0.96 g cm^{-3}) and a more rigid structure as the chains can fit together more closely. It is used to make containers and pipes.

ELECTROLYSIS OF SODIUM CHLORIDE

Chlorine is a powerful oxidizing agent with a standard electrode potential of $+1.36\,V$. Apart from fluorine, very few chemical oxidizing agents are powerful enough to oxidize chloride ions to chlorine so the production depends on using electrons themselves.

Chlorine gas is formed during the electrolysis of molten sodium chloride in the industrial production of sodium metal; however, the main source of chlorine is the electrolysis of aqueous sodium chloride (brine) as this requires less energy. Sodium chloride is a cheap raw material, which is readily available, and the process also produces sodium hydroxide and hydrogen, both of which are important industrial products. There are two main methods by which the electrolysis is achieved – the mercury cell and the diaphragm cell.

THE MERCURY CELL

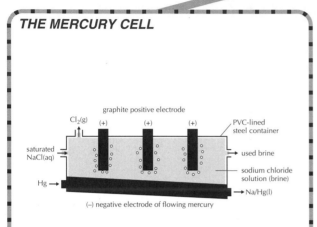

The negative electrode (cathode) is made of flowing mercury. Although sodium is above hydrogen in the electrochemical series sodium is preferentially discharged as it forms an alloy (known as an amalgam) with the mercury.

$$Na^+(aq) + e + Hg(l) \rightarrow Na/Hg(l)$$

The mercury then flows out of the electrolysis cell into a separate chamber where it reacts with water to produce hydrogen and sodium hydroxide solution. The mercury is recycled back into the electrolytic cell.

$$Na/Hg(l) + H_2O(l) \rightarrow Na^+(aq) + OH^-(aq) + \tfrac{1}{2}H_2(g) + Hg(l)$$

The cell itself is made of PVC-lined steel and the positive electrode (anode) where the chlorine is formed is made of graphite.

$$2Cl^-(aq) \rightarrow Cl_2(g) + 2e$$

THE DIAPHRAGM CELL

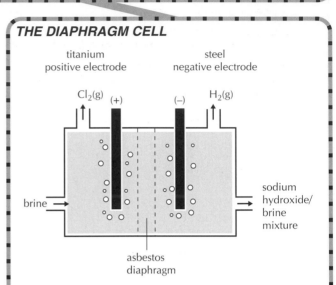

In the diaphragm cell the positive electrode is made of titanium and the negative electrode is made of steel. Hydrogen is formed at the negative electrode and chlorine at the positive electrode.

negative electrode $\quad 2H_2O(l) + 2e \rightarrow H_2(g) + 2OH^-(aq)$
positive electrode $\quad 2Cl^-(aq) \rightarrow Cl_2(g) + 2e$

The diaphragm, made of asbestos, allows the sodium chloride solution to flow between the electrodes but separates the chlorine and hydrogen gas and helps to prevent the OH^- ions flowing towards the positive electrode. The sodium hydroxide solution formed accumulates in the cathode compartment and is piped off. The resulting solution contains about 10% sodium hydroxide and 15% unused sodium chloride by mass. It is concentrated by evaporation and the sodium chloride crystallizes out leaving a 50% solution of sodium hydroxide.

A more modern version of the diaphragm cell (known as an ion exchange membrane cell) uses a partially permeable ion exchange membrane rather than asbestos. The membrane is made of a fluorinated polymer and is permeable to positive ions but not negative ions.

ENVIRONMENTAL IMPACT OF THE CHLOR-ALKALI INDUSTRY

In many parts of the world the mercury cell has been replaced by the diaphragm or ion exchange membrane cell. The ion exchange membrane cell is much cheaper to run due to the development of modern polymers. The main reason why the use of the mercury cell has been discontinued is environmental. In theory all the mercury is recycled but in practice some leaks into the environment and can build up in the food chain to toxic levels.

The chlorine produced has many important industrial uses, among them chlorinated organic solvents, water purification, pesticides, feedstock for inorganic chemicals (e.g. hydrochloric acid), and bleaching paper. Concern is mounting over the use of chlorinated organic compounds. Several have been shown to be carcinogenic and the C–Cl bond can break homolytically in the presence of ultraviolet light at higher altitudes to form chlorine radicals, which can contribute to ozone depletion.

IB QUESTIONS – OPTION E –
CHEMICAL INDUSTRIES

1. (a) Aluminium is manufactured by the electrolysis of alumina dissolved in molten cryolite.

    (i)   Explain the function of the cryolite. [1]

    (ii) Give an ionic equation for the reaction at the positive electrode (anode) during the electrolysis. [1]

    (iii) Explain with the aid of an equation why the positive electrode slowly disappears. [1]

    (b) Explain how the production of pure alumina from bauxite takes advantage of the amphoteric nature of aluminium oxide. [2]

    (c) Give **two** properties **and** related uses which make aluminium an important metal in today's world. [2]

    (d) Despite aluminium being the most abundant metal in the earth's crust, it is frequently recycled. Give **two** reasons which favour recycling. [2]

2. (a) Explain why crude oil contains small amounts of sulfur. [1]

    (b) Why must the sulfur be removed from crude oil before further refining takes place? [1]

    (c) One of the chemical processes used in the refining of crude oil is *cracking*. Give a balanced equation for the thermal cracking of $C_{10}H_{22}$ and explain why the process is important. [2]

    (d) Crude oil can also be refined by *re-forming*. One re-forming process is *isomerisation*. State what takes place during this process and what use is made of the product. [2]

    (e) A second type of re-forming is *cyclisation* and *aromatisation*. One such example is the conversion of hexane to benzene using a catalyst at 500 °C and a pressure of 20 atmospheres.

    (i)   Write a balanced equation for this reaction. [1]

    (ii) For which important industrial process is the inorganic product from this reaction used as a feedstock? [1]

3. The graph below (not to scale) indicates the variation of K_c with temperature for an industrial process:

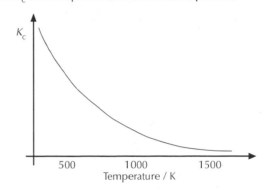

    (a) Based on the graph, explain whether the reaction is exothermic or endothermic. [2]

    (b) Industrially, this process is carries out at 750 K. Explain why it is not carried out at a much higher or much lower temperature. State how a catalyst could increase the rate of this reaction at 750 K. [3]

---

4. The structures of two important polymers are given below:

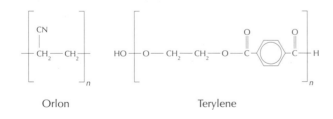

    Orlon   Terylene

    (a) Deduce and draw the structural formulas of the monomers used to make these polymers. [3]

    (b) State the conditions necessary for the manufacture of high density polythene. How does the mechanism for the manufacture of high density polythene differ from the mechanism for the manufacture of low density polythene? [3]

5. Some metals can be isolated from their oxides simply by heating (e.g. mercury) while other metals cannot be isolated by heating alone but can be obtained by heating the oxides in the presence of carbon (e.g. lead). With reference to information on Ellingham Diagrams in the Data Booklet:

    (a) describe and explain the conditions under which mercury can be obtained from its oxide. [3]

    (b) explain why lead cannot be obtained just by heating the oxide alone but can be obtained by heating in the presence of carbon and specify the conditions under which this occurs. [4]

    (c) account for the fact that aluminium cannot be isolated from its oxide even when the oxide is heated with carbon. [3]

Energy sources

DESIRABLE CHARACTERISTICS OF ENERGY SOURCES

Energy is released when a system moves from a higher energy state to a lower energy state. Energy may be stored as potential energy, for example in chemical bonds or water in dams, or be in the form of kinetic energy, for example wave movement or flowing water. When one form of energy is converted into a different form some of the energy available to do work is always lost during the process. For an energy source to be useful it should have the following characteristics:

- it should be easily accessible
- it should release energy at a reasonable rate (not too slow or too fast)
- it should be cheap and plentiful
- if possible its use should not be detrimental to the environment or health.

DIFFERENT SOURCES OF ENERGY

Apart from tidal energy and nuclear energy the Sun is the ultimate source of most of the different forms of energy on Earth.

Fossil fuels These are coal, oil, and natural gas and are essentially non-renewable energy sources. They contain mainly carbon and hydrogen atoms and the energy is released when they are oxidized to carbon dioxide and water.

Nuclear energy In **nuclear fission** large atomic nuclei are split into smaller nuclei. In the process matter is converted into energy. This process has been harnessed in nuclear power stations and the energy released converted into electricity. Nuclear fission is capable of producing large amounts of energy. The disadvantages include controlling the reactions and all the problems associated with radioactive materials.

Nuclear fusion is potentially an even more powerful source of energy. It involves the combination of small nuclei to make larger nuclei and is the source of the Sun's energy. It has been used destructively in hydrogen bombs but the problems associated with harnessing the energy in a controlled way have not been overcome and as yet it is of no commercial use.

Other energy sources These are generally renewable sources which are as yet not harnessed very efficiently. They include wind power, wave power, and hydroelectric power, which use wind and/or water to drive a generator. Tidal energy harnesses the gravitational force between the Earth and the Moon but can only be generated when there is tidal movement. Geothermal energy utilizes the heat stored in the interior of the Earth, which is generated by gravitational forces and through natural radioactivity.

Biomass Fuel produced by biological processes. The energy from the Sun is harnessed through photosynthesis.

$$6CO_2(g) + 6H_2O(l) \rightarrow C_6H_{12}O_6(s) + 6O_2(g)$$

When plant materials (e.g. wood) are burned the reverse process takes place, releasing the energy.

The fermentation of starch and sugars from plant material can form ethanol; an alternative fuel.

Biomass is a renewable source but cannot produce nearly enough energy to sustain the energy demands of modern society.

Electrochemical cells (batteries) The difference in the redox potential between two half-reactions is utilized in electrochemical cells to produce a flow of electrons (electricity). They are useful as a portable source of energy and as stores for limited amounts of energy.

Solar energy Solar energy is non-polluting and freely available during daylight hours. It can be harnessed by solar heating panels, which capture and store the energy directly, or through photovoltaic cells, which convert the solar energy into electricity.

POLLUTION CAUSED BY ENERGY SOURCES

There are many different forms of pollution. Wind power may cause little, if any, chemical pollution but windmills can be noisy and unsightly. Similarly the building of dams to provide hydroelectric power can destroy large areas of land and alter the local ecology. Many energy sources also cause thermal pollution by heating the surrounding water which lowers the dissolved oxygen content.

The most polluting fuels are fossil fuels and nuclear fuels. Radioactive materials can escape from nuclear power stations into the air and water and spent radioactive waste can remain radioactive for thousands of years. Coal and oil both contain sulfur, which was once present in the amino acids of the plants and animals from which they are made. This burns to form sulfur dioxide which is converted into sulfur trioxide and then sulfuric acid – a constituent of acid rain.

$$S(s) + O_2 \rightarrow SO_2(g)$$
$$2SO_2(g) + O_2(g) \rightleftharpoons 2SO_3(g)$$
$$SO_3(g) + H_2O(l) \rightarrow H_2SO_4(aq)$$

At the high temperatures produced when fossil fuels burn oxides of nitrogen can also be formed. These can react to form nitrogen dioxide, NO_2. This is converted into nitric acid, which is also a constituent of acid rain. Nitrogen oxides contribute to photochemical smog. They break down in the presence of sunlight to form oxygen radicals and then secondary pollutants, such as ozone and peroxyacylnitrates (PANs). Coal and oil also form particulates in the form of fly ash and soot from coal and carbon particles from the incomplete combustion of gasoline (petrol) and diesel. Incomplete combustion also produces carbon monoxide and petroleum fuels evaporate to produce potentially carcinogenic volatile hydrocarbons. Natural gas is much cleaner but like all fossil fuels increases the concentration of carbon dioxide in the atmosphere when it is combusted. Carbon dioxide is a greenhouse gas and contributes to global warming.

Fossil fuels (1)

FORMATION OF FOSSIL FUELS

Coal is fossilized plant material. Most coal was formed during the carboniferous period (286–360 million years ago). The action of pressure and heat through geological forces converted the plant material in stages from peat to lignite to bituminous soft coal and finally to hard coal (anthracite). At each stage the percentage of carbon increased. Coal contains between 80 and 90% carbon by mass.

Crude oil was formed from the remains of marine organisms, mainly during the Paleozoic period some 600 million years ago. Thick sediments built up on top of the organic layers and under the action of high pressure and biochemical activity crude oil was formed. The oil migrated through rocks due to earth movements and collected in traps. Crude oil is a complex mixture of straight-chain, branched, cyclic, and aromatic hydrocarbons, although it consists mainly of alkanes.

Natural gas was formed at the same time as crude oil and the two are often found together, although it may occur on its own or with coal. It consists mainly of methane (85–95%) with varying amounts of ethane, propane, butane, and other gases, such as hydrogen sulfide.

CALORIFIC VALUES OF FOSSIL FUELS

In a plentiful supply of oxygen fossil fuels burn to produce carbon dioxide and water. The relative strengths of the O–H and C=O bonds account for the exothermic nature of the reactions. Weight for weight natural gas is the best fossil fuel.

Energy value / kJ kg^{-1}	Natural gas: 55 000	Crude oil: 42 000	Coal: 29 000

The enthalpies of combustion of fuels can be determined experimentally using a water calorimeter. A known mass of the fuel is burned in oxygen and the heat used to raise the temperature of a known quantity of water and the surrounding apparatus. By knowing the specific heat capacity of water and the water equivalent of the apparatus the heat evolved by the fuel can be calculated. This calculation is identical to the example for a food given in 14. Option C – Human Biochemistry. Enthalpies of combustion can also be calculated theoretically using enthalpies of formation. For example, the standard enthalpies of formation of methane, carbon dioxide, and liquid water are –75.6, –394 and –286 kJ mol^{-1} respectively.

From the energy cycle:

$$\text{CH}_4(g) + 2\text{O}_2(g) \xrightarrow{\Delta H_c^{\ominus}(\text{CH}_4)} \text{CO}_2(g) + 2\text{H}_2\text{O}(l)$$

$$\Delta H_f^{\ominus}(\text{CH}_4) \quad \Delta H_f^{\ominus}(\text{CO}_2) \quad 2\Delta H_f^{\ominus}(\text{H}_2\text{O})$$

$$\text{C(s)} + 2\text{H}_2(g) + 2\text{O}_2(g)$$

It can be seen that $\Delta H_c^{\ominus}(\text{CH}_4) = \Delta H_f^{\ominus}(\text{CO}_2) + 2\Delta H_f^{\ominus}(\text{H}_2\text{O}) - \Delta H_f^{\ominus}(\text{CH}_4)$

$$= -394 + (2 \times -286) - (-75.6) = -890.4 \text{ kJ mol}^{-1}$$

Since one mole of methane has a mass of 16.0 g this is equivalent to $-890.4 \times \frac{1000}{16.0} = -55\,600$ kJ kg^{-1}.

THE COMPOSITION AND CHARACTERISTICS OF CRUDE OIL FRACTIONS

The fractional distillation of oil is covered in 15. option E – Chemical Industries. The number of carbon atoms, boiling ranges, and the uses of the different fractions are summarized in the table:

Fraction	Carbon chain length	Boiling range / °C	Main uses
Refinery gas	1–4	<30	Used as fuel on site, gaseous cooking fuel, and as feedstock for chemicals, e.g. methane is used to provide hydrogen gas for the Haber process.
Gasoline and naphtha	5–10	40–180	Gasoline (petrol) for cars. Feedstock for organic chemicals (by steam cracking).
Kerosine	11–12	160–250	Fuel for jet engines; domestic heating; cracked to provide extra gasoline (petrol).
Gas oil (diesel oil)	13–25	220–350	Diesel engines and industrial heating; cracked to produce extra gasoline (petrol).
Residue	>20	>350	Fuel for large furnaces; vacuum distilled to make lubricating oils and waxes. Residue of bitumen and asphalt used to surface roads and waterproof roofs.

Fossil fuels (2)

OCTANE RATING

In an efficient internal combustion engine a spark ignites the fuel–air mixture just before the piston reaches 'top dead centre' so that the full force of the explosion pushes the piston down just as it reaches the top of the cylinder. Under the conditions of high temperature and pressure the reaction may start before the spark and the engine will be less efficient. This is known as pre-ignition or **knocking**. The more straight-chain the alkane the higher the tendency for knocking.

Fuels are classified according to their **octane number**. Generally the more branched the alkane the higher the octane number. Pure heptane is assigned an octane number of zero and an isomer of octane, 2,2,4-trimethylpentane, has an octane rating of 100. Thus gasoline (petrol) with an octane rating of 95 will burn as efficiently as a mixture of 95% 2,2,4-trimethylpentane and 5% heptane. In the past tetraethyllead $Pb(C_2H_5)_4$ was added to petrol to raise the octane rating. Lead-free gasoline (petrol) contains added aromatic hydrocarbons, such as benzene, and more branched hydrocarbons obtained through cracking.

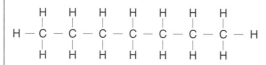

heptane: octane rating 0

2,2,4-trimethylpentane: octane rating 100

COAL GASIFICATION AND LIQUEFACTION

Before the advent of plentiful supplies of natural gas coal was routinely turned into 'coal gas'. As supplies of natural gas diminish, interest in coal gasification may revive. Coal gas (also known as synthesis gas, water gas, or town gas) contains a mixture of hydrogen and carbon monoxide and is made by heating coal in the presence of steam.

$$C(s) + H_2O(g) \rightarrow CO(g) + H_2(g)$$

Reacting coal gas with more hydrogen in the presence of a heated catalyst converts it into SNG (substitute or synthetic natural gas).

$$CO(g) + 3H_2(g) \rightarrow CH_4(g) + H_2O(g)$$

SNG can also be made by heating crushed coal in steam at 700 °C using potassium hydroxide as a catalyst.

$$2C(s) + 2H_2O(g) \rightarrow CH_4(g) + CO_2(g)$$

SNG is a cleaner gas (as it removes pollution due to sulfur dioxide) which is easier to transport but the process is less efficient as it uses up some 30% of the available energy during the conversion process.

In Germany in the 1930s and in South Africa, where coal is abundant, coal has been converted into a liquid fuel. As the price of oil increases this process may become more important economically. The method is known as the Fischer–Tropsch process. Synthesis gas is reacted with more steam to increase the proportion of hydrogen in the mixture.

$$CO(g) + H_2O(g) \rightarrow CO_2(g) + H_2(g)$$

The hydrogen and carbon monoxide are then passed into a fluidized bed reactor containing iron or cobalt catalysts to produce a mixture of hydrocarbons that can be separated by fractional distillation.

Relative advantages and disadvantages of fossils fuels

Fossil fuel	Advantages	Disadvantages
Coal	1. Present in large quantities and distributed throughout the world. 2. Can be converted into synthetic liquid fuels and gases. 3. Feedstock for organic chemicals. 4. Has the potential to yield vast quantities of energy compared with renewable sources and safer than nuclear power. 5. Longer lifespan (350 years?) compared with oil or gas.	1. Contributes to acid rain and global warming. 2. Not so readily transported (no pipelines). 3. Coal waste (slag heaps) lead to ground acidity and visual and chemical pollution. 4. Mining is dangerous – cave-ins, explosions, and long term effect of coal dust on miners. 5. Dirty (produces dust, smoke, and particulates).
Oil	1. Easily transported in pipelines or by tankers. 2. Convenient fuel for use in cars, lorries, etc. 3. Feedstock for organic chemicals.	1. Contributes to acid rain and global warming. 2. Limited lifespan (30–50 years?) and uneven distribution world-wide. 3. Risk of pollution associated with transportation by tankers.
Natural gas	1. Clean fuel. 2. Easily transported in pipelines and pressurized containers. 3. Does not contribute to acid rain. 4. Releases a higher quantity of energy per kg than coal or oil.	1. Contributes to global warming. 2. Limited lifespan (30 years?) and uneven distribution world-wide. 3. Greater risk of explosions due to leaks.

Nuclear energy (1)

NUCLEAR REACTIONS

In a normal chemical reaction valence shell electrons are rearranged as bonds are broken and new bonds formed. There is no change in the nucleus and no new elements are formed. In nuclear reactions the nucleus itself rearranges. During nuclear fission a large nucleus breaks down to form two or more smaller nuclei. Nuclear fusion involves the combination of two light nuclei to form a heavier nucleus. In both cases the total mass of the products is less than the mass of the initial nucleus or nuclei. During nuclear reactions mass is converted into energy according to Einstein's equation $E = mc^2$ (where c is the velocity of light). Thus in a nuclear reaction new elements are formed and the energy change is potentially much greater than in a chemical reaction.

HALF-LIFE $t_{\frac{1}{2}}$

It is impossible to state when an individual unstable isotope will decay as it occurs randomly. However, when a large number of atoms are together in a sample of the isotope the rate of decay depends on the amount of atoms present. The time taken for any specified amount to decrease by exactly one half remains constant and, unlike chemical reactions, is independent of pressure and temperature. This time is known as the half-life. Note that the half-life for a particular isotope is defined as the time taken to decay to one half of the **mass** of the **original isotope**. In a chemical reaction half-life is defined as the time taken for the **concentration** of a reactant to decrease to one half of its initial value. The half-life of $^{32}_{15}P$ is 14.3 days. After 14.3 days a 1.0 g sample of $^{32}_{15}P$ will have decayed to 0.50 g and after a further 14.3 days only 0.25 g of $^{32}_{15}P$ will be remaining. Questions are often set the other way round. (For example, calculate the half-life of ^{131}I if a sample of ^{131}I is found to contain $\frac{1}{32}$ of the original amount after 40.30 days.)

$$1 \xrightarrow{t_{\frac{1}{2}}} \frac{1}{2} \xrightarrow{t_{\frac{1}{2}}} \frac{1}{4} \xrightarrow{t_{\frac{1}{2}}} \frac{1}{8} \xrightarrow{t_{\frac{1}{2}}} \frac{1}{16} \xrightarrow{t_{\frac{1}{2}}} \frac{1}{32}$$

The total number of half-lives is five so the half-life = $\frac{40.30}{5}$ = 8.06 days.

THE NATURE OF α, β, AND γ RADIATION

When an unstable radioactive isotope decays naturally it can emit three different types of radiation from the nucleus. α particles are helium nuclei. They have a positive charge and will be deflected towards the negative plate when an external electric field is applied. β particles (also known as β^- particles) are electrons. They will experience greater deflection in an electric field as they have much less mass. Since they are negatively charged β particles will be attracted to the positive plate. γ radiation is high energy electromagnetic radiation. Since it has no charge it is unaffected by an electric or magnetic field. It is the most penetrating of the three types of radiation as it has no mass.

Name	Type	Mass / amu	Relative charge	Penetrating power
alpha particle α	helium nucleus $^4_2He^{2+}$	4	+2	A few centimetres of air. Stopped by paper, skin, or clothing.
beta particle β	electron $^0_{-1}e$	$\frac{1}{1840}$	-1	A few metres of air. Stopped by thin (1 mm thick) aluminium.
gamma radiation γ	high energy electromagnetic radiation	0	0	A few kilometres of air. Stopped by 10 cm of lead or several metres of concrete.

NUCLEAR EQUATIONS

Nuclear equations must balance. The total mass numbers and the nuclear charge numbers (atomic numbers) must be equal on both sides of the equation. During α particle emission the new element will have a mass of four less than the original element and an atomic number that is two less.

$$\text{e.g. } {}^{238}_{92}U \rightarrow {}^{234}_{90}Th + {}^4_2He$$

During β emission when an electron is ejected from the nucleus at a velocity approaching the speed of light the new element will have the same mass number but the atomic number will have increased by one.

$$\text{e.g. } {}^{14}_6C \rightarrow {}^{14}_7N + {}^0_{-1}e$$

Nuclear reactions also occur artificially when nuclei are bombarded with other small particles, such as α particles or neutrons. In each case the total mass numbers and nuclear charges on both sides of the equation must still balance.

$$\text{e.g. } {}^{235}_{92}U + {}^1_0n \rightarrow {}^{144}_{56}Ba + {}^{90}_{36}Kr + 2{}^1_0n$$

Other small particles that may be involved in nuclear reactions include protons ^1_1p and positrons $^0_{+1}e$. Positrons are positive electrons, sometimes called β^+ particles to distinguish them from electrons (β^- particles).

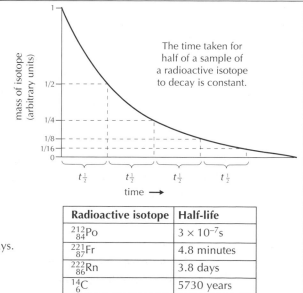

mass of isotope (arbitrary units)

The time taken for half of a sample of a radioactive isotope to decay is constant.

$t_{\frac{1}{2}}$ $t_{\frac{1}{2}}$ $t_{\frac{1}{2}}$ $t_{\frac{1}{2}}$

time →

Radioactive isotope	Half-life
$^{212}_{84}Po$	3×10^{-7} s
$^{221}_{87}Fr$	4.8 minutes
$^{222}_{86}Rn$	3.8 days
$^{14}_6C$	5730 years
$^{238}_{92}U$	4.5×10^9 years

Nuclear energy (2)

NUCLEAR POWER

A nuclear power station essentially contains two main components: the **reactor** to produce the heat from a nuclear reaction and a **turbine** to drive a generator to produce electricity. The essential difference between a nuclear power station and a conventional power station is simply the method used to provide the heat. The nuclear reactor uses a fuel of uranium or plutonium and is housed in a concrete container which acts as a **shield**. The uranium used is ^{235}U. This reacts with neutrons to form smaller nuclei and more neutrons. A typical reaction is:

$$^{1}_{0}n + ^{235}_{92}U \rightarrow ^{141}_{56}Ba + ^{92}_{36}Kr + 3^{1}_{0}n + energy$$

Because more neutrons are produced than are used a chain reaction is possible. The mass of the products is less than the reactants and the mass defect is converted into energy. A one kilogram difference in mass equates to $1 \times (3 \times 10^8)^2 = 9 \times 10^{16}$ J kg^{-1} or 9×10^{13} kJ kg^{-1}. Compare the energy obtained from a conventional fuel, such as coal (2.9×10^4 kJ kg^{-1}). Natural uranium only contains a small percentage of ^{235}U, most of it is ^{238}U. In a breeder reactor neutrons react with ^{238}U to form plutonium -239 which is fissionable.

$$^{238}_{92}U + ^{1}_{0}n \rightarrow ^{239}_{92}U \rightarrow ^{0}_{-1}e + ^{239}_{93}Np \rightarrow ^{0}_{-1}e + ^{239}_{94}Pu$$

$$^{1}_{0}n + ^{239}_{94}Pu \rightarrow ^{90}_{38}Sr + ^{147}_{56}Ba + 3^{1}_{0}n + energy$$

The reactors contains several different components. The **fuel rods** contain the fissionable material, such as $^{235}_{92}U$ or $^{239}_{94}Pu$. The neutrons are controlled by **moderators**. These are made of an inert material, such as water or graphite. Their function is to slow down the neutrons, making them more likely to collide with the fissionable nuclei. As the nuclear reaction proceeds the chain reaction needs to be controlled. This is done by **control rods** made of cadmium or boron. These can be lowered or raised as necessary and their function is to absorb excess neutrons.

A **coolant** is used to control and extract the heat produced. In a pressurized water reactor, water at high pressure is used as the primary coolant. Other coolants used include liquid sodium, air, heavy water ($^{2}_{1}H_{2}O$), and carbon dioxide. The primary coolant is itself cooled by a secondary loop of water. This is used to provide steam to turn the blades of a turbine. The secondary loop is necessary to avoid the transfer of radioactivity to the water used to power the generator and to the cooling water around the condenser.

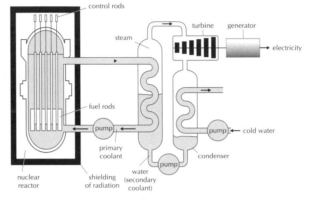

Electricity generation in a nuclear power plant

control rods
steam
turbine   generator
electricity
fuel rods
pump
pump ← cold water
primary coolant
condenser
pump
nuclear reactor
shielding of radiation
water (secondary coolant)

POWER FROM NUCLEAR FUSION

Nuclear fusion offers the possibility of an almost unlimited source of energy since the main fuel, deuterium, is abundant in sea water. The essential reaction is:

$$^{2}_{1}H + ^{2}_{1}H \rightarrow ^{3}_{2}He + ^{1}_{0}n + energy$$

However, the problems of controlling the reaction have yet to be overcome. Essentially the intensely hot reaction mixture, known as plasma, has to be contained and maintained for long enough to fuse the nuclei together. Since this involves temperatures approaching forty million degrees (4×10^7 °C) the problems are considerable. Current efforts centre around confining the plasma within a magnetic field.

SAFETY IN NUCLEAR POWER PLANTS

Arguments against the use of nuclear energy include the risk of terrorist attack, an accident, the disposal of waste, and the promotion of less polluting forms of alternative energy. The worst recorded accident at a nuclear power plant took place in Chernobyl in the Ukraine in 1986. This accident heightened the concerns about the safety of nuclear power stations.

1. **The possibility of a meltdown**
   A meltdown occurs when a nuclear reactor becomes out of control and essentially becomes a nuclear bomb. A nuclear power station requires the slow release of energy so the fuel contains much less of the fissionable isotope ($^{235}_{92}U$) than a nuclear bomb. The neutrons emitted by the fissionable material in a power station are absorbed by the non-fissionable isotope ($^{238}_{92}U$) so in theory cannot build up enough momentum to establish a spontaneously explosive chain reaction.

2. **Escape of radioactive material**
   This can occur as the fuel is being transported or while it is being used. In the Chernobyl disaster a fire ignited the graphite moderator and a cloud of radioactive gas spread across much of Europe. Ordinary materials, such as the

surrounding air and clothes worn by workers, also have the potential to transfer low level waste outside the plant. Concern has also been expressed about the risk of plutonium (particularly from the disused reactors in the former USSR) falling into the hands of terrorists.

3. **Escape of the coolant**
   The fast breeder reactor uses sodium metal as the coolant. The escape of liquid sodium and its subsequent reaction with water would have potentially disastrous consequences. Other coolants, such as heavy water and carbon dioxide, could also transfer radioactivity quickly over a wide area if they escaped.

4. **Nuclear waste**
   There are considerable problems associated with the disposal of high level nuclear waste from spent fuels rods. These must be stored for very long periods (hundreds if not thousands of years) before they become relatively harmless. Current methods include vitrifying the waste in glass and burying it deep underground or in ocean trenches. Humans have no experience of storing such materials safely for this length of time. The possibilities of disruption by earthquakes or the slow seepage of the waste into the water table make this one of the strongest arguments against using nuclear energy.

Solar energy (1)

CONVERSION OF SOLAR ENERGY INTO OTHER FORMS OF ENERGY

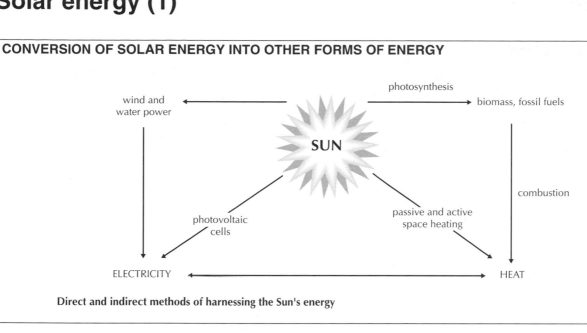

Direct and indirect methods of harnessing the Sun's energy

THE ROLE OF PHOTOSYNTHESIS

Approximately 5.6×10^{21} kJ of energy reach the surface of the planet from the Sun each year. About 0.06% of this is used to store energy in plants. This is achieved through photosynthesis – a complex process summarized by the reaction of carbon dioxide and water to form carbohydrates in the presence of chlorophyll.

$$6CO_2(g) + 6H_2O(l) \rightarrow C_6H_{12}O_6(s) + 6O_2(g)$$

The process is endothermic, requiring 2816 kJ of energy per mole of glucose. The green pigment chlorophyll interacts with light energy from the Sun and uses this energy to drive the process. The products of photosynthesis are used as food to provide energy for animals through the reverse process – respiration. They can also be converted into ethanol by fermentation or simply be burned to provide heat. Wood is mainly cellulose – a polymer made up of repeating glucose units.

CONVERSION OF BIOMASS INTO ENERGY

The energy stored in biomass can be released in a variety of ways.

1. Direct combustion

    e.g. $C_6H_{12}O_6(s) + 6O_2(g) \rightarrow 6CO_2(g) + 6H_2O(l)$

    $\Delta H = -2816$ kJ mol^{-1}

2. Combustion of waste materials derived from plants, such as animal dung. In many hot countries animal dung is dried and used as a fuel for heating and cooking. Garbage consisting of animal and vegetable waste is burned in incinerators in several cities to provide heat and electricity. This has the added advantage of reducing the amount of waste that has to be dumped in landfill sites.

3. Biogas – the anaerobic decay of organic matter by bacteria produces a mixture of mainly methane and carbon dioxide known as biogas. The manure from farm animals can generate enough methane to provide for the heating, cooking, and refrigeration needs of rural communities.

4. Fermentation to produce ethanol. Carbohydrates can be fermented by enzymes in yeast.

$$C_6H_{12}O_6(s) \rightarrow 2C_2H_5OH(l) + 2CO_2(g)$$

The ethanol can then be burned to produce energy.

$$C_2H_5OH(l) + 3O_2(g) \rightarrow 2CO_2(g) + 3H_2O(l)$$
$$\Delta H = -1371 \text{ kJ mol}^{-1}$$

By combining ethanol with gasoline a fuel called gasohol can be produced which can be used by unmodified cars, making them less reliant on the supply and cost of pure gasoline (petrol).

Biofuels are renewable, readily available, and relatively non-polluting. However, disadvantages include the fact that they are widely dispersed, they take up land where food crops can be grown, and they remove nutrients from the soil.

Solar energy (2)

USE OF SOLAR ENERGY FOR SPACE HEATING

The direct heat of the Sun can be stored in materials. By designing well-insulated homes with windows facing a sunny aspect the energy from the Sun can be used to supplement the heating system for the house. Solar panels with water running through them can utilize the high specific heat capacity of water and increase the efficiency of the storage. In many sunny countries this can provide enough energy to provide hot water for washing.

The advantages of passive heating include relatively low installation and running costs but it depends upon a plentiful supply of sunlight and needs a back-up energy supply. In active space heating the energy is absorbed using black bed collectors placed on the roof or against the side of a house that faces the Sun. The heated air or water in the collectors is then pumped into storage tanks containing water or rocks. When it is required pumps and fans distribute the stored heat. The installation is more expensive than passive heating and energy available to do work is lost each time the energy is converted into a different form.

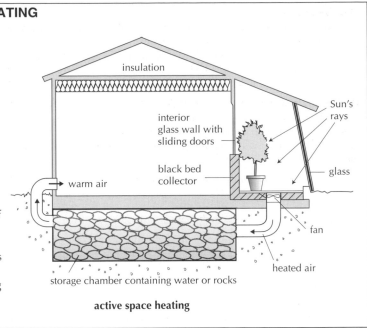

active space heating

CONVERSION OF SOLAR ENERGY INTO ELECTRICITY

The heat from the Sun can be concentrated using parabolic mirrors to focus the Sun's rays. A large bank of mirrors is required and they need to be controlled by computers to track the Sun. Currently the biggest solar power plant is in the Mojave Desert, California. The Sun's rays are used to heat oil or liquid sodium. The heated liquid is then pumped to a heat exchanger which in turn converts water to steam. The set-up is relatively quick to build, has a free and limitless fuel supply and is not polluting. However, large areas of mirrors are required and they work best in deserts where there is space and cloudless skies. The surfaces need to be able to withstand extremes of temperature and need constant attention to keep them clean.

On a smaller scale electricity can also be produced directly from the Sun by the photoelectric effect. Photovoltaic cells are used to power orbiting satellites and space stations and on Earth to power electronic instruments, such as calculators. They utilize the fact that when sunlight falls onto certain materials it produces a flow of electrons. Most are made from the semiconductors silicon or germanium. The best cells are made from gallium arsenide but these are even more expensive. They have no moving parts and electricity is generated indefinitely upon exposure to light. However, they have low efficiency (typically 10–20%), are expensive and require large surfaces.

Electrochemical energy

BATTERIES

A battery is a general term for an electrochemical cell in which chemical energy is converted into electrical energy. The electrons transferred in the spontaneous redox reaction taking place in the voltaic cell produce the electricity. Batteries are a useful way to store and transport relatively small amounts of energy. Some batteries (primary cells) can only be used once whereas secondary cells can be recharged.

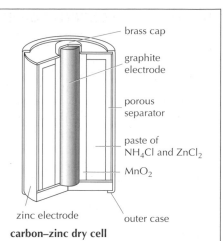

carbon–zinc dry cell

Carbon–zinc dry cell The common dry cell contains a paste of ammonium chloride and manganese(IV) oxide as the electrolyte. The positive electrode is made of graphite (often with a brass cap) and the zinc casing is the negative electrode. A porous separator acts as a salt bridge. When the cell is working electrons flow from the zinc to the graphite.

Oxidation (– electrode) $Zn \rightarrow Zn^{2+} + 2e$

Reduction (+ electrode) $MnO_2 + NH_4^+ + e \rightarrow MnO(OH) + NH_3$

The ammonia then reacts with the zinc ions to form the complex ion $[Zn(NH_3)_4]^{2+}$. This prevents a build up of ammonia gas.

A carbon–zinc dry cell produces 1.5 V when new. In addition to the disadvantage of having to replace the battery when it wears out, it has a poor shelf-life and the acidic ammonium chloride can corrode the zinc casing.

Alkaline dry cell Alkaline dry cells also produce a voltage of 1.5 V but have a longer shelf-life and are often used to power emergency lighting during a power failure. They also maintain operating voltage better at high current load. The negative electrode is a paste of powdered zinc and the electrolyte is potassium hydroxide. The positive electrode is manganese(VI) oxide.

Oxidation (– electrode) $Zn + 2OH^- \rightarrow Zn(OH)_2 + 2e$

Reduction (+ electrode) $MnO_2 + 2H_2O + e \rightarrow Mn(OH)_3 + OH^-$

Lead–acid battery The lead–acid battery is used in automobiles and is an example of a secondary cell. Usually it consists of six cells in series producing a total voltage of 12 V. The electrolyte is an aqueous solution of sulfuric acid. The negative electrodes are made of lead and the positive electrodes are made of lead(IV) oxide.

Oxidation (– electrode) $Pb + SO_4^{2-} \rightarrow PbSO_4 + 2e$

Reduction (+ electrode)

$PbO_2 + 4H^+ + SO_4^{2-} + 2e \rightarrow PbSO_4 + 2H_2O$

The overall reaction taking place is thus:
$Pb + PbO_2 + 4H^+ + 2SO_4^{2-} \rightarrow 2PbSO_4 + 2H_2O$

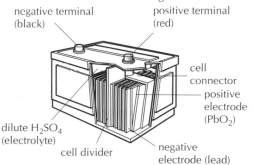

The reverse reaction takes place during charging. This can be done using a battery charger or through the alternator as the automobile is being driven. As sulfuric acid is used up during discharging the density of the electrolyte can be measured using a hydrometer to give an indication of the state of the battery. The disadvantages of lead–acid batteries are that they are heavy and both lead and sulfuric acid are potentially polluting.

Lead–acid battery – during recharging hydrogen and oxygen are evolved from the electrolysis of dilute H_2SO_4 so it needs topping up occasionally with distilled water.

FACTORS AFFECTING THE VOLTAGE AND POWER FROM A BATTERY

The **voltage** of a cell is essentially the difference in electrode potential between the two half-cells. It primarily depends only on the nature of the chemical components of the two half-cells. By combining two cells in series the voltage can be doubled. Three cells in series will triple the voltage of a cell, etc. The **power** of a cell is the rate at which it can deliver energy and is measured in joules per second. It is affected both by the size of the cell and the physical quantities of the materials present.

FUEL CELLS

A fuel cell utilizes the reaction between oxygen and hydrogen to produce water. Unlike combustion the energy is given out not as heat but as electricity. As reactants are used up more are added so a fuel cell can give a continuous supply of electricity. They are used in spacecraft as they do not need recharging. The electrolyte is aqueous sodium hydroxide. It is contained within the cell using porous electrodes which allow the passage of water, hydrogen, and oxygen.

Oxidation (– electrode)

$H_2 + 2OH^- \rightarrow 2H_2O + 2e$

Reduction (+ electrode)

$O_2 + 2H_2O + 4e \rightarrow 4OH^-$

hydrogen–oxygen fuel cell

The advantage of a fuel cell is that it does not pollute, as water is the only product. Currently they are very expensive to produce.

Storage of energy and photovoltaics

ENERGY STORAGE

Small quantities of energy can be stored in batteries but it is impracticable to use them to store large quantities. According to the second law of thermodynamics some energy capable of doing work is always lost during the process of converting one form of energy into another so all storage schemes have a built-in inefficiency. However, it can also be inefficient to keep stopping and starting power stations as demand varies.

One solution is to use the excess energy produced at times of low demand to pump water to a high level reservoir. When demand for electricity exceeds capacity the water can then be used to produce hydroelectric power. The advantages are that it makes use of cheap off-peak electricity, it can respond to increased demand rapidly, and it is clean and relatively efficient. Disadvantages include the initial high cost of construction and that it is limited to suitable locations.

Another solution is to use the surplus energy to produce hydrogen from water by electrolysis. This process requires much energy and is only about 60% efficient. The energy can then be released when hydrogen is burned. Although the burning of hydrogen produces a large amount of heat per kg of fuel, hydrogen is a gas and occupies a large volume. It cannot be liquefied easily as its boiling point is –253 °C at 1 atm pressure. This means it has to be stored and transported in large pressurized containers which add considerably to the weight. A further problem is controlling the safe burning of hydrogen, as a hydrogen–air mixture is potentially very explosive.

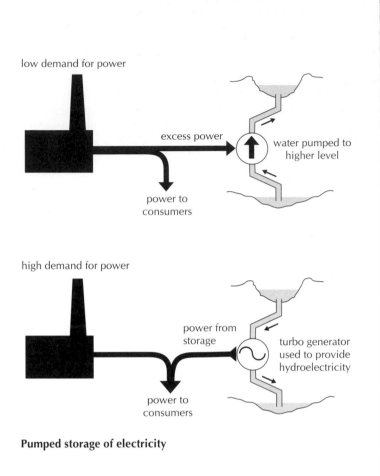

Pumped storage of electricity

PHOTOVOLTAIC CELLS

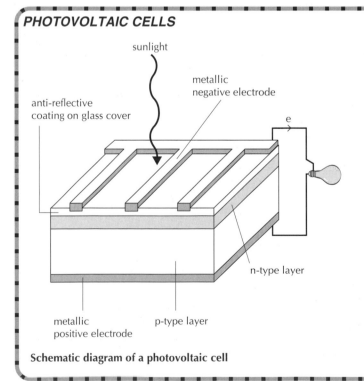

Schematic diagram of a photovoltaic cell

The purification of silicon and the electrical conductivity of silicon including doping are discussed in detail in 16. option E – Chemical Industries. n-type semiconductors are doped using a group 5 element, such as arsenic, and p-type semiconductors are doped with a group 3 element, such as gallium. Photovoltaic cells are made from sheets of n- and p-type semiconductors in close contact with each other. The electrons diffuse from the negative n-type to the p-type and the positive holes diffuse in the other direction. This generates a potential difference at the junction between the two types of semiconductor. This voltage accelerates the electrons released by the energy of the Sun. Connecting a conductor between the two layers will cause the freed electrons to flow through the external circuit as they move from the n-type to the p-type and hence produce a direct current.

Research into making more efficient photovoltaic cells is continuing and although still expensive the cost of producing electricity in this way is decreasing. Apart from cost and the relative inefficiency of the process a big disadvantage is that large areas of land are required to produce electricity in the quantities available from conventional power stations.

Nuclear stability

NEUTRON TO PROTON RATIOS

It can be seen that for the first few elements in the Periodic Table the number of protons and neutrons in the nucleus is approximately equal. As the atomic number increases the number of neutrons becomes greater than the number of protons. Some isotopes of elements are stable whereas others are radioactive. If a graph of number of protons against number of neutrons is plotted it is found that the stable isotopes all lie within a narrow band. Radioactive isotopes usually lie outside this band and get nearer to it as they decay. The upper limit of the stability band is $^{206}_{82}Pb$. Elements with an atomic number greater than 82 will be radioactive and the majority decay by α emission.

e.g. $^{210}_{84}Po \rightarrow ^{206}_{82}Pb + ^{4}_{2}He$

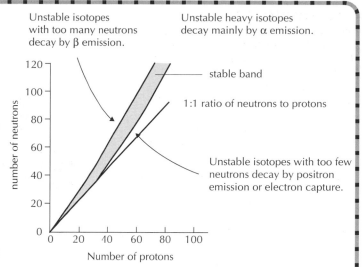

Unstable isotopes with too many neutrons decay by β emission.

Unstable heavy isotopes decay mainly by α emission.

stable band

1:1 ratio of neutrons to protons

Unstable isotopes with too few neutrons decay by positron emission or electron capture.

During α decay a $^{4}_{2}He^{2+}$ particle is emitted and the atomic number of the new element will be two less than the original element and the mass number will be four less. Lighter elements with isotopes that have too many neutrons lie to the left of the band of stability. By emitting a β particle the new element will have an atomic number that has increased by one but the mass will remain unchanged. This decreases the neutron:proton ratio and brings the element closer to the band of stability.

e.g. $^{14}_{6}C \rightarrow ^{14}_{7}N + ^{0}_{-1}e$

Artificially produced isotopes can lie to the right of the stable band. They regain stability by emitting a positron (β^{+} particle) or by electron capture when an electron in the lowest energy level (n = 1) 'falls' into the nucleus.

e.g. $^{30}_{15}P \rightarrow ^{30}_{14}Si + ^{0}_{+1}e$ (positron emission)

$^{30}_{15}P + ^{0}_{-1}e \rightarrow ^{30}_{14}Si$ (electron capture)

CALCULATING THE ENERGY RELEASED IN A NUCLEAR REACTION

The mass numbers and atomic numbers balance in a nuclear equation. However, if precise values are used for the masses it can be seen that a small decrease in mass takes place during the reaction. The amount of energy that is formed by this decrease in mass can be calculated using Einstein's equation $E = mc^2$.

e.g.

$$^{235}_{92}U + ^{1}_{0}n \rightarrow ^{90}_{36}Kr + ^{144}_{56}Ba + 2^{1}_{0}n$$

relative mass $\underbrace{235.0439 + 1.0087}_{236.0526}$ $\underbrace{89.9470 + 143.8810 + 2.0174}_{235.8454}$

The relative mass loss = 236.0526 – 235.8454 = 0.2072 which is approximately 0.1% of the initial uranium. If the masses are measured in grams then a loss of 0.2072 g (2.072×10^{-4} kg) is equivalent to $2.072 \times 10^{-4} \times (2.998 \times 10^{8})^{2} = 1.862 \times 10^{13}$ J (or 1.862×10^{10} kJ). This amount of energy is the theoretical amount that one mole of uranium-235 (235 g) can produce if it all reacted according to the above equation.

MASS DEFECT AND NUCLEAR BINDING ENERGY

Protons and neutrons in the nucleus of an atom are held together by very strong forces. A measure of these forces is known as the **binding energy**. This can be defined as the energy that must be supplied to one mole of the atoms to break down the nuclei into separate neutrons and protons *or* the energy released when separate neutrons and protons combine to form one mole of the atomic nuclei. It can be calculated from the **mass defect**.

The mass defect is the difference in the combined mass of all the separate protons and neutrons compared to the actual mass of the nucleus. Strictly speaking it also includes the mass of the electrons too but these are so small they can be ignored. Consider a helium nucleus. It contains two neutrons (relative mass = 2 × 1.0087) and two protons (relative mass = 2 × 1.0078) giving a total relative mass of 4.0330. However the actual relative atomic mass of helium is 4.0039. The relative mass defect is thus 0.0291. By using Einstein's equation this converts to a binding energy of approximately 2.7×10^{9} kJ mol^{-1}. A graph of binding energy per nucleon against mass number shows that atomic nuclei with a mass number of about 56 (e.g. iron nucleus) have the maximum binding energy and are thus the most stable. Nuclei to the left or right of this maximum will undergo nuclear change in such a way that as they approach the maximum energy will be released. This explains nuclear fusion, whereby small atoms combine to form heavier nuclei, and nuclear fission – the splitting of heavy nuclei to form two or more lighter nuclei.

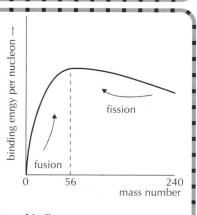

How binding energy varies with mass number

RATE OF RADIOACTIVE DECAY

The decay of radioactive isotopes is a first order reaction, i.e.

rate = $k[A]$ where k is the rate constant and [A] is the concentration at time t.

If the rate is expressed as $-\dfrac{d[A]}{dt}$

then $-\dfrac{d[A]}{dt} = k[A]$

Integration of this expression gives $\ln\left(\dfrac{[A]_o}{[A]}\right) = kt$

where $[A]_o$ is the initial concentration when $t = 0$ and ln is the natural logarithm.

Since radioactive decay depends on the nucleus it does not matter how the isotope is chemically bonded or whether it is present as the free element. The rate of decay will be the same and it is more usual to refer to [A] as the amount of isotope present rather than concentration. This can be expressed in moles or as mass.

At $t_{\frac{1}{2}}$ the amount of A will be half the initial amount. Thus $[A]_o/[A]$ will equal 2. The integrated expression then becomes:

$$\ln 2 = Kt_{\frac{1}{2}} \text{ or } t_{\frac{1}{2}} = \frac{0.693}{k}$$

From this expression it can be seen that the half-life of radioactive decay is independent of the amount of isotope present, as stated earlier in this Option. The integrated form of the rate equation and the expression for the half-life are both given in the IB Data Booklet. They can be used to solve problems involving the change in activity over a period of time.

Worked examples

1   The half-life of 226-radium is 1622 years. Calculate how long it will take for a sample of 226-radium to decay to 10% of its original radioactivity.

   Step 1 Use $t_{\frac{1}{2}}$ to find the rate constant from the equation $kt_{\frac{1}{2}} = 0.693$.

   $k = 0.693/1622 = 4.27 \times 10^{-4}$ year^{-1}

   Step 2 Insert the value for k into the integrated form of the rate equation $kt = \ln ([A]_o/[A])$.

   $4.27 \times 10^{-4} \times t = \ln (100\%/10\%)$
   => $t = 5392$ years

   It will take 5392 years to decay to 10% of its original activity.

2   A piece of old wood was found to give 10 counts per minute per gram of carbon when subjected to ^{14}C analysis. New wood has a count of 15 cpm g^{-1}. The half-life of ^{14}C is 5570 years. Calculate the age of the old wood.

   Step 1 Use $t_{\frac{1}{2}}$ to find the rate constant.

   $k = 0.693/5570 = 1.24 \times 10^{-4}$ years^{-1}

   Step 2 Insert the value of k into the integrated form of the rate equation.

   $1.24 \times 10^{-4} \times t = \ln ([A]_o/[A]) =$
   $\ln\left(\dfrac{^{14}C \text{ content in new wood}}{^{14}C \text{ content in old wood}}\right) = \ln\left(\dfrac{15}{10}\right) = \ln 1.5$

   => $t = 3270$ years

   The wood is 3270 years old.

DIFFERENT TYPES OF NUCLEAR WASTE

There are various types of nuclear waste. Some are radioisotopes from research laboratories or hospitals. Some are spent fuel rods from nuclear power stations and some are materials that have come into contact with radioactive material and become contaminated themselves. Essentially nuclear waste can be divided into **high level waste** and **low level** waste. Low level waste includes items such as rubber gloves, paper towels, and protective clothing that have been used in areas where radioactive materials are handled. The level of activity is low and the half-lives of the radioactive isotopes are generally short. High level waste has high activity and generally the isotopes have long half-lives so the waste will remain active for a long period. Most high level waste comes from spent fuel rods or the reprocessing of spent nuclear fuel.

STORAGE AND DISPOSAL OF NUCLEAR WASTE

Low level waste Different methods are used to dispose of low level waste. Although many governments have now banned the practice some is simply discharged straight into the sea where it becomes diluted. Since the decay produces heat it is better to store it in vast tanks of cooled water called 'ponds' where it can lose much of its activity. Before it is then discharged into the sea it is filtered through an ion exchange resin which removes strontium and caesium, the two elements responsible for much of the radioactivity. Other methods of disposal include keeping the waste in steel containers inside concrete-lined vaults.

High level waste During reprocessing of spent fuel about 96% of the uranium is recovered for re-use. About 1% is plutonium which is a valuable fuel. The remaining 3% is high level liquid waste. One method used to treat this is to vitrify it. The liquid waste is dried in a furnace and then fed into a melting pot together with glass making material. The molten material is then poured into stainless steel tubes where it solidifies. Air flows round the containers to keep them cool. Because of the high activity and long half-lives some of the waste will remain active for hundreds if not thousands of years. The problem is how to store it safely for this length of time. Currently the best solution seems to be burying it in deep remote places that are geologically stable such as disused mines or in granite rock. The concern is that the radioactive material may eventually leach into the water table and then into drinking water.

IB QUESTIONS – OPTION F –
FUELS AND ENERGY

1. **(a)** Two possible reactions of coal are given below with their associated enthalpy changes per mole of product:

$$C(s) + \tfrac{1}{2}O_2(g) \rightarrow CO(g) \qquad \Delta H = -111 \text{ kJ}$$

$$C(s) + 2H_2(g) + \tfrac{1}{2}O_2(g) \rightarrow CH_3OH(l) \qquad \Delta H = -239 \text{ kJ}$$

   (i) From this information, calculate the heat of reaction for:

   $$CO(g) + 2H_2(g) \rightarrow CH_3OH(l) \qquad \textbf{[2]}$$

   (ii) Give a balanced equation for the complete combustion of methanol. **[1]**

   (iii) Use information provided in Table 2 of the Data Booklet to calculate the amount of heat required to raise the temperature of 500 kg of water at 25.0° C to water at 100.0° C. **[1]**

   (iv) Use information provided in Table 13 of the Data Booklet to calculate the mass of methanol that must be burnt completely to produce the amount of heat required in **(iii)**. **[2]**

2. **(a)** When biomass, such as animal waste, decomposes in the absence of oxygen, *biogas* is formed. Name the main gas present in biogas. **[1]**

   (b) When wood and crop residues are burnt in a limited amount of oxygen, a mixture of gases known as *producer gas* is formed. Name **one** combustible gas present in producer gas. **[1]**

   (c) Name **two** substances hazardous to health which are produced when wood is burnt in an enclosed space. **[2]**

   (d) Why is biomass likely to become more important as a fuel in the future? **[1]**

3. **(a)** In the context of nuclear reactions, explain the meaning of *fission*. **[1]**

4. **(a)** Explain why germanium and silicon are semi-conductors while carbon (in the form of diamond) is a non-conductor of electricity. **[2]**

   (b) What changes occur in the electrical conductivity of germanium when small amounts of

   (i) gallium

   (ii) arsenic

   are incorporated in the germanium lattice? Explain why the changes occur. **[4]**

(b) Explain why a fission reaction results in the release of a large quantity of energy. **[1]**

(c) The main reaction in a uranium reactor is fission of ^{235}U. A side reaction is caused when neutrons react with ^{238}U. Complete the following equation by filling in the symbols and values in the table below: **[4]**

$$^{238}_{a}U + ^{b}_{c}X \rightarrow ^{239}_{92}U \rightarrow ^{239}_{94}Y + 2^{d}_{e}Z$$

X		b	
Y		c	
Z		d	
a		e	

(d) A diagram of a nuclear power plant to produce electricity is shown below:

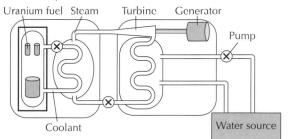

The water which produces the steam to drive turbines in a nuclear power plant is not heated directly. Explain why more than one heat exchange loop is used and name **one** substance used in the primary cooling loop. **[2]**

(e) ^{14}C can be used for dating organic remains. Whilst a plant or animal is alive the amount of ^{14}C remains constant. At death the amount of ^{14}C decreases by a first order reaction with a half-life of 5730 years. In a living sample the ^{14}C: ^{12}C ratio is 1.2×10^{-12}. If an object is found with a ^{14}C: ^{12}C ratio of 1.5×10^{-13}, how old is the object? **[2]**

5. **(a)** State the trend in the neutron to proton ratio for stable nuclei with increasing atomic number, and give a reason for this behaviour. **[3]**

   (b) Explain why elements above atomic number 83 with mass number 209 do not exist as stable isotopes. **[1]**

   (c) If a radionuclide has a lower n to p ratio than for a stable nuclei, give a balanced nuclear equation to represent what happens. **[1]**

Analytical techniques and the principles of spectroscopy

INFORMATION FROM DIFFERENT ANALYTICAL TECHNIQUES

This option builds on some of the techniques that have already been mentioned in 11. Topic 20 – Organic Chemistry. Techniques such as ultraviolet/visible spectroscopy, infrared spectroscopy, mass spectrometry, and nuclear magnetic resonance spectroscopy are extremely useful tools to chemists. They are becoming ever more refined and some of them (e.g. mass spectrometry) can be used on extremely small samples. This has revolutionized many branches of chemistry, for example combinatorial chemistry in the search for new drugs. This is only possible because very small amounts of thousands of new compounds can be analysed accurately and efficiently.

The main uses for these techniques are structural determination, the analysis of the different composition of compounds, and to determine purity. Before analysis can usually take place it is important to separate any mixture into its individual components – hence the need for chromatography. Often information is not obtained from a single technique but from a combination of several of them. Some examples are:

*   Ultraviolet and visible spectroscopy – assaying of metal ions, organic structural determination, and detection of drug metabolites.
*   Infrared spectroscopy – organic structural determination, information on the strength of bonds, information about the secondary structure of proteins, measuring the degree of unsaturation of oils and fats, and determining the level of alcohol in the breath.
*   Mass spectrometry – organic structural determination, isotopic dating (e.g. ^{14}C dating).
* ^1H nuclear magnetic resonance – organic structural determination, body scanning.
*   Gas chromatography – mass spectrometry (GC–MS) – drug testing in the blood and urine, food testing, and forensic science.

THE ELECTROMAGNETIC SPECTRUM

The electromagnetic spectrum has already been briefly described in 2. Topic 2 – Atomic Theory. You should be familiar with the relationship $c = \lambda f$ and know the different regions of the spectrum.

The electromagnetic spectrum

Wavelength / m	10^{-10}	10^{-9}	10^{-8}	10^{-7}	10^{-6}	10^{-5}	10^{-4}	10^{-3}	10^{-2}	10^{-1}	10^{0}	10^{1}	10^{2}	10^{3}
Frequency / MHz		3×10^{10}		3×10^{8}		3×10^{6}		3×10^{4}		3×10^{2}		3		
Type of radiation	X-rays γ-rays		ultraviolet	visible		infrared		microwaves					radio waves	
Type of transition	inner electron		outer electron			molecular vibrations		molecular rotations					nuclear spin	

◄————————————————Increasing energy————————————————

ABSORPTION SPECTRA AND EMISSION SPECTRA

Spectroscopy can be divided into two main types. Emission spectroscopy involves the analysis of light emitted by excited atoms or molecules as they return to their ground state. The atomic emission spectrum of hydrogen is a good example of this. Many analytical techniques involve absorption spectroscopy. When radiation is passed through a sample some of the energy is absorbed by the sample to excite an atom or molecule to an excited state. The spectrometer analyses the transmitted energy relative to the incident energy. Since the energy levels are quantized only radiation with a frequency corresponding to the difference in the energy levels will be absorbed. The relationship between energy and frequency is given by:

$E = hf$ where h is Planck's constant, 6.626×10^{-34} J s.

The greater the energy difference between the levels the higher the frequency (or the shorter the wavelength) of the light absorbed. The most energetic absorptions are atomic electronic transitions which involve bond breaking and ionization. Absorptions in the ultraviolet and visible region are due to atomic and molecular transitions in which electrons become excited to higher levels. Molecular vibrations (stretching and bending) occur in the infrared region and molecular rotations in the microwave region. The weakest transitions of all involve nuclear spin. These occur in the radio wave region and form the basis of nuclear magnetic resonance spectroscopy.

HL **Visible and ultraviolet spectroscopy (1)**

FACTORS AFFECTING THE COLOUR OF TRANSITION METAL COMPLEXES

Transition metals are defined as elements having an incomplete d sub-level in one or more of their oxidation states. Compounds of Sc^{3+} which have no d electrons and of Cu^+ and Zn^{2+} which both have complete d sub-shells are colourless. This strongly suggests that the colour of transition metal complexes is related to an incomplete d level. The actual colour is determined by four different factors.

1. The nature of the transition element. For example $Mn^{2+}(aq)$ and $Fe^{3+}(aq)$ both have the configuration $[Ar]3d^5$. $Mn^{2+}(aq)$ is pink whereas $Fe^{3+}(aq)$ is yellow.
2. The oxidation state. $Fe^{2+}(aq)$ is green whereas $Fe^{3+}(aq)$ is yellow.
3. The identity of the ligand. $[Cu(H_2O)_6]^{2+}$ (sometimes shown as $[Cu(H_2O_4)^{2+}]$), is blue, $[Cu(NH_3)_4(H_2O)_2]^{2+}$ (sometimes shown as $[Cu(NH_3)_4]^{2+}$), is blue/violet whereas $[CuCl_4]^{2-}$ is yellow (green in aqueous solution).
4. The stereochemistry of the complex. The colour is also affected by the shape of the molecule or ion. In the above example $[Cu(H_2O)_6]^{2+}$ is octahedral whereas $[CuCl_4]^{2-}$ is tetrahedral. However for the IB only octahedral complexes in aqueous solution will be considered.

SPLITTING OF THE d ORBITALS

In the free ion the five d orbitals are degenerate. That is they are all of equal energy.

Note that three of the orbitals (d_{xy}, d_{yz} and d_{zx}) lie *between* the axes whereas the other two ($d_{x^2-y^2}$ and d_{z^2}) lie *along* the axes. Ligands act as Lewis bases and donate a non-bonding pair of electrons to form a co-ordinate bond. As the ligands approach the metal along the axes to form an octahedral complex the non-bonding pairs of electrons on the ligands will repel the $d_{x^2-y^2}$ and d_{z^2} orbitals causing the five

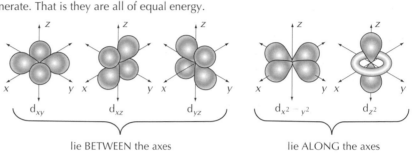

lie BETWEEN the axes   lie ALONG the axes

d orbitals to split, three to lower energy and two to higher energy. The difference in energy between the two levels corresponds to the wavelength of visible light.

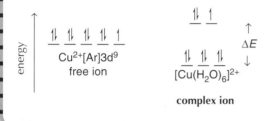

When white light falls on the aqueous solution of the complex the colour corresponding to ΔE is absorbed and the transmitted light will be the complementary colour. For example, $[Cu(H_2O)_6]^{2+}$ absorbs red light so the compound appears blue. The amount that the d orbitals are split will determine the exact colour. Changing the transition metal changes the number of protons in the nucleus which will affect the levels. Similarly changing the oxidation state will affect the splitting as the number of electrons in the level is different. Different ligands will also cause different amounts of splitting depending on their electron density.

COMPLEMENTARY COLOURS

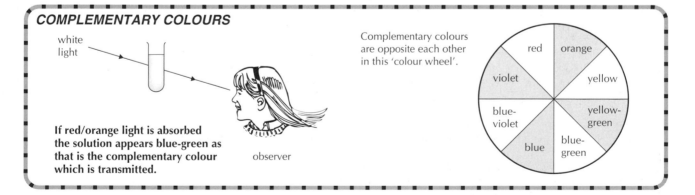

white light

If red/orange light is absorbed the solution appears blue-green as that is the complementary colour which is transmitted.

observer

Complementary colours are opposite each other in this 'colour wheel'.

red | orange
violet | yellow
blue-violet | yellow-green
blue | blue-green

THE SPECTROCHEMICAL SERIES

Ligands can be arranged in order of their ability to split the d orbitals.

$$I^- < Br^- < Cl^- < OH^- < H_2O < NH_3 < CN^-$$

This order is known as the spectrochemical series. Iodide ions cause the smallest splitting and cyanide ions the largest splitting. The energy of light absorbed increases when ammonia is substituted for water in Cu^{2+} complexes as the splitting increases, i.e. in going from $[Cu(H_2O)_6]^{2+}$ to $[Cu(NH_3)_4(H_2O)_2]^{2+}$. This means that the wavelength of the light absorbed decreases and this is observed in the colour of the transmitted light which changes from blue to a blue-violet colour.

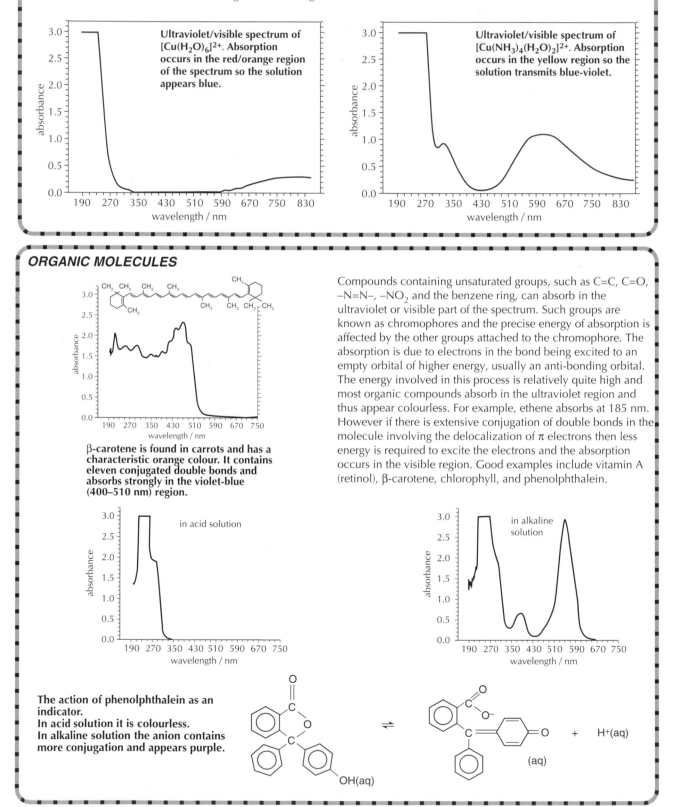

Ultraviolet/visible spectrum of $[Cu(H_2O)_6]^{2+}$. Absorption occurs in the red/orange region of the spectrum so the solution appears blue.

Ultraviolet/visible spectrum of $[Cu(NH_3)_4(H_2O)_2]^{2+}$. Absorption occurs in the yellow region so the solution transmits blue-violet.

ORGANIC MOLECULES

Compounds containing unsaturated groups, such as C=C, C=O, –N=N–, $–NO_2$ and the benzene ring, can absorb in the ultraviolet or visible part of the spectrum. Such groups are known as chromophores and the precise energy of absorption is affected by the other groups attached to the chromophore. The absorption is due to electrons in the bond being excited to an empty orbital of higher energy, usually an anti-bonding orbital. The energy involved in this process is relatively quite high and most organic compounds absorb in the ultraviolet region and thus appear colourless. For example, ethene absorbs at 185 nm. However if there is extensive conjugation of double bonds in the molecule involving the delocalization of π electrons then less energy is required to excite the electrons and the absorption occurs in the visible region. Good examples include vitamin A (retinol), β-carotene, chlorophyll, and phenolphthalein.

β-carotene is found in carrots and has a characteristic orange colour. It contains eleven conjugated double bonds and absorbs strongly in the violet-blue (400–510 nm) region.

in acid solution

in alkaline solution

The action of phenolphthalein as an indicator.
In acid solution it is colourless.
In alkaline solution the anion contains more conjugation and appears purple.

Visible and ultraviolet spectroscopy (3)

BEER–LAMBERT LAW

The Beer–Lambert law states that:

$$\log_{10}\frac{I_o}{I} = \varepsilon l c$$

where: I_o is the intensity of the incident radiation and I is the intensity of the transmitted radiation.
ε is the molar absorption co-efficient (a constant for each absorbing substance).
l is the path length of the absorbing solution (usually 1.0 cm) and c is the concentration.

The intensity of the light transmitted will also depend on the wavelength and it is usual to carry out experiments involving the Beer–Lambert law at the wavelength of maximum absorption λ_{max}. Most spectrometers measure $\log_{10} I_o / I$ directly as absorbance. Often knowing the values of ε and λ_{max} is enough to identify a particular substance.

DETERMINING THE CONCENTRATION OF AN UNKNOWN SOLUTION

Most dilute solutions obey the Beer–Lambert law and it can be used to determine the concentration of an unknown solution. Examples include the amount of iron in a sample of blood, the percentage of copper in brass (by first converting it into a copper(II) ion solution), or investigating the reaction kinetics of a reaction involving coloured species. The method is essentially the same in each case and is illustrated by the following example.

A student was asked to plan an experiment to determine the concentration of Cu^{2+} in an unknown solution. She was provided with a separate solution of 1.00 mol dm^{-3} copper(II) sulfate.

Concentration of $CuSO_4(aq)$ / mol dm^{-3}	Absorbance at 720 nm
0.100	0.362
0.150	0.498
0.200	0.798
0.250	0.901
0.300	1.002
0.500	1.751

Using a spectrometer the student measured the absorbance of a diluted sample of the copper(II) sulfate solution at different wavelengths to obtain the value of 720 nm for λ_{max}. She then carefully diluted the 1.00 mol dm^{-3} solution to give separate solutions with a range of known concentrations. The absorbance at 720 nm for each of these diluted solutions was then measured.

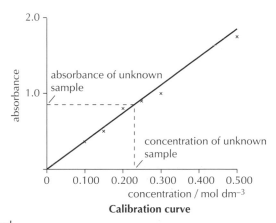

Calibration curve

The student used these results to plot a calibration curve using the line of best fit. The absorbance of the unknown solution at 720 nm was then measured. The value was found to be 0.850. From the graph the unknown concentration was determined to be 0.232 mol dm^{-3}.

VIBRATION OF BONDS

When molecules vibrate they absorb energy. This energy lies in the frequency range $1.2 \times 10^{14} - 1.2 \times 10^{13}$ s^{-1} (Hz), i.e. wavelengths of $2.5 \times 10^{-6} - 2.5 \times 10^{-5}$ m which is in the infrared region of the electromagnetic spectrum. For simple diatomic molecules made up of different atoms, such as HCl, there is only one form of vibration. This is stretching where the atoms alternatively move further apart then closer together. Different molecules absorb at different frequencies as the energy needed to excite a vibration depends on the bond enthalpy. Weaker bonds require less energy.

Note that infrared absorptions are usually given in cm^{-1}. Frequency and wavelength are related by the equation $c = \lambda f$ where c is the velocity of light in a vacuum. Since c is a constant the reciprocal of wavelength is a direct measure of frequency. This reciprocal of wavelength ($1/\lambda$) is known as the **wavenumber** and has the units cm^{-1}. Hence an absorption of 2886 cm^{-1} corresponds to a wavelength of 3.465×10^{-6} m. The longer the wavelength the lower the energy and the smaller the value of $1/\lambda$ in cm^{-1}. Conversely the higher the wavenumber the higher the energy.

Molecule	Bond enthalpy / kJ mol^{-1}	Absorption / cm^{-1}
H–Cl	431	2886
H–Br	366	2559
H–I	299	2230

CHANGE IN BOND POLARITY

Not all vibrations absorb infrared radiation. For absorption there must be a change in the dipole moment (bond polarity) as the vibration occurs. Thus diatomic gas molecules containing only one element, such as H$_2$, Cl$_2$, and O$_2$, do not absorb infrared radiation.

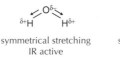

Vibrations of H$_2$O, SO$_2$, and CO$_2$

asymmetrical stretching IR active / symmetrical stretching IR active / symmetrical bending IR active

asymmetrical stretching IR active / symmetrical stretching IR active / symmetrical bending IR active

Stretching involves a change in dipole: IR active

Stretching involves no change in dipole: IR inactive

O$^{\delta-}$—C$^{\delta+}$—O$^{\delta-}$ asymmetrical stretching IR active

O$^{\delta-}$—C$^{\delta+}$—O$^{\delta-}$ symmetrical stretching IR active

O$^{\delta-}$—C$^{\delta+}$—O$^{\delta-}$ symmetrical bending two modes at right angles IR active

For more complex molecules only those vibrations which result in a dipole change will be infrared active. For example, the symmetrical stretch in carbon dioxide will be infrared inactive whereas the asymmetric stretch and the bending are both infrared active as they result in a dipole change.

Stretching and bending are the main modes of vibration but bending can be sub-divided into rocking, scissoring, twisting and wagging as exemplified by the –CH$_2$– group.

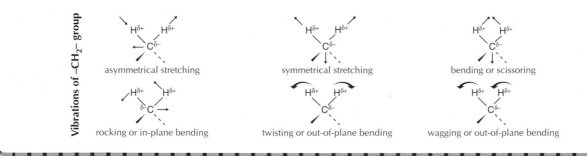

Vibrations of –CH$_2$– group

asymmetrical stretching / symmetrical stretching / bending or scissoring

rocking or in-plane bending / twisting or out-of-plane bending / wagging or out-of-plane bending

Infrared spectroscopy (2)

THE OPERATING PRINCIPLES OF A DOUBLE BEAM INFRARED SPECTROMETER

Traditional infrared spectrometers work by scanning wavelengths from about 2.5×10^{-6} m (4000 cm^{-1}) to 2.5×10^{-5} m (400 cm^{-1}). By using a rotating mirror the beam of monochromatic radiation is alternately passed through the sample and a reference. A photomultiplier converts photons of radiation into an electrical current. The spectrum is generated by comparison of the currents produced by the sample and the reference beams. Modern spectrometers pass all the wavelengths through the sample at the same time and a mathematical technique known as Fourier transformation is used to automatically analyse the transmission at each wavelength. Modern spectrometers are also linked to computers which store the spectra of known compounds. This enables an unknown compound to be identified by exact matching with the spectrum of a known compound.

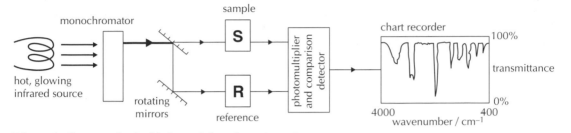

Schematic diagram of a double beam infrared spectrometer

USE OF INFRARED SPECTROSCOPY

Infrared spectrum of hexane C$_6$H$_{14}$

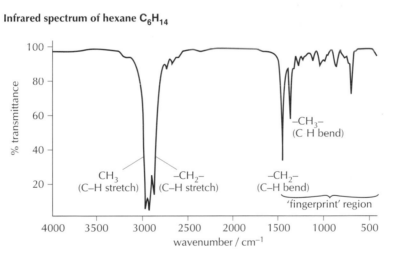

Note that infrared spectra normally show percentage transmittance, rather than absorption, from a base of 100% transmittance.

Because of all the different vibrations possible most molecules actually have quite complex spectra and can be identified from the 'fingerprint' region, that is the region between about 1400–400 cm^{-1}. However, for simple laboratory use there are several main absorptions which can be used to identify particular functional groups. The most common of these have been mentioned previously in 11. Topic 20 – Organic Chemistry and are the –OH, C=O, C–H, C=C, and C–O bonds. The precise region at which these absorb is determined by the neighbouring atoms as these will influence the bond enthalpy. The main use of infrared spectroscopy is thus to confirm the presence of particular functional groups. The intoximeter for determining the amount of alcohol in the breath (see 14. option B – Medicines and Drugs) provides a good example of how it can also be used to give quantitative data on the amount of sample present.

Nuclear magnetic resonance (NMR) spectroscopy (1)

NUCLEAR SPIN

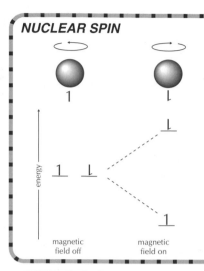

Unlike most other branches of spectroscopy NMR involves the interaction of **nuclei**, not electrons, with electromagnetic radiation. NMR concerns nuclei with uneven mass numbers, e.g. ^1H, ^{13}C, ^{19}F, and ^{31}P. All nuclei that possess spin behave as small magnets. In the absence of a magnetic field the two orientations of spin have the same energy. In the presence of a magnetic field the two orientations have a small difference in energy as they line up with or against the magnetic field. Nuclei can absorb energy as they move from the lower to the higher state. This small difference in energy corresponds to radio frequencies in the region of 10^7 to 10^8 s^{-1} (Hz).

To obtain an NMR spectrum the sample, either a pure liquid or in solution, together with a reference is placed in a cylindrical tube. The tube is then placed between the poles of a powerful electromagnet. A probe coil connected to a radio-frequency generator and receiver surrounds the sample. The spectrum is obtained either by varying the radio frequency or more commonly by varying the strength of the magnetic field.

THE MAIN FEATURES OF ^1H NMR SPECTRA

1. **The number of different absorptions (peaks)**
   Each proton in a particular chemical environment absorbs at a particular frequency. The number of peaks thus gives information as to the number of different chemical environments occupied by the protons.

2. **The area under each peak**
   The area under each absorption is proportional to the number of hydrogen atoms in that particular chemical environment. Normally each area is integrated and the heights of the integrated traces can be used to obtain the ratio of the number of hydrogen atoms in each environment.

3. **The chemical shift**
   Because spinning electrons create their own magnetic field the surrounding electrons of neighbouring atoms can exert a shielding effect. The greater the shielding the lower the frequency for the resonance to occur. The 'chemical shift' (δ) of each absorption is measured in parts per million (ppm) relative to a standard. The normal standard is tetramethylsilane (TMS) which is assigned a value of 0 ppm. This has been mentioned previously in 11. Topic 20 – Organic Chemistry and a table of chemical shifts is to be found in the IB Data Booklet.

4. **Splitting pattern**
   In ^1H NMR spectroscopy the chemical shift of protons within a molecule is slightly altered by protons bonded to adjacent carbon atoms. This spin–spin coupling shows up in high resolution ^1H NMR as splitting patterns. If the number of adjacent equivalent protons is equal to n then the peak will be split into (n + 1) peaks.

Tetramethylsilane as a reference sample

tetramethylsilane

The advantages of using tetramethylsilane $Si(CH_3)_4$ as the standard reference are:

- All the protons are in the same environment so it gives a strong single peak.
- It is not toxic and is very unreactive (so does not interfere with the sample).
- It absorbs upfield well away from most other protons.
- It is volatile (has a low boiling point) so can easily be removed from the sample.

INTERPRETING AN ^1H NMR SPECTRUM

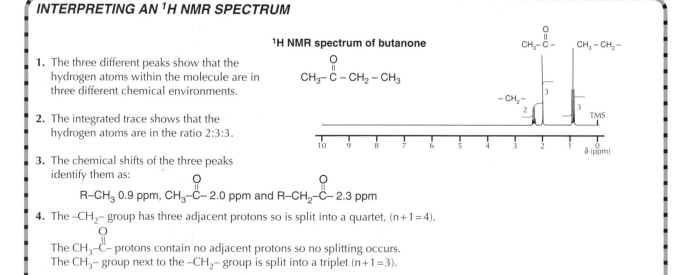

1. The three different peaks show that the hydrogen atoms within the molecule are in three different chemical environments.

2. The integrated trace shows that the hydrogen atoms are in the ratio 2:3:3.

3. The chemical shifts of the three peaks identify them as:

   R–CH$_3$ 0.9 ppm, CH$_3$–C(=O)– 2.0 ppm and R–CH$_2$–C(=O)– 2.3 ppm

4. The –CH$_2$– group has three adjacent protons so is split into a quartet, (n+1=4).

   The CH$_3$–C(=O)– protons contain no adjacent protons so no splitting occurs.
   The CH$_3$– group next to the –CH$_2$– group is split into a triplet (n+1=3).

Nuclear magnetic resonance (NMR) spectroscopy (2)

(HL)

SPIN–SPIN COUPLING

Splitting patterns are due to spin–spin coupling. For example, if there is one proton (n = 1) adjacent to a methyl group then it will either line up with the magnetic field or against it. The effect will be that the methyl protons will thus experience one slightly stronger and one slightly weaker external magnetic field resulting in an equal splitting of the peak. This is known as a doublet (n + 1 = 2).

This atom splits the methyl absorption into a doublet

If there is a $-CH_2-$ group (n = 2) adjacent to a methyl group then there are three possible energy states available.

1. Both proton spins are aligned with the field
2. One is aligned with the field and one against it (2 possible combinations)
3. Both are aligned against the field.

These two H atoms split the methyl absorption into a triplet

This results in a triplet with peaks in the ratio of 1:2:1.

The pattern of splitting can always be predicted using Pascal's triangle to cover all the possible combinations.

Number of adjacent protons (n)	Splitting pattern	Type of splitting
0	1	singlet
1	1 1	doublet
2	1 2 1	triplet
3	1 3 3 1	quartet
4	1 4 6 4 1	quintet

Thus a methyl group (n = 3) next to a proton will result in the absorption for that proton being split into a quartet with peaks in the ratio 1:3:3:1.

These three H atoms split the absorption due to the single proton into a quartet

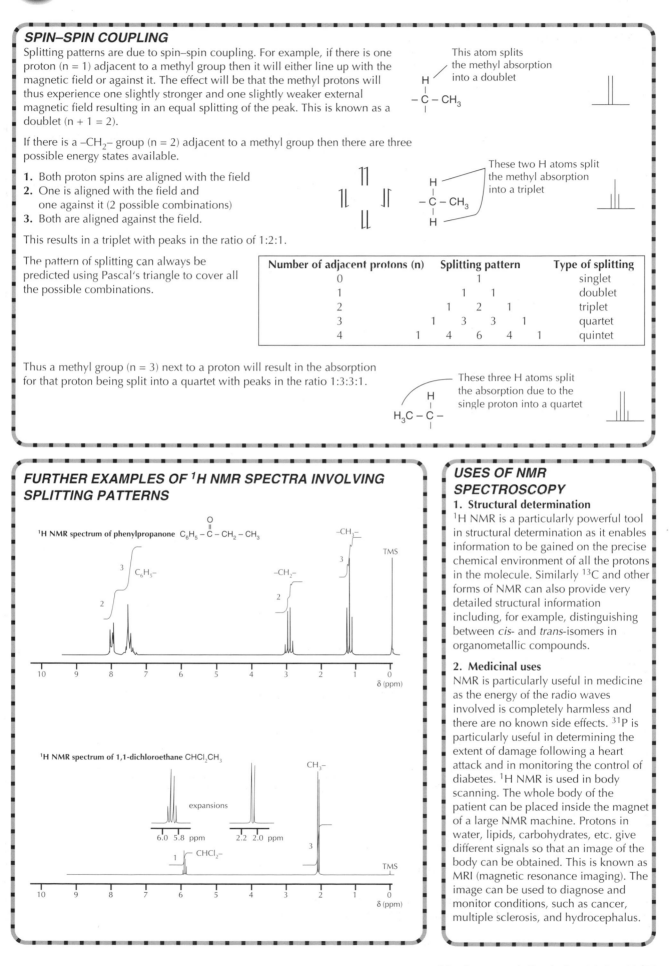

FURTHER EXAMPLES OF ^1H NMR SPECTRA INVOLVING SPLITTING PATTERNS

^1H NMR spectrum of phenylpropanone $C_6H_5 - \overset{\overset{O}{\|}}{C} - CH_2 - CH_3$

^1H NMR spectrum of 1,1-dichloroethane $CHCl_2CH_3$

USES OF NMR SPECTROSCOPY

1. Structural determination

^1H NMR is a particularly powerful tool in structural determination as it enables information to be gained on the precise chemical environment of all the protons in the molecule. Similarly ^{13}C and other forms of NMR can also provide very detailed structural information including, for example, distinguishing between *cis*- and *trans*-isomers in organometallic compounds.

2. Medicinal uses

NMR is particularly useful in medicine as the energy of the radio waves involved is completely harmless and there are no known side effects. ^{31}P is particularly useful in determining the extent of damage following a heart attack and in monitoring the control of diabetes. ^1H NMR is used in body scanning. The whole body of the patient can be placed inside the magnet of a large NMR machine. Protons in water, lipids, carbohydrates, etc. give different signals so that an image of the body can be obtained. This is known as MRI (magnetic resonance imaging). The image can be used to diagnose and monitor conditions, such as cancer, multiple sclerosis, and hydrocephalus.

Modern analytical chemistry 147

HIGH RESOLUTION MASS SPECTROMETRY

The principles of how a mass spectrometer works and the use of fragmentation patterns in low resolution mass spectrometry have already been covered in 2. Topic 2 – Atomic Theory and 11. Topic 20 – Organic Chemistry. Mass spectra can be obtained on extremely small samples, e.g. less than 10^{-6} g. This makes it an extremely useful and sensitive technique. In high resolution mass spectrometry it is possible to identify individual samples from the value of the molecular ion, M^+ alone. For example, consider a compound having a relative molecular mass of 78 which contains C, H, N, and O. The precise values for the relative atomic masses of C, H, N, and O are 12.0000, 1.0078, 14.0031 and 15.9949 respectively.

The accurate relative molecular masses for all ten possible species containing C, H, N, and O with a relative molecular mass of 78 can be calculated.

Formula	M_r	Formula	M_r
CH_2O_4	77.995	C_4NO	77.998
CH_4NO_3	78.019	$C_4H_2N_2$	78.022
$CH_6N_2O_2$	78.043	C_5H_2O	78.011
$C_2H_6O_3$	78.032	C_5H_4N	78.034
C_3N_3	78.009	C_6H_6	78.047

Thus if the high resolution spectrum gives a peak for M^+ with an m/z value of 78.032 the substance can be identified positively as having the molecular formula $C_2H_6O_3$.

THE IMPORTANCE OF ISOTOPES IN MASS SPECTROMETRY

In any sample of an organic compound there is a probability of 1.1% that any one carbon atom will be a carbon-13 isotope hence there will be a small peak at $(M + 1)^+$ due to the presence of ^{13}C. By measuring the ratio of the $(M + 1)^+$ peak to the molecular ion, M^+, the actual number of carbon atoms in the compound can be calculated.

Similarly the presence of chlorine or bromine gives rise to additional peaks. The relative natural abundance of ^{35}Cl and ^{37}Cl is 3:1. For molecular ions containing one chlorine atom the peaks will be separated by 2 m/z units and their relative intensities will be 3:1. Compounds containing two chlorine atoms will have peaks separated by 2 and 4 m/z units. Bromine also contains two isotopes, ^{79}Br and ^{81}Br. Since these occur in approximately equal amounts bromine-containing compounds will also contain two peaks separated by 2 m/z units of approximately equal intensity.

THE MASS SPECTRUM OF CHLOROBENZENE C_6H_5Cl

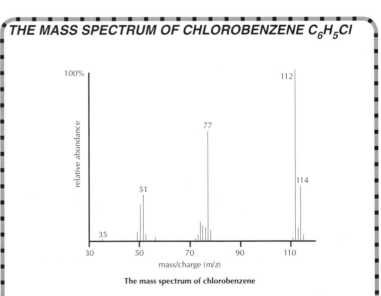

The mass spectrum of chlorobenzene

The signal at 112 is due to $C_6H_5{}^{35}Cl^+$ and the signal at 114 is due to $C_6H_5{}^{37}Cl^+$. The small signals at 113 and 115 are due to one atom of ^{13}C replacing one of ^{12}C. Since there are six carbon atoms in the molecule there is a $6 \times 1.1\%$ chance of having one ^{13}C in a particular ion so the peak at 113 should be 6.6% the height of the peak at 112. Similarly the peak at 115 will be 6.6% of the intensity of the peak at 114. The peak at 77 does not have a corresponding peak at 79 so chlorine must have been lost and this peak corresponds to $C_6H_5{}^+$. Similarly the peaks at 50 and 51 are due to $C_4H_2{}^+$ and $C_4H_3{}^+$ respectively.

HL Combination of different analytical techniques to determine structure

The determination of the organic structure of an unknown compound is usually achieved by combining the information from several different analytical techniques. This is illustrated by the following worked example for **Compound X**.

ELEMENTAL ANALYSIS

Compound X was found to contain 48.63% carbon, 8.18% hydrogen, and 43.19% oxygen by mass.

From this information the empirical formula of **Compound X** can be deduced as $C_3H_6O_2$.

Element	Amount / mol	Simplest ratio
C	48.63/12.01 = 4.05	3
H	8.18/1.01 = 8.10	6
O	43.19/16.00 = 2.70	2

INFRARED SPECTROSCOPY

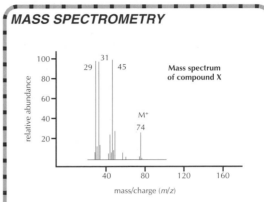

IR spectrum of compound X

Information available from the infrared spectrum:
- Absorption at 2980 cm^{-1} due to presence of C–H in **Compound X**.
- Absorption at 1725 cm^{-1} due to presence of C=O in **Compound X**.
- Absorption at 1200 cm^{-1} due to presence of C–O in **Compound X**.
- Absence of broad absorption at 3300 cm^{-1} indicates **Compound X** does not contain O–H.

MASS SPECTROMETRY

Mass spectrum of compound X

Information available from the mass spectrum:
- Since M$^+$ occurs at 74 the relative molecular mass of **Compound X** = 74.
- From this and the empirical formula it can be deduced that the molecular formula of **Compound X** is $C_3H_6O_2$.
- Fragment at 45 due to (M – 29)$^+$ so **Compound X** may contain C_2H_5– and/or CHO–.
- Fragment at 31 due to (M – 43)$^+$ so **Compound X** may contain C_2H_5O–.
- Fragment at 29 due to (M – 45)$^+$ so **Compound X** may contain HOOC– or H–C(=O)–O–.
- Peak at 75 due to the presence of ^{13}C.

¹H NMR SPECTROSCOPY

Information available from the ¹H NMR spectrum:
- Number of separate peaks is three so **Compound X** contains hydrogen atoms in three different chemical environments.
- From the integration trace the hydrogen atoms are in the ratio of 3:2:1 for the peaks at 1.3, 4.2, and 8.1 ppm respectively. Since there are six hydrogen atoms in the molecule this is the actual number of protons in each environment.

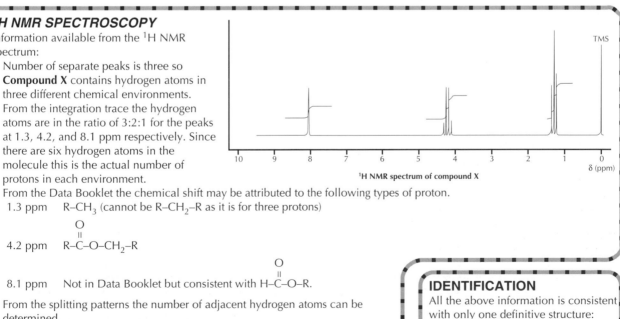

¹H NMR spectrum of compound X

- From the Data Booklet the chemical shift may be attributed to the following types of proton.

  1.3 ppm   R–CH$_3$ (cannot be R–CH$_2$–R as it is for three protons)

  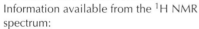
  4.2 ppm   R–C(=O)–O–CH$_2$–R

  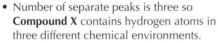
  8.1 ppm   Not in Data Booklet but consistent with H–C(=O)–O–R.

- From the splitting patterns the number of adjacent hydrogen atoms can be determined.

  1.3 ppm   triplet   two adjacent hydrogen atoms
  4.2 ppm   quartet   three adjacent hydrogen atoms
  8.1 ppm   singlet   no adjacent hydrogen atoms

IDENTIFICATION

All the above information is consistent with only one definitive structure:

Compound X is: H–C(=O)–O–CH$_2$CH$_3$
ethyl methanoate

Modern analytical chemistry 149

Chromatography (1)

INTRODUCTION

There are several ways in which mixtures can be separated into their individual components, for example filtration, crystallization, distillation, and solvent extraction. Chromatography can be used to separate mixtures containing very small amounts of the individual components. Coupled with other techniques, such as mass spectrometry, it can be used to separate and identify complex mixtures both quantitatively and qualitatively. It can also be used to determine how pure a substance is.

There are several different types of chromatography. They include paper, thin layer (TLC), column (LC), gas–liquid (GLC), and high performance liquid chromatography (HPLC). In each case there are two phases: a **stationary phase** that stays fixed and a **mobile phase** that moves. Chromatography relies upon the fact that in a mixture the components have different tendencies to adsorb onto a surface or dissolve in a solvent. This provides a way of separating them.

ADSORPTION AND PARTITION

Adsorption involves a solid stationary phase with a moving liquid phase (LC, HPLC, and at times TLC). The rate at which the solute moves through the solid phase depends on the equilibrium between its solubility in the moving liquid phase and its adsorption to the solid phase. The more tightly the component is adsorbed the slower it will elute.

Partition involves a stationary liquid phase and a mobile gaseous or liquid phase (paper, TLC, and GLC). It is dependent on the relative solubility of each component in the two phases if both are liquid. If the mobile phase is gaseous then the rate of movement depends upon the volatility of the components. The more soluble or volatile the component in the mobile phase the faster it will elute.

PAPER CHROMATOGRAPHY

Most people are familiar with a simple example of paper chromatography. When black ink is placed on blotting paper the different colours in the ink move at different rates so it quickly separates into several colours. Paper consists largely of cellulose fibres. These contain a large number of hydroxyl groups making the paper quite polar. Water molecules hydrogen bond to these groups so that a sheet of 'dry' paper actually contains about 10% water. It is this water that acts as the stationary phase. The mobile phase is a solvent, either water itself or a polar organic liquid such as ethanol, propanone, or ethanoic acid. Normally a small amount of the mixture is spotted onto the paper about 1.0 cm from its base. The paper is then suspended in a small quantity of the solvent (known as the eluent) in a closed container. (The container is closed in order to saturate the atmosphere and prevent evaporation of the solvent from the paper to give better and faster separation.) As the solvent rises up the paper the components in the mixture partition between the two phases depending on their relative solubility. As the solvent nears the top of the paper a mark is made to record the level and the paper then removed and dried. Coloured components can be seen with the naked eye. Other components can be made visible by staining (e.g. with iodine) or by irradiating with an ultraviolet lamp.

As with all equilibria the partition of a solute between the two phases is constant at a fixed temperature, thus the solute will always move the same fraction of the distance moved by the solvent. Each solute will have a particular retention factor (R_f) for a given eluent. It is obtained by measuring the distance from the original spot both to the centre of the particular component and to the solvent front.

$$R_f = \frac{\text{distance moved by solute}}{\text{distance moved by solvent (eluent)}} = \frac{x}{y}$$

Substances can be identified by their R_f values. If two substances have similar R_f values in one solvent the paper can be turned 90° and a different solvent used. This is known as two way chromatography and is often used to separate amino acids (see 14. option C – Human Biochemistry).

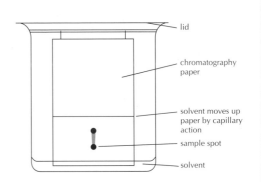

lid

chromatography paper

solvent moves up paper by capillary action

sample spot

solvent

paper chromatography

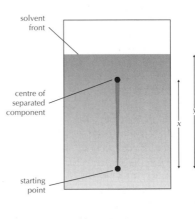

solvent front

centre of separated component

starting point

chromatogram

THIN LAYER CHROMATOGRAPHY (TLC)

This is very similar to paper chromatography but uses a thin layer of solid, such as alumina (Al_2O_3) or silica (SiO_2), on an inert support, such as glass. When absolutely dry it works by adsorption but, like paper, silica and alumina have a high affinity for water therefore the separation occurs more by partition, with water as the stationary phase. The separated components can be recovered pure by scraping off the section containing the component and dissolving it in a suitable solvent. Pregnancy tests may use TLC to detect pregnanediol in urine.

COLUMN CHROMATOGRAPHY (LC)

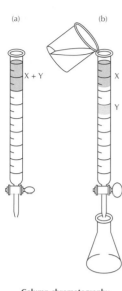

Column chromatography

(a) Mixture containing X and Y is placed on prepared column.
(b) More solvent is added as X and Y elute. Y elutes faster than X and can be collected first.

Column chromatography is usually used to separate the components of a mixture for further use rather than for identification. The stationary phase is alumina or silica gel ($SiO_2.xH_2O$). The column is set up by packing the dry stationary phase on top of a piece of glass wool in a long column with a tap at the end and then saturating it with the eluting solvent. The sample is added at the top and as it moves down the column more eluent is added. After some of the components have been eluted it is possible to change the solvent to elute the more tightly held components.

HIGH PERFORMANCE LIQUID CHROMATOGRAPHY (HPLC)

Column chromatography uses gravity to allow the solvent to elute. High performance liquid chromatography works in much the same way but the mobile phase is forced through under pressure. The stationary phase is commonly composed of silica particles with long chain alkanes adsorbed onto their surface. The separation is very efficient so long columns are not needed. The separated components are usually detected by ultraviolet spectroscopy. Like GLC the results are recorded onto a chart showing the different retention times. HPLC can be used for identification as well as for separation. It is particularly useful for non-volatile components or components that decompose near their boiling point. It can also be used to separate enantiomers (chiral separation) using columns containing optically active material.

GAS–LIQUID CHROMATOGRAPHY (GLC)

Schematic diagram of a gas–liquid chromatograph

This method is used to separate and identify the components present in mixtures of volatile liquids that do not decompose at temperatures at or near their boiling points. The stationary phase consists of a liquid (e.g. a long chain alkane) coated onto a solid support in a long, thin capillary tube. The mobile phase is an inert gas such as nitrogen or helium. The sample is injected through a self-sealing cap into an oven for vaporization. The sample is then carried by the inert gas into the column which is coiled and fitted into an oven. At the end of the column the separated components exit into a detector. This is usually a flame ionization detector connected to a chart recorder.

Each component will have a separate retention time and the area under the peak will be proportional to the amount of component present. GLC can thus detect the alcohol concentration in a sample of blood and can be used to provide evidence in court cases involving drunken driving. It is possible to programme the temperature of the oven to increase during the operation to speed up the elution of the less volatile components.

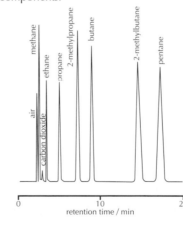

Separation of the components of a sample of natural gas by gas–liquid chromatography (GLC). The less volatile the component, the greater the retention time.

GAS CHROMATOGRAPHY – MASS SPECTROMETRY (GC–MS)

GC–MS is an incredibly powerful tool used extensively in forensic science and in medicine and other analytical laboratories. Because the mass spectrometer is so sensitive it can be used to detect and measure the concentration of extremely small samples of substances. It is used to detect traces of banned drugs in athletes, toxins in food samples, and the different components in mixtures of crude oil.

The technique works by combining the operation of gas liquid chromatography with mass spectrometry. As the separate components elute from the chromatograph they pass straight into a mass spectrometer. The mass spectrometer is connected to a computer which contains a library of the spectra of all known compounds. The computer matches the spectra and gives a print-out of all the separate components and their concentrations. A similar technique (HPLC–MS) combines high performance liquid chromatography with mass spectrometry.

1. **(a)** Infrared spectroscopy is a powerful tool for identifying organic compounds. State what occurs at the molecular level during the absorption of infrared (ir) radiation and identify the change that is necessary for ir absorption to occur. Discuss why infrared studies are particularly helpful in the characterisation of organic molecules. **[4]**

   (b) Use information in Table 18 of the Data Booklet to list the absorption regions expected for:

   (i) ethanoic acid. **[2]**

   (ii) methyl methanoate. **[2]**

   (c) Identify the absorption listed in **(b)** which could be used to distinguish between these two compounds. Explain why the other absorptions could not be used. **[2]**

   (d) Identify the absorption listed in **(b)** which has the highest energy and calculate its wavelength in cm. **[2]**

2. **(a)** In the use of paper chromatography, a retention factor, R_f is determined. Define R_f **[1]**

   (b) A drop of green dye is placed 2 cm from the bottom of a strip of filter paper. The filter paper is suspended in a graduated cylinder with 1 cm of the paper immersed in a water–alcohol solvent. After 30 minutes, the green spot is no longer present and there is a yellow spot and a blue spot.

   (i) Describe how the R_f value of the blue spot could be determined. **[2]**

   (ii) Account for the difference in the R_f values of the yellow and blue dyes. **[2]**

   (iii) What is the significance of an R_f value of 1.0? **[2]**

3. The structure of carotene is given below:

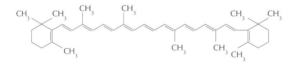

   (a) What feature of the carotene molecule accounts for the fact that it absorbs light in the visible region of the spectrum? **[2]**

   (b) The extinction coefficient of the carotene absorption spectrum is 130 000 $dm^3 mol^{-1} cm^{-1}$ at its absorption maximum of 450 nm. Using the Beer-Lambert formula in the Data Booklet calculate the concentration of carotene required to absorb 99% of the incident light at that wavelength as it passes through a 1.0 cm cell. **[2]**

   (c) Carotene usually occurs mixed with other complex organic molecules. A technique often employed with such mixtures is to analyse them by combining gas phase chromatography with mass spectrometry. Explain briefly why these two techniques complement each other so well. **[2]**

4. The ^1H NMR spectra of an unknown compound is given below:

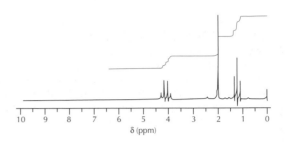

   (a) Why is tetramethylsilane used in ^1H NMR spectroscopy? **[1]**

   (b) What word is used to describe the multiplicity of the peaks centred at 4.1 ppm? **[1]**

   (c) What is the ratio of the number of hydrogen atoms responsible for the chemical shifts centred at 1.2, 2.0 and 4.1 ppm respectively? **[1]**

   (d) (i) From the Data Booklet, identify two general structures which contain protons in the correct environment to produce the chemical shift centred at 4.1 ppm. **[1]**

   (ii) From consideration of the rest of the spectra only one of these general structures is possible. Identify which one and explain your reasoning. **[3]**

   Below is the infra-red spectrum of the same compound:

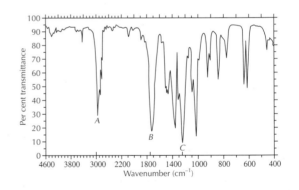

   (e) Identify which particular vibrations are responsible for the peaks labelled *A*, *B* and *C*. **[3]**

   (f) The mass spectrum of the same compound shows a molecular ion peak at 88 m/z. The fragmentation pattern shows prominent peaks at 73 m/z and 59 m/z amongst others. Identify the ions responsible for these peaks. **[3]**

   (g) Give the name and structural formula of the compound. **[2]**

Stereoisomerism

Structural isomers share the same molecular formula but have different structural formulas. That is, their atoms are bonded in different ways. Stereoisomers have the same structural formula but differ in their spatial arrangement. There are two types of stereoisomerism: geometrical isomerism and optical isomerism.

GEOMETRICAL ISOMERISM

Geometrical isomerism occurs when rotation about a bond is restricted or prevented. The classic examples of geometric isomers occur with asymmetric non-cyclic alkenes. A *cis*- isomer is one in which the substituents are on the same side of the double bond. In a *trans*- isomer the substituents are on opposite sides of the double bond. For example consider *cis*-but-2-ene and *trans*-but-2-ene.

cis-   *trans-*

When there is a single bond between two carbon atoms free rotation about the bond is possible. However, the double bond in an alkene is made up of a σ and a π bond. The π bond is formed from the combination of two p orbitals, one from each of the carbon atoms. These two p orbitals must be in the same plane to combine. Rotating the bond would cause the π bond to break so no rotation is possible.

Cis- and *trans*- isomerism will always occur in alkenes when the two groups attached to each of the two carbon atoms are different.

π bond prevents rotation

cis-   *trans-*

Geometrical isomerism can also occur in disubstituted cycloalkanes. The rotation is restricted because the C–C single bond is part of a ring system. Examples include 1,2-dichlorocyclopropane and 1,3-dichlorocyclobutane.

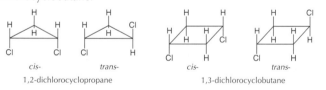

cis-   *trans-*   *cis-*   *trans-*
1,2-dichlorocyclopropane   1,3-dichlorocyclobutane

PHYSICAL AND CHEMICAL PROPERTIES OF GEOMETRICAL ISOMERS

The chemical properties of geometric isomers tend to be similar but their physical properties are different. For example, the boiling point of *cis*-1,2-dichloroethene is 60.3 °C whereas *trans*-1,2-dichloroethene boils at the lower temperature of 47.5 °C. Sometimes there can be a marked difference in both chemical and physical properties. This tends to occur when there is some sort of chemical interaction between the substituents. *cis*-but-2-ene-1,4-dioic acid melts with decomposition at 130–131 °C. However, *trans*-but-2-ene,1,4-dioic acid does not melt until 286 °C. In the *cis*- isomer the two carboxylic acid groups are closer together so that intramolecular hydrogen bonding is possible between them. In the *trans*- isomer they are too far apart to attract each other so there are stronger intermolecular forces of attraction between different molecules, resulting in a higher melting point. The *cis*-isomer reacts when heated to lose water and form a cyclic acid anhydride. The *trans*- isomer cannot undergo this reaction.

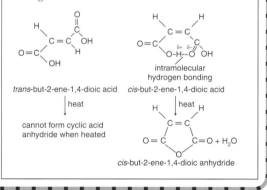

trans-but-2-ene-1,4-dioic acid   *cis*-but-2-ene-1,4-dioic acid

intramolecular hydrogen bonding

cannot form cyclic acid anhydride when heated   *cis*-but-2-ene-1,4-dioic anhydride

ENANTIOMERS

Optical isomerism has already been introduced in 11. Topic 11 – Organic Chemistry. You should be familiar with the terms **plane-polarized light** and **asymmetric** or **chiral** carbon atom. The similarities and differences in their physical and chemical properties have also been discussed.

The optical activity of enantiomers can be detected and measured by an instrument called a polarimeter. It consists of a light source, two polarizing lenses, and between the lenses a tube to hold the sample of the enantiomer.

When light passes through the first polarizing lens (polarizer) it becomes plane-polarized. That is, it is vibrating in a single plane. With no sample present the observer will see the maximum intensity of light when the second polarizing lens (analyser) is in the same plane. Rotating the analyser by 90° will cut out all the light. When the sample is placed between the lenses the analyser must be rotated by θ degrees, either clockwise (dextrorotatory) or anticlockwise (laevorotatory) to give light of maximum intensity. The two enantiomers rotate the plane of plane-polarized light by the same amount but in opposite directions. If both enantiomers are present in equal amounts the two rotations cancel each other out and the mixture appears to be optically inactive. Such a mixture is known as a **racemic mixture** or **racemate**.

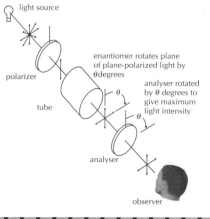

Free radical substitution reactions

CHLORINATION OF METHANE

In the presence of ultraviolet light alkanes undergo substitution reactions with halogens. This can be exemplified by the reaction between methane and chlorine. In the dark no reaction occurs but in ultraviolet light chloromethane and hydrogen chloride are formed. If excess chlorine is present then further substitution can occur.

$$CH_4(g) + Cl_2(g) \xrightarrow{UV} CH_3Cl(g) + HCl(g)$$
$$\text{chloromethane}$$

For the reaction to proceed the Cl–Cl and C–H bonds must break.

$$Cl_2 \rightarrow Cl^\bullet + Cl^\bullet \quad \Delta H^{\ominus} = +242 \text{ kJ mol}^{-1}$$
$$CH_4 \rightarrow CH_3^\bullet + H^\bullet \quad \Delta H^{\ominus} = +415 \text{ kJ mol}^{-1}$$

Heating does not provide sufficient energy to break these bonds but ultraviolet light does provide sufficient activation energy to split the chlorine molecule homolytically to form chlorine radicals. This stage of the mechanism is called **initiation**.

These free radicals contain an unpaired electron and are highly reactive. When they come into contact with a methane molecule they combine with a hydrogen atom to produce hydrogen chloride and a methyl radical. Since a new radical is produced this stage is called **propagation**. The movement of a single electron can be shown using a curly arrow with half an arrowhead.

$$H_3C \overset{\frown}{-} H \quad \overset{\frown}{} Cl^\bullet \rightarrow H_3C^\bullet + H{-}Cl$$

The methyl free radical is also extremely reactive and reacts with a chlorine molecule to form the product and regenerate another chlorine radical. This is a further propagation step. This enables a chain reaction to occur as the process can repeat itself.

$$CH_3^\bullet + Cl_2 \rightarrow CH_3{-}Cl + Cl^\bullet$$

In theory a single chlorine radical may cause up to 10 000 molecules of chloromethane to be formed. **Termination** occurs when two radicals react together.

$$\left. \begin{array}{l} Cl^\bullet + Cl^\bullet \rightarrow Cl_2 \\ CH_3^\bullet + Cl^\bullet \rightarrow CH_3Cl \\ CH_3^\bullet + CH_3^\bullet \rightarrow C_2H_6 \end{array} \right\} \quad \text{termination}$$

Further substitution can occur when chlorine radicals react with the substituted products. For example:

$$Cl{-}\underset{\underset{H}{|}}{\overset{\overset{H}{|}}{C}}{-}H + Cl^\bullet \rightarrow Cl{-}\underset{\underset{H}{|}}{\overset{\overset{H}{|}}{C}}{-}^\bullet + HCl$$

$$\text{then } Cl{-}\underset{\underset{H}{|}}{\overset{\overset{H}{|}}{C}}{-}^\bullet + Cl_2 \rightarrow Cl{-}\underset{\underset{H}{|}}{\overset{\overset{H}{|}}{C}}{-}Cl + Cl^\bullet$$

The overall mechanism is called **free radical substitution**. [Note that in this mechanism hydrogen radicals H• are not formed.]

REACTION OF CHLOROALKANES WITH OZONE

The use of chlorinated organic compounds is being curtailed for two main reasons. Some of the compounds are potentially carcinogenic (cancer forming) and they also have the potential to lower the concentration of ozone in the ozone layer.

FREE RADICAL SUBSTITUTION OF METHYLBENZENE

Methylbenzene (toluene) behaves like a substituted alkane when it reacts with chlorine in ultraviolet light.

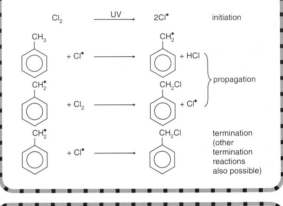

The mechanism is identical to the reaction of methane with chlorine. The initial product is chloromethylbenzene although in excess chlorine further substitution can also occur.

Ozone is continually being formed and depleted in the ozone layer between about 12 km and 50 km above the surface of the Earth. The strong double bond in oxygen is broken by high energy ultraviolet light from the Sun to form radicals. One oxygen radical can then react with an oxygen molecule to form ozone.

$$O{=}O(g) \xrightarrow{UV \text{ (high energy)}} 2O^\bullet(g)$$
$$O^\bullet(g) + O_2(g) \rightarrow O_3(g)$$

The bonds in ozone are weaker so UV light of less energy will break them. When they are broken the reverse process happens and the ozone breaks down back to an oxygen molecule and an oxygen radical. The radical can then react with another ozone molecule to form two oxygen molecules.

$$O_3(g) \xrightarrow{UV \text{ (lower energy)}} O_2(g) + O^\bullet(g)$$
$$O_3(g) + O^\bullet(g) \rightarrow 2O_2(g)$$

Overall the rate of production of ozone is equal to the rate of ozone destruction – this is known as a steady state. However, compounds containing C–Cl bonds can alter this steady state. Chlorinated hydrocarbons tend to be rather unreactive until they reach the stratosphere where the ultraviolet light causes the C–Cl bond to break homolytically. The chlorine radicals can then react with ozone molecules by a chain reaction. This causes a substantial lowering of the ozone concentration resulting in 'holes' in the ozone layer, particularly over the polar regions.

$$RCH_2Cl(g) \rightarrow RCH_2^\bullet(g) + Cl^\bullet(g) \quad \text{(radical initiation)}$$
$$Cl^\bullet(g) + O_3(g) \rightarrow ClO^\bullet(g) + O_2(g) \quad \text{(propagation}$$
$$ClO^\bullet(g) + O_3(g) \rightarrow Cl^\bullet(g) + 2O_2(g) \quad \text{of radicals)}$$
$$\text{Overall} \quad 2O_3(g) \rightarrow 3O_2(g)$$

ELECTROPHILIC ADDITION TO SYMMETRIC ALKENES

Ethene readily undergoes addition reactions. With hydrogen bromide it forms bromoethane.

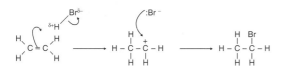

The reaction can occur in the dark which suggests that a free radical mechanism is not involved. The ethene molecule has a region of high electron density above and below the plane of the molecule due to the π bond. Hydrogen bromide is a polar molecule due to the greater electronegativity of bromine compared with hydrogen. The hydrogen atom (which contains a charge of δ+) from the H–Br is attracted to the π bond and the H–Br bond breaks, forming a bromide ion. At the same time the hydrogen atom adds to one of the ethene carbon atoms leaving the other carbon atom with a positive charge. A carbon atom with a positive charge is known as a **carbocation**. The carbocation then combines with the bromide ion to form bromoethane. Because the hydrogen bromide molecule is attracted to a region of electron density it is described as an **electrophile** and the mechanism is described as **electrophilic addition**.

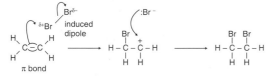

Electrophilic addition also takes place when bromine adds to ethene in a non-polar solvent to give 1,2-dibromoethane. Bromine itself is non-polar but as it approaches the π bond of the ethene an induced dipole is formed by the electron cloud.

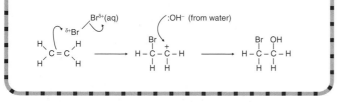

Evidence for this mechanism is that when bromine water is reacted with ethene the main product is 2-bromoethanol not 1,2-dibromoethane. This suggests that hydroxide ions from the water add to the carbocation in preference to bromide ions.

SYMMETRIC AND ASYMMETRIC ALKENES

Asymmetric alkenes contain different groups attached to the carbon atoms of the C=C bond.

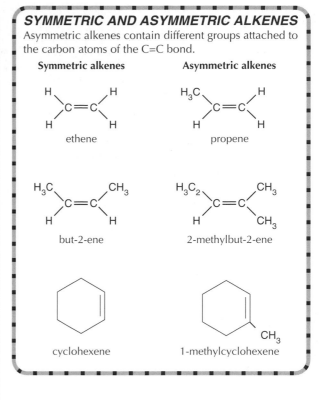

Electrophilic addition reactions (2)

MARKOVNIKOV'S RULE

When hydrogen halides add to asymmetric alkenes two products are possible depending upon which carbon atom the hydrogen atom bonds to. For example, the addition of hydrogen bromide to propene could produce 1-bromopropane or 2-bromopropane.

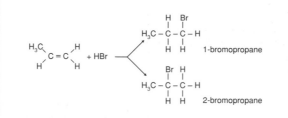

Markovnikov's rule enables you to predict which isomer will be the major product. It states that the hydrogen halide will add to the carbon atom that already contains the most hydrogen atoms bonded to it. Thus in the above example 2-bromopropane will be the major product.

EXPLANATION OF MARKOVNIKOV'S RULE

Markovnikov's rule enables the product to be predicted but it does not explain why. It can be explained by considering the nature of the possible intermediate carbocations formed during the reaction.

When hydrogen ions react with propene two different carbocations can be formed.

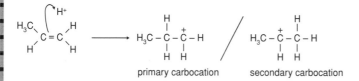

primary carbocation   secondary carbocation

The first one has the general formula RCH_2^+ and is known as a **primary carbocation**. The second one has two R– groups attached to the positive carbon ion R_2CH^+ and is known as a **secondary carbocation**. A **tertiary carbocation** has the general formula R_3C^+. The R-groups (alkyl groups) tend to push electrons towards the carbon atom they are attached to which tends to stabilize the positive charge on the carbocation. This is known as a **positive inductive effect**. This effect will be greatest with tertiary carbocations and smallest with primary carbocations.

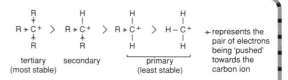

Thus in the above reaction the secondary carbocation will be preferred as it is more stable than the primary carbocation. This secondary carbocation intermediate leads to the major product, 2-bromopropane.

Understanding this mechanism enables you to predict what will happen when an interhalogen adds to an asymmetric alkene even though no hydrogen atoms are involved. Consider the reaction of iodine chloride ICl with but-1-ene. Since iodine is less electronegative than chlorine the iodine atom will act as the electrophile and add first to the alkene.

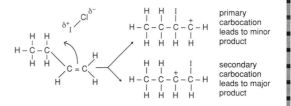

The major product will thus be 2-chloro-1-iodobutane.

ⓗ Electrophilic substitution reactions (1)

Benzene does not readily undergo addition reaction since an additional 150 kJ mol⁻¹ of energy would be required to overcome the energy of delocalization. Instead it undergoes electrophilic substitution reactions. Electrophiles are attracted to the region of high electron density above and below the plane of the molecule due to the delocalized π bond.

NITRATION OF BENZENE

Benzene reacts with a mixture of concentrated nitric acid and concentrated sulfuric acid when warmed at 50 °C to give nitrobenzene and water. Note that the temperature should not be raised above 50 °C otherwise further nitration to dinitrobenzene will occur.

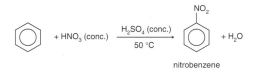

nitrobenzene

The electrophile is the **nitryl cation NO₂⁺** (also called the nitronium ion). The concentrated sulfuric acid acts as a catalyst. Its function is to protonate the nitric acid which then loses water to form the electrophile. In this reaction nitric acid is acting as a base in the presence of the more acidic sulfuric acid.

$$H_2SO_4 + HNO_3 \rightleftharpoons H_2NO_3^+ + HSO_4^-$$
$$\downarrow$$
$$H_2O + NO_2^+$$

The NO₂⁺ is attracted to the delocalized π bond and attaches to one of the carbon atoms. This requires considerable activation energy as the delocalized π bond is partially broken. The positive charge is distributed over the remains of the π bond in the intermediate. The intermediate then loses a proton and energy is evolved as the delocalized π bond is reformed. The proton can recombine with the hydrogen sulfate ion to regenerate the catalyst.

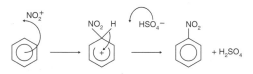

Although it is more correct to draw the intermediate as a partially delocalized π bond it can sometimes be convenient to show benzene as if it does contain alternate single and double carbon to carbon bonds. In this model the positive charge is located on a particular carbon atom.

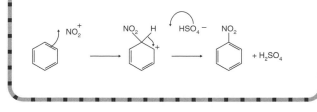

CHLORINATION AND ALKYLATION OF BENZENE

Benzene can also undergo electrophilic substitution with chlorine. This only happens in the presence of a **halogen carrier**, such as anhydrous aluminium chloride or iron.

chlorobenzene

The halogen carrier is a catalyst (sometimes known as a Friedel–Crafts catalyst after its discoverers). It functions as a Lewis acid by accepting a pair of electrons from the chlorine. The electrophile is the resulting chlorine ion, Cl⁺.

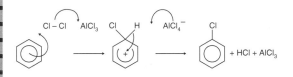

Halogen carriers can also be used to attract the electron pair in the C–Cl bond in halogenoalkanes R–Cl. The resulting electrophile will now be the positive alkyl ion R⁺. The resulting substituted product will be an alkylbenzene. For example, methylbenzene can be prepared by warming benzene with chloromethane in the presence of anhydrous aluminium chloride.

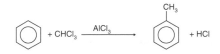

Mechanism:

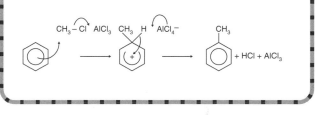

HL Electrophilic substitution reactions (2)

SUBSTITUTION REACTIONS OF METHYLBENZENE

The reaction between methylbenzene and chlorine provides a good example of how altering the conditions can completely alter the products. In the presence of ultraviolet light the reaction proceeds by a free radical substitution mechanism and chloromethylbenzene is formed. In this reaction methylbenzene is behaving as a substituted alkane. In the presence of a halogen carrier it behaves as a substituted arene and the substitution takes place on the aromatic ring. The organic products are a mixture of 2-chloromethylbenzene and 4-chloromethylbenzene.

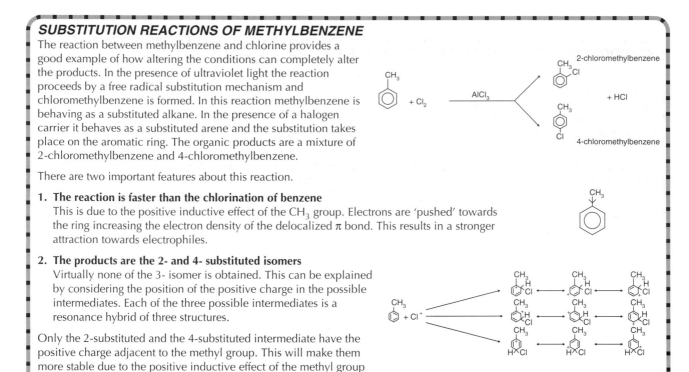

There are two important features about this reaction.

1. **The reaction is faster than the chlorination of benzene**
   This is due to the positive inductive effect of the CH_3 group. Electrons are 'pushed' towards the ring increasing the electron density of the delocalized π bond. This results in a stronger attraction towards electrophiles.

2. **The products are the 2- and 4- substituted isomers**
   Virtually none of the 3- isomer is obtained. This can be explained by considering the position of the positive charge in the possible intermediates. Each of the three possible intermediates is a resonance hybrid of three structures.

Only the 2-substituted and the 4-substituted intermediate have the positive charge adjacent to the methyl group. This will make them more stable due to the positive inductive effect of the methyl group and thus more likely to be formed.

DIRECTING EFFECTS OF SUBSTITUENTS IN SUBSTITUTED BENZENE COMPOUNDS

The methyl group has a positive inductive effect and this both increases the rate of substitution and directs towards the 2- and 4-positions. Many other groups are electron withdrawing and these can be sub-divided into two groups depending on whether the atom bonded to the benzene ring contains a non-bonding pair of electrons or not.

Electron withdrawing with no non-bonded pair

Examples include:

Because the electron density on the ring is reduced the substitution reaction will be slower relative to benzene. The substitution will direct towards the 3-position since none of the intermediate structures will be destabilized by having the positive charge adjacent to the electron withdrawing group. For example, the nitration of nitrobenzene requires the nitrating mixture of the two concentrated acids to be refluxed with nitrobenzene rather than just heated to 50 °C to produce 1,3-dinitrobenzene.

Electron withdrawing with a non-bonded pair

Examples include:

phenol   aminobenzene   chlorobenzene

These substituents make the compound considerably more reactive (faster) towards electrophilic substitution than benzene. The delocalization in the ring is now able to include the non-bonding pair of electrons on the substituent. Whereas six electrons are delocalized over six atoms in benzene eight electrons are delocalized over seven atoms in phenol and phenylamine. This increases the electron density and makes them much more attracted to electrophiles. This can be exemplified by the reaction of phenol with chlorine. The reaction is so fast that no halogen carrier is necessary and the trisubstituted product is formed immediately.

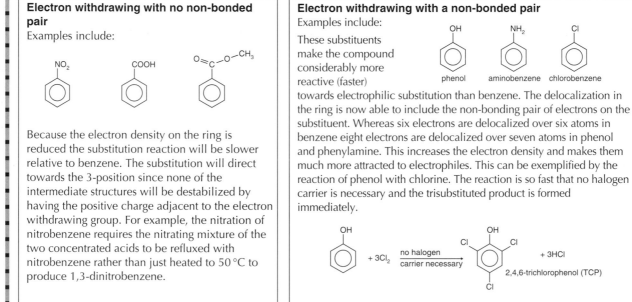

+ 3Cl₂ → no halogen carrier necessary → + 3HCl

2,4,6-trichlorophenol (TCP)

These substituents are 2- and 4-directing as the intermediate can now be considered as a resonance hybrid of four structures thus increasing its stability. In the 3-position the positive charge cannot be positioned on the substituent.

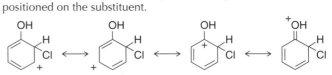

The four possible locations of the positive charge in the 2-substituted intermediate of phenol.

HL Nucleophilic addition reactions

ADDITION OF HYDROGEN CYANIDE TO ALDEHYDES AND KETONES

Aldehydes and ketones are both carbonyl compounds as they contain the $>$C=O group. The double bond in carbonyl compounds is similar to the C=C double bond in alkenes in that it is made up of a σ bond and a π bond and the region around the bond is planar with bond angles of 120°.

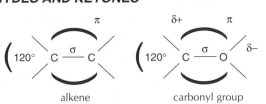

alkene   carbonyl group

The essential difference is that oxygen is much more electronegative than carbon so that the bond is polar. Whereas alkenes undergo electrophilic addition, carbonyl compounds undergo **nucleophilic addition reactions**.

A typical example is the reaction of ethanal with hydrogen cyanide. Hydrogen cyanide is weakly acidic and dissociates to form the cyanide ion which acts as the nucleophile. The intermediate anion is then protonated to form the product, 2-hydroxypropanenitrile.

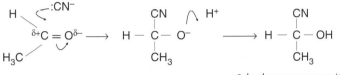

2-hydroxy propanenitrile

This reaction is useful in synthesis as it is a method by which an extra carbon atom can be introduced into a carbon chain. Nitriles can be hydrolysed in the presence of acid to form carboxylic acids so this product can easily be converted into 2-hydroxypropanoic acid (lactic acid).

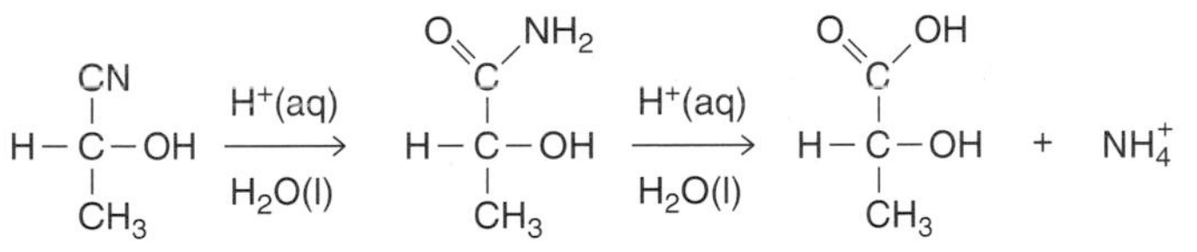

2-hydroxypropanoic acid

Note that 2-hydroxypropanoic acid contains an asymmetric carbon atom so the product will be a racemic mixture of the two enantiomers.

enantiomers of 2-hydroxypropanoic acid

REACTION OF 2,4-DINITROPHENYLHYDRAZINE WITH ALDEHYDES AND KETONES

With nucleophiles containing the $-NH_2$ group the addition can be followed by the elimination of water. For example, the reaction of hydrazine with propanone. The initial mode of attack is still nucleophilic addition but then the carbon atom forms a double bond with the nitrogen atom and a molecule of water is lost. The overall reaction is known either as an **addition–elimination reaction** or more simply as a **condensation reaction**.

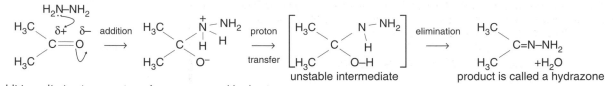

addition–elimination reaction of propanone and hydrazine

If 2,4-dintritrophenylhydrazine (2,4-DNP) is used instead of hydrazine the product is known as a 2,4-dinitrophenylhydrazone.

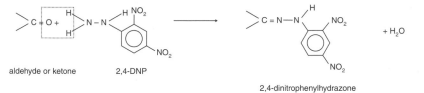

All aldehydes and ketones form red or orange crystalline solid 2,4-dinitrophenylhydrazones each of which has a characteristic melting point. These derivatives of 2,4-DNP are easy to make and purify so provide a convenient way of distinguishing between different aldehydes and ketones, although in a well-equipped laboratory chemists are now more likely to use a combination of spectroscopic techniques (e.g. ^1H NMR and mass spectrometry).

**Melting points of 2, 4-dinitrophenyl-
hydrazones of aldehydes and ketones**

M. pt / °C		M. pt / °C	
methanal	166	propanone	126
ethanal	168	butanone	116
propanal	155	pentan-3-one	156
butanal	126	pentan-2-one	144
benzaldehyde	237	cyclohexanone	162

Elimination reactions

ELIMINATION REACTIONS OF HALOGENOALKANES

The reactions of halogenoalkanes with hydroxide ions provide an example of how altering the reaction conditions can cause the same reactants to produce completely different products. [Note that another good example is the reaction of methylbenzene with chlorine.] With dilute sodium hydroxide solution the OH⁻ ion acts as a nucleophile and substitution occurs to produce an alcohol, e.g.

$$HO^-: \quad R–Br \longrightarrow R–OH + Br^-$$

However with hot alcoholic sodium hydroxide solution (i.e. sodium hydroxide dissolved in ethanol) **elimination** occurs and an alkene is formed, e.g.

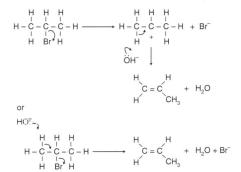

In this reaction the hydroxide ion reacts as a base. The elimination of HBr can proceed either by a carbocation or as a concerted process, e.g.

or

In the presence of ethanol there will also be some ethoxide ions present. Ethoxide is a stronger base than hydroxide so the equilibrium lies to the left but some ethoxide ions will be present and these may be the actual species acting as the base.

$$HO^- + C_2H_5OH \rightleftharpoons H_2O + C_2H_5O^-$$

ELIMINATION REACTIONS OF ALCOHOLS

Alcohols can be dehydrated (elimination of water) by heating with concentrated sulfuric acid at 180 °C. In practice phosphoric acid is often used in place of sulfuric acid as fewer side reactions take place.

The strong acid acts as a catalyst by protonating the oxygen atom in the alcohol. Water is then lost to form a carbocation. The carbocation then donates a proton to form the double bond and regenerate the catalyst.

Tertiary alcohols are more readily dehydrated than primary and secondary alcohols because the intermediate tertiary carbocation is more stable.

Ⓗ Nucleophilic substitution reactions

Nucleophilic substitition has already been introduced in 11. Topic 20 – Organic Chemistry. Primary halogenoalkanes react by an S_N2 mechanism and tertiary halogenoalkanes react by an S_N1 mechanism. Tertiary halogenoalkanes react faster than primary halogenoalkanes and iodoalkanes react faster than chloroalkanes.

EFFECT OF CHANGING THE NUCLEOPHILE ON THE RATE OF SUBSTITUTION

The effectiveness of a nucleophile depends on its electron density. Anions tend to be more reactive than the corresponding neutral species. For example, the rate of substitution with the hydroxide ion is faster than with water. Among species with the same charge a less electronegative atom carrying a non-bonded pair of electrons is a better nucleophile than a more electronegative one. Thus ammonia is a better nucleophile than water. This is because the less electronegative atom can more easily donate its pair of electrons as they are held less strongly.

$$CN^- > OH^- > NH_3 > H_2O$$

order of reactivity of common nucleophiles

INDUCTIVE AND STERIC EFFECTS OF SUBSTITUENTS

The inductive and steric effects of substituents can help to explain why primary and tertiary halogenoalkanes undergo nucleophilic substitution reactions by different mechanisms. In both cases it is helpful to consider the intermediate or transition state formed.

Tertiary halogenoalkanes form a tertiary carbonium ion intermediate during the S_N1 process.

e.g.

$$H_3C-\overset{\overset{\displaystyle CH_3}{|}}{\underset{\underset{\displaystyle CH_3}{|}}{C}}-Br \longrightarrow H_3C\rightarrow \overset{\overset{\displaystyle CH_3}{\downarrow}}{\underset{\underset{\displaystyle CH_3}{\uparrow}}{C^+}} + Br^-$$

All three of the R-groups bonded to the carbocation will have a positive inductive effect. This means the positive charge is stabilized and much more likely to be formed than a primary carbocation.

The S_N2 mechanism for primary halogenoalkanes involves five groups surrounding the central carbon atom in the transition state. With primary halogenoalkanes two of these groups are hydrogen atoms so there is room for the other three groups. Tertiary halogenoalkanes cannot proceed by this mechanism as there is no room to accommodate five bulky groups around the carbon atom. This is an example of steric hindrance.

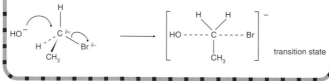

HYDROLYSIS OF HALOGENATED BENZENE COMPOUNDS

Substitution reactions involving the ionic S_N1 mechanism are faster than those proceeding via the S_N2 mechanism. Thus the hydrolysis of a tertiary chloroalkane with water or hydroxide ions is faster than the hydrolysis of a primary chloroalkane under the same conditions. Chlorobenzene is very much slower and does not readily undergo nucleophilic substitution reactions at all.

$$H_3C-\overset{\overset{\displaystyle CH_3}{|}}{\underset{\underset{\displaystyle CH_3}{|}}{C}}-Cl \;>\; H_3C-\overset{\overset{\displaystyle H}{|}}{\underset{\underset{\displaystyle CH_3}{|}}{C}}-Cl \;>\; H-\overset{\overset{\displaystyle H}{|}}{\underset{\underset{\displaystyle CH_3}{|}}{C}}-Cl \;\gg\; \text{(chlorobenzene)}$$

There are two accepted explanations for the inertness of halogenated benzene compounds towards hydrolysis.

1. A non-bonding pair of p electrons on the chlorine atom can interact with the delocalized π bond on the benzene ring increasing the strength of the C–Cl bond and making it harder to be broken. The presence of the delocalized electrons may also reduce the polarity of the C–Cl bond.
2. The high electron density of the delocalized π bond repels the nucleophile and hinders its access to the δ+ carbon atom attached to the chlorine atom.

Acid–base reactions

ACIDIC PROPERTIES OF SUBSTITUTED CARBOXYLIC ACIDS

Although alcohols and carboxylic acids both contain an –OH group carboxylic acids are weak acids whereas alcohols are not. The electron withdrawing carbonyl group $\diagdown C = O$ adjacent to the –OH group weakens the normally strong O–H bond so that the carboxyl group can lose a proton.

$$H_3C-C\overset{O}{\underset{O-H}{\diagup}} + H_2O \rightleftharpoons H_3C-C\overset{O}{\underset{O}{\diagup}} + H_3O^+ \quad pK_a = 4.76$$

The acidity of carboxylic acids can also be explained by considering the conjugate base formed. In the carboxylate anion the negative charge is delocalized over three atoms as both C–O bond lengths are identical. This stabilizes the ion and makes it less likely to attract a proton as the charge density of the negative charge is reduced. Compare this with the strongly basic ethoxide ion where the negative charge is localized on the oxygen atom.

$$C_2H_5-O-H + H_2O \rightleftharpoons C_2H_5-O^- + H_3O^+ \quad pK_a = \text{approx. } 16$$

Electron withdrawing atoms such as chlorine can further delocalize the negative charge on the anion increasing the acid strength.

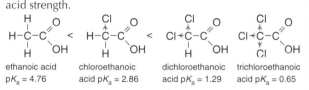

| ethanoic acid $pK_a = 4.76$ | chloroethanoic acid $pK_a = 2.86$ | dichloroethanoic acid $pK_a = 1.29$ | trichloroethanoic acid $pK_a = 0.65$ |

Groups with a positive inductive effect such as alkyl groups will decrease the ability of the charge to delocalize and the acid will be weaker.

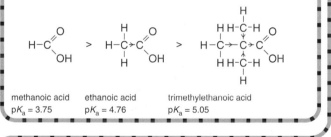

| methanoic acid $pK_a = 3.75$ | ethanoic acid $pK_a = 4.76$ | trimethylethanoic acid $pK_a = 5.05$ |

RELATIVE BASICITIES OF AMMONIA, AMINES AND AMIDES

Ammonia is a weak base with a pK_b of 4.75 and forms salts with strong acids.

$$NH_3 + H_2O \rightleftharpoons NH_4^+ + OH^-$$

$$NH_3 + HCl \rightarrow NH_4^+Cl^-$$

Amines are more basic than ammonia because the positive inductive effect of the alkyl group 'pushes' electrons towards the nitrogen atom increasing the electron density of the non-bonding pair of electrons. Thus aminoethane (ethylamine) is more basic than aminomethane (methylamine) as the positive

| ammonia $pK_b = 4.75$ | methylamine $pK_b = 3.36$ | ethylamine $pK_b = 3.27$ | diethylamine $pK_b = 3.08$ |

ACIDIC PROPERTIES OF PHENOLS

Phenol is weakly acidic because the negative charge in the conjugate phenoxide ion can be delocalized over the benzene ring. This makes it less likely to attract a proton than the ethoxide ion or hydroxide ion where delocalization cannot occur.

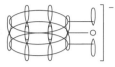

Phenol has a pK_a value of 10.00 making it a weaker acid than carboxylic acids. It is too weak to react with carbonates to give carbon dioxide but it will react with sodium to give hydrogen and form salts with sodium hydroxide.

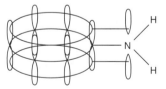

Substituted phenols containing electron withdrawing groups are much more acidic as the negative charge can be further delocalized. For example, 2-nitrophenol has a pK_a value of 7.21. In 2,4,6-trinitrophenol the effect is so great that the substituted phenol is nearly a strong acid with a pK_a value of 0.42.

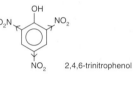

2,4,6-trinitrophenol

inductive effect of an ethyl group is greater than a methyl group. Secondary amines are more basic still.

However, aminobenzene (phenylamine), although still basic with a pK_b of 9.48, is considerably less basic than ammonia. The non-bonding pair of electrons on the nitrogen atom is delocalized with the π electrons in the ring and so is less available to donate to a proton.

All amines can be obtained from their salts by reacting with sodium hydroxide.

$$RNH_3^+ + OH^- \rightarrow RNH_2 + H_2O$$

Amides are non-basic. For example, ethanamide, CH_3CONH_2 with a pK_b value of 15.1 does not form salts with strong acids. This is because the non-bonding pair of electrons on the nitrogen atom delocalizes with the carbonyl group and cannot donate to a proton.

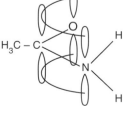

HL Summary of reaction mechanisms

The previous pages have detailed the seven different organic reaction mechanisms required for this Option. To avoid confusion it is helpful if you have an overview of all seven mechanisms and learn at least one example for each type.

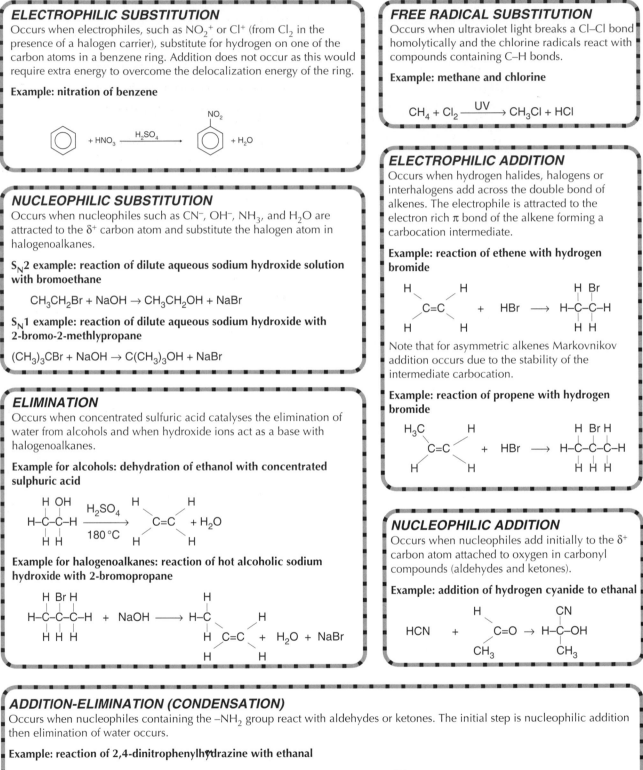

ELECTROPHILIC SUBSTITUTION

Occurs when electrophiles, such as NO_2^+ or Cl^+ (from Cl_2 in the presence of a halogen carrier), substitute for hydrogen on one of the carbon atoms in a benzene ring. Addition does not occur as this would require extra energy to overcome the delocalization energy of the ring.

Example: nitration of benzene

NUCLEOPHILIC SUBSTITUTION

Occurs when nucleophiles such as CN^-, OH^-, NH_3, and H_2O are attracted to the δ^+ carbon atom and substitute the halogen atom in halogenoalkanes.

S_N2 **example: reaction of dilute aqueous sodium hydroxide solution with bromoethane**

$$CH_3CH_2Br + NaOH \rightarrow CH_3CH_2OH + NaBr$$

S_N1 **example: reaction of dilute aqueous sodium hydroxide with 2-bromo-2-methlypropane**

$$(CH_3)_3CBr + NaOH \rightarrow C(CH_3)_3OH + NaBr$$

ELIMINATION

Occurs when concentrated sulfuric acid catalyses the elimination of water from alcohols and when hydroxide ions act as a base with halogenoalkanes.

Example for alcohols: dehydration of ethanol with concentrated sulphuric acid

Example for halogenoalkanes: reaction of hot alcoholic sodium hydroxide with 2-bromopropane

FREE RADICAL SUBSTITUTION

Occurs when ultraviolet light breaks a Cl–Cl bond homolytically and the chlorine radicals react with compounds containing C–H bonds.

Example: methane and chlorine

$$CH_4 + Cl_2 \xrightarrow{UV} CH_3Cl + HCl$$

ELECTROPHILIC ADDITION

Occurs when hydrogen halides, halogens or interhalogens add across the double bond of alkenes. The electrophile is attracted to the electron rich π bond of the alkene forming a carbocation intermediate.

Example: reaction of ethene with hydrogen bromide

Note that for asymmetric alkenes Markovnikov addition occurs due to the stability of the intermediate carbocation.

Example: reaction of propene with hydrogen bromide

NUCLEOPHILIC ADDITION

Occurs when nucleophiles add initially to the δ^+ carbon atom attached to oxygen in carbonyl compounds (aldehydes and ketones).

Example: addition of hydrogen cyanide to ethanal

ADDITION-ELIMINATION (CONDENSATION)

Occurs when nucleophiles containing the –NH_2 group react with aldehydes or ketones. The initial step is nucleophilic addition then elimination of water occurs.

Example: reaction of 2,4-dinitrophenylhydrazine with ethanal

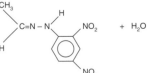

1. Benzene can be converted into phenylethene (styrene) by the following series of reactions:

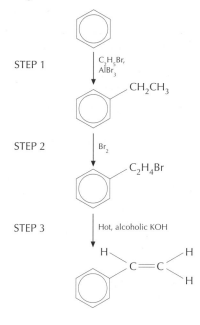

(a) Step 1: (i) Name the mechanism for this reaction. [1]

(ii) Give mechanistic equations for the reaction clearly indicating the role of the aluminium bromide. [4]

(b) Step 2: (i) What reaction condition is necessary for Step 2? [1]

(ii) State the name of the reaction mechanism and outline the steps involved. [4]

(c) Step 3: (i) Name and explain the mechanism of the reaction described in Step 3. [3]

(ii) In Step 3 a different reaction takes place if a warm dilute aqueous solution of potassium hydroxide is used instead of a hot, alcoholic one. Give the structural formula of the major organic product formed under these conditions, and state the name of the mechanism. [2]

2. When hydrogen bromide is added to 2-methylpropene two possible isomeric products could be formed.

(a) Give the structural formulas and the names of the two possible products. [2]

(b) State Markovnikov's rule for the addition of hydrogen halides to asymmetric alkenes. [1]

(c) According to the rule which isomer will be the major product in the above reaction? [1]

(d) Outline the mechanism for the reaction. [2]

(e) Explain clearly **why** only one of the possible products is formed. [3]

(f) Give the structural formula of the major product formed when iodine chloride, ICl, is added to 2-methylpropene. [1]

3. Compound A, an organic acid, has the molecular formula $C_5H_8O_2$. This compound can exist as two geometric isomers, *cis*-A and *trans*-A, each containing two methyl groups.

(a) Give the structural formulas of the two geometric isomers. [2]

(b) When hydrogen gas is added to the *cis*-isomer, a racemic mixture of a carboxylic acid B is obtained. The same racemic mixture is also obtained when hydrogen gas is added to the *trans*-isomer. Write the equation for the hydrogenation of *cis*-A or *trans*-A isomer to form B. [1]

(c) Name B. [1]

(d) Explain what is meant by a racemic mixture. [1]

(e) Under certain conditions, the racemic mixture could be separated into two optically active enantiomers, C and D. Give the structural formulas of the enantiomers C and D. [2]

(f) Describe the similarities and the difference in the physical properties of enantiomers. [2]

Study methods

This book has been written to provide you with all the information you need to gain the highest grade in Chemistry whether at SL or at HL. It is not intended as a 'teach yourself' book and is not a substitute for a good teacher nor for the practical work to support the theory. There is no magic solution which will compensate for a lack of knowledge or understanding but there are some pieces of advice that should ensure that you achieve to the best of your ability.

DURING THE COURSE

The IB course for both SL and HL is scheduled to last for two years, although some schools do attempt to cover the whole course in one year. There is a tendency for some students to take it easy in the first year as the final exams seem a long way off. Don't be tempted to do this as it will be hard to catch up later. Equally do not try to simply learn all the information given about each topic. The exam does not particularly test recall, more how to apply your knowledge in different situations. Although there are some facts that must be learned, much of Chemistry is logical and knowledge about the subject tends to come much more from understanding than from 'rote learning'. During each lesson concentrate on trying to understand the content. A good teacher will encourage you to do this by challenging you to think. At the end of the lesson or in the evening go over your notes, add to them or rearrange them to ensure you have fully understood everything. Read what this book has to say on the subject and read around the topic in other books to increase your understanding. If there are parts you do not understand ask your teacher to explain them again. You can also benefit much by talking and working through problems with other students. You will only really know if you understand something if you have to explain it to someone else. You can test your understanding by attempting the problems at the end of each topic in this book.

Some of the early parts of the course involve basic calculations. Some students do find these hard initially. Persevere and see if you can identify exactly what the difficulty is and seek help. Most students find that as the course progresses and more examples are covered their confidence to handle numerical problems increases considerably. If you ensure that you do understand everything during the course then you will find that by the time it comes to the exam, learning the essential facts to support your understanding is much easier.

MATHEMATICAL SKILLS

One big advantage of the IB is that all students study maths so the mathematical skills required for Chemistry should not present a problem. Essentially they concern numeracy rather than complex mathematical techniques. Make sure that you are confident in the following areas.

- Perform basic functions: addition, subtraction, multiplication and division.
- Carry out calculations involving means, decimals, fractions, percentages, ratios, approximations and reciprocals.
- Use standard notation (e.g. 1.8×10^5).
- Use direct and indirect proportion.
- Solve simple algebraic equations.
- Plot graphs (with suitable scales and axes) and sketch graphs.
- Interpret graphs, including the significance of gradients, changes in gradient, intercepts, and areas.
- Interpret data presented in various forms (e.g. bar charts, histograms, pie charts etc.).

USING YOUR CALCULATOR

Most calculators are capable of performing functions far beyond the demands of the course. When simple numbers are involved try to solve problems without using your calculator (you will need to do this for real in Paper 1). Even when the numbers are more complex try to estimate approximately what the answer will be before using the calculator. This should help to ensure that you do not accept and use a wrong answer because you failed to realize that you pushed the wrong buttons. Don't just give the 'calculator answer' but record the answer to the correct number of significant figures.

For HL students and SL students studying Option A make sure you know how to use your calculator to work out problems involving logarithms for pH and pK_a calculations. The examples given below are for a TI-83.

To convert a hydrogen ion concentration of 1.8×10^{-5} mol dm^{-3} into pH.

$pH = -\log_{10} 1.8 \times 10^{-5}$

To obtain the value press the following keys in sequence.

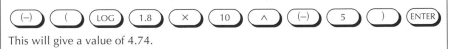

This will give a value of 4.74.

To convert a pK_a value of 3.75 into a K_a value

Press

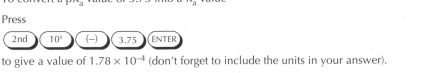

to give a value of 1.78×10^{-4} (don't forget to include the units in your answer).

The final examinations

PREPARING FOR THE EXAMINATIONS

Hopefully for much of the course the emphasis has been on enjoying learning and understanding Chemistry rather than always worrying about grades. Towards the end of the course, however, it does make sense to prepare yourself for the final exam. Examiners are human and mark positively (i.e. they look to give credit rather than penalize mistakes). You have to help them by being clear in your answers and addressing the particular question(s) asked.

- **Know what it is you have to know** Ask your teacher for a copy of the current programme for the core and the two options you are taking. Higher Level students should also have a copy of the Additional Higher Level material. Go through the programme carefully and make sure you recognize and have covered all the points listed for each topic and sub-topic.
- **Understand what depth you have to cover** Each sub-topic on the programme has an objective next to it. Objective 1 is the lowest and implies you just have to define or state the information. Objective 2 means you have to apply your knowledge of the topic in a straightforward situation. Objective 3 is the highest level and means that you will have to recognize the problem and select the appropriate method to solve it.
- **Be familiar with key action verbs** Each statement in the programme and each question in the exam will contain a key action verb. A list of all these verbs and their precise meaning can be found in the programme. If a question asks you to *describe* a reaction then a very different answer is required than if the question had asked you to *explain* a reaction. Examiners can only award marks for the correct answers to the question asked for. Not paying careful attention to the correct action verb may cost you marks unnecessarily.
- **Practice with past papers** Most schools will give their students a 'mock' or 'trial' exam. This is helpful as it enables you to judge the correct amount of time to spend on each question. Make your mistakes in the mock exam and learn from them. Of course the IB questions are different each year but they do tend to follow a similar pattern. It helps to have seen similar questions before and know what level to expect.
- **Organize your notes** As you review your work it is often helpful to rewrite your notes. Concentrate on just the key points – they should trigger your memory. This book already contains the important points in a fairly condensed form. Condense them even more to make your own set of review notes. Each time you review each topic try to condense the notes even more. By the time you are ready to take the exam all you personal review notes should ideally fit onto a single page!
- **Be familiar with using the IB Data Booklet** You should get into the habit of using this throughout the course so that you are completely familiar with its contents and how to use them by the time of the exams.
- **Know the format of the exam papers** Both HL and SL students take three exam papers. Papers 1 and 2 examine the core (and additional HL) material. Paper 3 will be taken on the next day (or examination session if a weekend intervenes) and covers the options. Paper 1 is multiple choice and you are not allowed a calculator or the Data Booklet. A Periodic Table is provided. Paper 2 contains short answer questions in Section A that you must attempt and longer questions in Section B. SL students must choose one of the longer questions from a choice of three and HL students must choose two longer questions from a choice of four. Paper 3 contains questions on all the options. You are required to answer all the questions on two of the options.
- **Know the dates of the exams** Plan your review timetable carefully in advance. Remember that you will have exams in other subjects and that you may not have much time for a 'last minute' review.

TAKING THE EXAMINATIONS

- Try to ensure that the night before you are able to take some time to relax and get a good night's sleep.
- Take all you need with you to the examination room, i.e. pens, pencils, ruler, and a simple translating dictionary if English is not your first language. You will need your calculator for Paper 2 and Paper 3 – remember to include a spare battery.
- There is no reading time allowed for Paper 1. Work through the questions methodically. HL have 40 questions in 1 hour, SL have 30 questions in 45 minutes. If you get stuck on a question move on and then come back to it if you have time at the end. Make a note of those questions you are unsure about. You can then come back to these at the end rather than going through all of them again. Make sure you give one answer for each question. You are not penalized for wrong answers so if you run out of time make an educated guess rather than leave any questions unanswered.
- Use the five minutes reading time for Paper 2 wisely. You have to answer the questions on Section A so use this time to read through the questions on Section B thoroughly and decide which you will choose to answer.
- Read each question very carefully. Make a mental note of the key action verb so that you give the required answer.
- Try to write your answers within the required space. Write as legibly as you can. For questions involving calculations do not round up too early but make sure your final answer is given to the correct number of significant figures. Always include the correct units. If you do need extra space continue in a separate answer booklet.
- Attempt all the required number of questions. If you do not attempt a question you can receive no marks. For sequential numerical questions even if you get the first part wrong continue as you will not be marked wrong twice for the same mistake. For this reason it is essential that you show your working. Do not answer more questions than required. The examiner will simply mark the required number in the order they are written, not necessarily the best ones.
- Leave yourself time to read through what you have written to correct any mistakes.
- Ensure that you have filled in the front of the paper correctly, including stating the number of the optional questions answered, before leaving the examination room.

Internal assessment (1)

INTRODUCTION

You are expected to spend 40 hours (SL) or 60 hours (HL) during the two years engaged in work that can be assessed internally. This is essentially time spent in the laboratory and comprises 24% of the final marks. Chemistry is an experimental science and practical work is an important component of the course. Your teacher should devise a suitable practical programme for you to follow. Practical work can have many different aims; for example, to improve your skills at different techniques, to reinforce the theoretical part of the course and to give you experience of planning your own investigations. Hopefully it will make studying Chemistry much more challenging and rewarding. Through the practical course you are expected to understand and implement safe practice and also to respect the environment.

INTERNAL ASSESSMENT – THE FACTS

Your practical work will be assessed continually throughout the two years of the course. The assessment is exactly the same for both SL and HL. It is assessed according to eight different criteria. Each criterion carries a maximum of three marks to give a total maximum mark of 24. Some work may not be formally assessed at all and some may only be assessed for one or two of the criteria. However, your teacher must assess each of the eight criteria at least twice during the two years. Most teachers will assess more than this. Your final mark will be the average of the two assessments chosen by your teacher to represent each criterion and should reflect the standard you have reached by the end of the course. Each criterion is broken down into two or three different aspects. Your teacher assesses whether you have covered each aspect *completely, partially, or not at all* to arrive at the mark given for each criterion. Each school will send the work of a few students for moderation so that everyone is graded uniformly.

GRADING OF INTERNAL ASSESSMENT

You will only gain good marks if you try to address each aspect of each criterion completely. It is important that you understand fully what is required for each of the different aspects

CRITERION	ASPECTS		
Planning (a)	Defining the problem or research question	Formulating an hypothesis or prediction	Selecting the key variables
Planning (b)	Designing a method with appropriate apparatus/materials	Designing a method for the control of variables	Designing a method for the collection of data
Data collection	Observing and collecting raw data		Presenting raw data clearly
Data processing and presentation	Transforming and manipulating (processing) raw data		Presenting processed data
Conclusion and evaluation	Evaluating (interpreting) results Drawing conclusions	Evaluating the procedure	Modifying the procedure
Manipulative skills	Carrying out a range of techniques proficiently with due regard to safety		Following a variety of instructions
Personal skills (a)	Working within a team	Recognizing the contributions of others	Encouraging the contributions of others
Personal skills (b)	Approaching scientific investigations with self-motivation and perseverance	Approaching scientific investigations in an ethical manner	Approaching scientific investigations while paying due attention to the environmental impact

THE WAY IN WHICH MARKS FOR EACH CRITERION ARE AWARDED

The teacher will mark each aspect and determine whether, in their opinion, you have covered the aspect completely, partially, or not at all, then use the appropriate grid to arrive at the mark for each particular criterion marked.

(a) For those criteria with three aspects

	3	2	2	2	1	1	1	1	0	0
Completely	✔✔✔	✔✔	✔✔	✔		✔		✔		
Partially		✔		✔✔	✔✔✔	✔	✔✔		✔	
Not at all			✔			✔	✔	✔✔	✔✔	✔✔✔

(b) For those criteria with two aspects

	3	2	1	1	0	0
Completely	✔✔	✔	✔			
Partially		✔		✔✔	✔	
Not at all			✔		✔	✔✔

Internal assessment (2)

HOW TO MAXIMIZE YOUR INTERNAL ASSESSMENT MARKS

General points

- Ascertain before you undertake the investigation which criteria (if any) are being assessed.
- Check that you are clear about all the aspects to be assessed.
- Record all your work as you proceed in your log book or laboratory notebook.
- Record the title of the experiment (or piece of work), the date and the name(s) of any partner(s) you worked with.
- Record precise details of all equipment used, e.g. a balance weighing to + or – 0.001 g, a thermometer measuring from –10 to +110 °C to an accuracy of + or –0.1°C, a 25.00 cm^3 pipette measuring to + or –0.04 cm^3, etc.
- Record precise details of any chemicals used, e.g. copper(II) sulfate pentahydrate $CuSO_4.5H_2O(s)$.
- Record all measurements accurately to the correct number of significant figures and include all units.
- Record all observations. Include colour changes, solubility changes, whether heat was evolved or taken in, etc.
- Draw up a checklist to cover each criterion being assessed. As you write the laboratory account check that each aspect is addressed fully. (Some students give each aspect a sub-heading. Although this is not strictly necessary it does help to draw the aspect to the attention of the teacher).
- Your work may be hand-written (in ink) or word-processed. Ensure that it is neat, correct, and legible.
- Write clearly and succinctly.
- Hand your work in on time. Teachers are within their rights to refuse to mark work handed in late as you may benefit from using other students' marked assignments.
- Learn from your mistakes. In the early part of the course do not expect to get everything correct the first time you do it. Find out why you lost marks and improve your next presentation.
- Keep all your laboratory reports. At the end of the course some of them may need to be sent off for moderation.

Specific points for each criterion

Planning (a)

- Identify the research question/problem and state it clearly.
- State the hypothesis carefully. Do not just guess but try to arrive at the hypothesis logically and, if appropriate, with a calculation.
- Identify all the variables and state those over which you have control.

Planning (b)

- Give accurate and concise details about the apparatus and materials used.
- When defining and describing the method include the rationale behind the method and include other methods considered and the reasons why they were discarded.

Data collection

- Ensure all data is recorded. Pay particular attention to significant figures and make sure all units are stated.
- Record the level of uncertainty for each quantitative reading.
- Present your results clearly. Often it is better to use a table or a graph. If using a graph make sure the graph has a title and both axes are labelled clearly and that the correct scale is chosen to utilize most of the graph space.
- When carrying out an acid–base titration ensure that the indicator is clearly stated and the change in colour recorded to signify the end-point.

Data processing and presentation

- Ensure that you have used your data correctly to produce the required result.
- In quantitative experiments ensure that the limits of accuracy of each piece of apparatus have been stated and then summed to give the limits of accuracy with which you can state your result. Calculate it first in percentage terms then transform it into the + and – amount pertaining to your actual result.
- Include any other errors or uncertainties which may affect the validity of your result.

Conclusion and evaluation

- Include a valid conclusion. This should relate to the initial problem or hypothesis.
- Compare your result to the expected (Literature or Data Book) result.
- Calculate the percentage error from the expected value.
- Evaluate your method. State any assumptions that were made which may affect the result. Comment on the limitations of the method chosen.
- Suggest how the method chosen could be improved to obtain more accurate and reliable results.

Manipulative skills, personal skills (a), and personal skills (b)

These three criteria cannot be moderated from the written work submitted. They are assessed by your teacher on your performance when you are actually working in the laboratory. To gain high marks ensure:

- You show a high regard for safety in the laboratory.
- You show proficiency in a wide range of chemical techniques.
- You are enthusiastic in your approach.
- You can follow instructions and show initiative where necessary.
- You work well in a team. Listen to the views of others and respect their input. Encourage all members of the team to make suggestions.
- You show respect for the environment in the way you conduct your experiments and dispose of any residues.

Internal assessment (3)

ERROR AND UNCERTAINTY

The **error** is the difference between the result obtained and the generally accepted 'correct' result found in the data book or other literature. If the 'correct' result is available it should be recorded and the percentage error calculated and commented upon in your conclusion. Without the 'correct' value no useful comment on the error can be made.

Uncertainty occurs due to the limitations of the apparatus itself and the taking of readings from scientific apparatus. For example, during a titration there are generally four separate pieces of apparatus, each of which contributes to the uncertainty.

e.g. when using a balance that weighs to ± 0.001 g the uncertainty in weighing 2.500 g will equal

$$\frac{0.001}{2.500} \times 100 = 0.04\%$$

Similarly a pipette measures 25.00 $cm^3 \pm 0.04$ cm^3.

The uncertainty due to the pipette is thus

$$\frac{0.04}{25.00} \times 100 = 0.16\%$$

Assuming the uncertainty due to the burette and the volumetric flask is 0.50% and 0.10% respectively the overall uncertainty is obtained by summing all the individual uncertainties:

Overall uncertainty = 0.04 + 0.16 + 0.50 + 0.10 = 0.80% ≈ 1.0%.

Hence if the answer is 1.87 mol dm^{-3} the uncertainty is 1.0% or 0.0187 mol dm^{-3}.

The answer should be given as 1.87 ± 0.02 mol dm^{-3}.

Sometimes it is not possible to give precise percentage uncertainties. For example, in a titration the end-point taken could vary according to the person carrying out the titration. In such cases you should state the colour change taken (e.g. until a faint permanent pink colour was obtained). Any assumptions made which can add to the uncertainty (e.g. the specific heat capacity of the solution was taken to be the same as that for pure water) should be stated in the evaluation.

SIGNIFICANT FIGURES

Whenever a measurement of a physical quantity is taken there will be uncertainty in the reading. The measurement quoted should include the first figure that is uncertain. This should include zero if necessary. Thus a reading of 25.30 °C indicates that the temperature was taken with a thermometer that is accurate to + or –0.01 °C. If a thermometer accurate to only + or –0.1 °C was used the temperature should be recorded as 25.3 °C.

Zero can cause problems when determining the number of significant figures. Essentially zero only becomes significant when it comes *after* a non-zero digit (1, 2, 3, 4, 5, 6, 7, 8, 9).

000 123.4	0.000 1234	1.0234	1.2340
zero not a significant figure		zero is a significant figure	
values quoted to 4 sig. figs		values quoted to 5 sig figs	

Zeros after a non-zero digit but before the decimal point may or may not be significant depending on how the measurement was made. For example 123 000 might mean exactly one hundred and twenty three thousand or one hundred and twenty three thousand to the nearest thousand. This problem can be neatly overcome by using scientific notation.

$1.230\,00 \times 10^6$ quoted to six significant figures
1.23×10^6 quoted to three significant figures

Calculations

1. When adding or subtracting it is the number of decimal places that is important.

e.g.   7.10 g   +   3.10 g   =   10.20 g
     3 sig. figs   3 sig. figs   4 sig. figs

This answer can be quoted to four significant figures since the balance used in both cases was accurate to + or –0.01 g.

2. When multiplying or dividing it is the number of significant figures that is important. The value with the least number of significant figures used in the calculation determines how many significant figures should be used when quoting the answer.

e.g. When the temperature of 0.125 kg of water is increased by 7.2 °C the heat required =
0.125 kg × 7.2 °C × 4.18 kJ kg^{-1} °C = 3.762 kJ

Since the temperature was only recorded to two significant figures the answer should strictly be given as 3.8 kJ.

In practice the IB does not tend to penalize in exams if the number of significant figures in an answer differs by one from the correct number (unless the questions specially asks for them) but will penalize if they are grossly wrong.

Extended Essays (1)

WHAT IS AN EXTENDED ESSAY?

In order to fulfil the requirements of the IB all Diploma candidates must submit an Extended Essay in an IB subject of their own choice. The Essay is an in-depth study of a limited topic within a subject. The purpose of the Essay is to provide you with an opportunity to engage in independent research. Approximately forty hours should be spent in total on the Essay. Each Essay must be supervised by a competent teacher. The length of the Essay is restricted to a maximum of 4000 words and it is assessed according to a carefully worded set of criteria. The marks awarded for the Extended Essay are combined with the marks for the Theory of Knowledge course to give a maximum of three bonus points.

EXTENDED ESSAYS IN CHEMISTRY

Although technically any IB Diploma student can choose to write their Essay in Chemistry it does help if you are actually studying Chemistry as one of your six subjects! Most Essays are from students taking Chemistry at Higher Level but there have been some excellent Essays submitted by Standard Level students. All Essays must have a sharply focused Research Question. Essays may be just library-based or also involve individual experimental work. Although it is possible to write a good Essay containing no experimental work it is much harder to show personal input and rarely do such Essays gain high marks. The experimental work is best done in a school laboratory although the word 'laboratory' can be interpreted in the widest sense and includes the local environment. It is usually much easier for you to control, modify, or redesign the simpler equipment found in schools than the more sophisticated (and expensive) equipment found in university or industrial research laboratories.

CHOOSING THE RESEARCH QUESTION

Choosing a suitable Research Question is really the key to the whole Essay. Some supervisors have a list of ready-made topics. The best Essays are usually where you, the student, identify a particular area or chemistry problem that you are interested in and together with the supervisor formulate a precise and sharply focused Research Question. It must be focused. A title such as 'A study of analysis by chromatography' is far too broad to complete in 4000 words. A focused title might be 'An analysis of the red dyes present in different brands of tomato ketchup by thin layer chromatography'. It is more usual to choose a topic then decide which technique(s) might be used to solve the problem. An alternative way is to look at what techniques are available and see what problems they could address.

SOME DIFFERENT TECHNIQUES (TOGETHER WITH A RESEARCH QUESTION EXAMPLE) THAT CAN BE USED FOR CHEMISTRY EXTENDED ESSAYS

The list below shows some examples of how standard techniques or equipment available in a school laboratory can be used to solve Research Questions. Although one example has been provided for each technique many Research Questions will, of course, involve two or more of these techniques.

Redox titration
Do the fossils found in different strata of rocks contain different amounts of sulfur?

Extension of a standard practical
What gas is evolved when zinc is added to $CuSO_4(aq)$ and what factors affect its formation?

Acid–base titration
How do storage time and temperature affect the vitamin C content of fruit juices?

Chromatography
Do all strawberry jellies worldwide contain the same red dye(s)?

Calorimetry
Is it better to use cow dung as a fuel or as a fertilizer?

pH meter
Can different types of chewing gum affect the pH of the mouth and prevent tooth decay?

Steam distillation
Can mosquito repellants be extracted from the Papua New Guinea plant species *Genus ocimum*?

Electrochemistry
Relationship between concentration and ratio of O_2:Cl_2 evolved during the electrolysis of NaCl(aq).

Refinement of a standard practical
How can the yield be increased in the laboratory preparation of 1,3-dinitrobenzene?

Questionnaire
Does the way we are taught chemistry affect our later understanding of the subject?

Theoretical
Why do some substances have much greater cryoscopic constants than others?

Data logging probes
Does the nature of the catalyst affect the rate expression for the decomposition of H_2O_2?

Visible spectrometry
Do different underarm deodorants contain different amounts of aluminium?

Gravimetric analysis
What is the percentage of salt in a typical McDonald's Big Mac Hamburger?

Inorganic reactions
An investigation into the oxidation states of manganese – does Mn(V) exist?

Microscale/small scale
How can the residues from a typical college practical programme be reduced?

Extended Essays (2)

RESEARCHING THE TOPIC

Once the topic is chosen research the background to the topic thoroughly before planning the experimental work. Information can be obtained from a wide variety of sources: a library, the internet, personal contacts, questionnaires, newspapers, etc. Make sure that each time you record some information you make an accurate note of the source as you will need to refer to this in the bibliography. Treat information from the internet with care. If possible try to determine the original source. Articles in journals are more reliable as they have been vetted by experts in the field. Together with your supervisor plan your laboratory investigation carefully. Your supervisor should ensure that your investigation is safe, capable of producing results (even if they are not the expected ones) and lends itself to a full evaluation.

THE LABORATORY INVESTIGATION

Make sure you understand the chemistry that lies behind any practical technique before you begin. Keep a careful record of everything you do at the time that you are doing it. If the technique 'works' then try to expand it to cover new areas of investigation. If it does not 'work' (and most do not the first time) try to analyse what the problem is. Try changing some of the variables, such as increasing the concentration of reactants, changing the temperature or altering the pH. It may be that the equipment itself is faulty or unsuitable. Try to modify it. Use your imagination to design new equipment in order to address your particular problem (modern packaging materials from supermarkets can often be used imaginatively to great effect). Because of the time limitations it is often not possible to get reliable repeatable results but attempt to if you can. Remember that the written Essay is all that the external examiner sees so leave yourself plenty of time to write the Essay.

WRITING THE ESSAY

Before starting to write the Essay make sure you have read and understood the assessment criteria. Your school or supervisor will provide you with a copy. It may be useful to look at some past Extended Essays to see how they were set out. Almost all Essays are word-processed and this makes it easier to alter draft versions but they may be written by hand. You will not be penalized for poor English but you will be penalized for bad chemistry so make sure that you do not make simple word-processing errors when writing formulas, etc.

Start the Essay with a clear introduction and make sure that you set out the Research Question clearly and put it into context. The rest of the Essay should then be very much focused on addressing the Research Question. Some of the marks are gained simply for fulfilling the criteria (e.g. numbering the pages, including a list of contents, etc.). These may be mechanical but you will lose marks if you do not do them. Give precise details of any experimental techniques and set out the results clearly and with the correct units and correct number of significant figures. If you have many similar calculations then show the method clearly for one and set the rest out in tabular form. Numerical results should give the limits of accuracy and a suitable analysis of uncertainties should be included. Relate your results to the Research Question in your discussion and compare them with any expected results and with any secondary sources of information you can locate. State any assumptions you have made and evaluate the experimental method fully. Suggest possible ways in which the research could be extended if more time were available. Throughout the whole Essay show that you understand what it is that you are doing and demonstrate personal input and initiative.

When you have completed the essay write an abstract. This should state clearly the Research Question, the scope of the investigation and the conclusion in less than 300 words.

Give your supervisor at least one draft version for comment before completing the final version. Before handing in the final version go through the following checklist very carefully. If you can honestly tick 'yes' to every box then your Essay will be at least satisfactory and hopefully much better than this.

Extended Essays (3)

EXTENDED ESSAY CHECK LIST

In order to gain the maximum credit possible for your Essay it is crucial that you can answer YES to the following questions *before* you finally submit your Essay to your supervisor.

(Criteria **A–H** apply to all Extended Essays. Criteria **J**, **K**, and **L** apply just to Chemistry Extended Essays.)

A. Research Question

Is the Research Question sharply focused? [ ]

Is the Research Question clearly and precisely stated in the early part of the Essay? [ ]

(N.B. the abstract and the title do **NOT** count as part of the Essay.)

B. Approach to the Research Question

Has all *relevant* information been included and irrelevant information excluded? [ ]

Does the Essay address and develop the Research Question? [ ]

C. Analysis/interpretation

Have the relevant information/data/sources/evidence been analysed appropriately? [ ]

D. Argument/evaluation

Have you developed a convincing argument from the materials/information considered which completely addresses
the Research Question? [ ]

Is your argument well organized and expressed clearly? [ ]

Is your argument or evaluation fully substantiated? [ ]

E. Conclusion

Is your conclusion clearly stated and relevant to the Research Question? [ ]

Is the conclusion consistent with the argument presented in the Essay? [ ]

Does your conclusion indicate unresolved questions and new questions that have emerged from the research? [ ]

F. Abstract

Is the abstract less than 300 words? [ ]

Are the Research Question, the scope of the investigation, and the conclusion all stated clearly? [ ]

G. Formal presentation

Is the Essay less than 4000 words? [ ]

Is there a list of contents which is clearly set out? [ ]

Is illustrative material appropriate, well set out and used effectively? [ ]

Are all the pages numbered? [ ]

Does the bibliography include all, and only, those works which have been consulted? [ ]

Are the references set out in a consistent standard format which specifies: author(s), title, date of publication,
and publisher? [ ]

If there is an appendix, does it contain only information that is required in support of the text? [ ]

H. Holistic judgement

Have you demonstrated to the best of your ability the following qualities in your Essay: Initiative; Personal input;
Inventiveness; Insight; Depth of understanding; Flair? [ ]

J. Principles and ideas used to describe and explain the properties and behaviour of materials

Are you sure that the whole emphasis of your Essay is on Chemistry? [ ]

Have you included all the relevant principles relating to Chemistry? [ ]

Have you explained the relevant principles to show that you understand them fully and have applied them correctly? [ ]

K. Use of methods and sources appropriate to Chemistry

Have you shown that the methods you have used are appropriate to the investigation? [ ]

Have you shown that you have used the methods correctly? [ ]

Have you shown clearly how *you* personally have devised/adapted/modified the methods? [ ]

L. Reasoning surrounding the research and its limitations

Have you clearly analysed the uncertainties in your experimental data? [ ]

Have you stated and accounted for all the assumptions and approximations you have made? [ ]

Have you taken into account any inadequate experimental design(s) or systematic errors? [ ]

Have you verified your experimental results (where possible) with secondary sources? [ ]

Are all your explanations, confirmations, and refutations supported by sound arguments? [ ]

Answers to questions

STOICHIOMETRY
1. C   2. C   3. B   4. A   5. D   6. C   7. D   8. A   9. B
10. A   11. D   12. A

13 (a) 2-hydroxybenzoic acid,   (b) 19.6 g, (c) 69.9%.
14. (a) 4.00×10^{-2} moles,   (b) 4.00×10^{-2} moles,
(c) 362,   (d) $A_r = 133$, M = Cs.

ATOMIC THEORY
1. A   2. D   3. A   4. D   5. B   6. B   7. D   8. A   9. D
10. C   11. B   12. B   13. B   14. C   15. A

PERIODICITY
1. A   2. A   3. D   4. C   5. C   6. C   7. B   8. C   9. D
10. A   11. D   12. D   13. D   14. A   15. A   16. A

BONDING
1. A   2. B   3. A   4. C   5. D   6. A   7. B   8. A   9. D
10. D   11. C   12. A   13. B   14. D   15. B   16. A

STATES OF MATTER
1. A   2. D   3. C   4. D   5. B   6. B   7. C   8. D   9. A
10. A   11. B   12. C   13. D   14. C

ENERGETICS
1. D   2. D   3. A   4. C   5. C   6. B   7. C   8. D   9. A
10. B   11. C   12. C   13. A   14. B

KINETICS
1. B   2. B   3. D   4. D   5. D   6. B   7. D   8. B   9. D
10. D   11. A   12. B   13. A   14. D   15. A   16. C

EQUILIBRIUM
1. A   2. B   3. D   4. A   5. D   6. A   7. B   8. B   9. B
10. C   11. D   12. D   13. D

ACIDS AND BASES
1. A   2. B   3. D   4. C   5. A   6. B   7. D   8. C   9. B
10. B   11. D   12. A   13. C   14. C   15. B   16. D

OXIDATION AND REDUCTION
1. C   2. D   3. A   4. B   5. B   6. A   7. B   8. A   9. D
10. B   11. A   12. C   13. A   14. C

ORGANIC CHEMISTRY
1. D   2. B   3. C   4. C   5. C   6. D   7. A   8. D   9. A
10. C   11. A   12. A   13. B   14. D   15. B   16. D

Option A – Higher physical organic chemistry
1. (a)(i) C_5H_{12} [1],   (ii) $CH_3CH_2CH_2CH_2CH_3$,
$CH_3CH(CH_3)CH_2CH_3$, $CH_3C(CH_3)_3$ [3].   (iii) CH_3^+ [1]. Causes
a mass loss of 15 [1].   (iv) one from:
$CH_2^+/CH_3CH_2^+/CH_3CH_2CH_2^+/CH_3CH_2CH_2CH_2^+$ [1].
(v) A = $CH_3C(CH_3)_3$, 2,2-dimethylpropane [2], B =
$CH_3CH_2CH_2CH_2CH_3$, pentane [2].   (b) A = single peak [1], all
hydrogen atoms in same chemical environment [1], B = 3
different environments [1].   (c) both contain only C–C and
C–H bonds so their infrared spectra will be very similar [1].

2. (a) Rate = $k[H_2][NO]^2$ [3], overall order is three [1].
(b) Increase by a factor of 8 [1].   (c)(i) Rate = $k[O_3][NO]$ [3],
(ii) $k = 2.2 \times 10^7$ dm^3 mol^{-1} s^{-1} [2].
(iii) 7.13×10^{-4} mol dm^{-3} s^{-1} [1].

3. (a) $C_2H_5COOH(aq) + H_2O(l) \rightleftharpoons C_2H_5COO^-(aq) +$
$H_3O^+(aq)$ [1].   (b) $K_a = [C_2H_5COO^-][H^+]/[C_2H_5COOH]$ [1].
(c) pH = 2.78 [2], the equilibrium concentration of the acid is
still 0.200 mol dm^{-3} or $[C_2H_5COOH] - [H^+] =$
$[C_2H_5COOH]$ [1].   (d) 0.961 g [2].   (e) The solution is a
buffer [1]. $H^+ + C_2H_5COO^- \rightleftharpoons C_2H_5COOH$; $OH^- +$
$C_2H_5COOH \rightleftharpoons C_2H_5COO^- + H_2O$ [1]

4. (a) $C(CH_3)_3Cl + OH^- \rightarrow C(CH_3)_3OH + Cl^-$ [1].   (b) Rate =

$k[C(CH_3)_3Cl]$ [1].   (c) S_N1 [1], $C(CH_3)_3Cl \rightarrow C(CH_3)_3^+ + Cl^-$
[1], $C(CH_3)_3^+ + OH^- \rightarrow C(CH_3)_3OH$ [1].   (d)(i) Decrease [1],
1-chlorobutane is a primary halogenoalkane so the mechanism
is S_N2 which is slower [1].   (ii) Increase [1], the C–I bond is
weaker than the C–Cl bond so breaks more readily [1].

Option B – Medicines and drugs
1. (a) A nucleic acid containing genetic materials within a
protein coating [1].   (b) Any two from: they replicate quickly
giving a high viral population/they mutate/they work by
entering a host cell and any treatment is likely to also affect the
host cell [2].   (c)   [1]

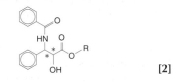

(d) Any two from: the nitrogen atoms become protonated by
the acid as they are bases to form a soluble salt/the –OH group
becomes protonated to form a soluble salt/hydrogen bonding
occurs between the Acyclovir and water [2].   (e) Some drugs
are decomposed by stomach acid. Water-soluble drugs are not
easily absorbed through the fatty tissue of the stomach wall [2].

2. (a) Increases resistance to the penicillinase enzyme **or** alters
the effectiveness of the 'penicillin' [1].   (b) Some bacteria may
remain unaffected [1].   (c) Bacteria can become resistant to
penicillins [1].   (d) Antibiotics that are effective against a wide
variety of bacteria [1].   (e) Penicillins are effective against
some, but not all, bacteria [1]. Penicillins work by interfering
with the chemicals needed by bacteria to form normal cell
walls [2].

3. (a) $C_7H_6O_3$ [1].   (b) Either circle ester group, pH = 7 **or**
circle carboxylic acid group pH = 2 – 6 [3].
(c) H_2N⟨ ⟩OH   [2].

(d) Aspirin: stomach bleeding/Reyes syndrome/allergic
reactions [1] paracetamol: overdose can permanently damage
liver and kidneys [1].

4. (a)(i) H_3N Cl   (ii) The non-bonding pairs of electrons
        Pt   on the N atoms can form co-
     Cl NH_3 [1]   ordinate bonds with the
                        platinum [1], Lewis acid/base [1]

(b) The different enantiomers can have different biological
properties [1], one form may be beneficial and the other form
harmful [1].   (c) A chiral auxilliary is itself an enantiomer [1].
It is bonded to the reacting molecule to create the
stereochemical conditions necessary to follow a certain
geometric path [1]. Once the desired enantiomer is formed the
auxilliary is removed [1].
(d)

[2]

Option C – Human biochemistry
1. (a) ketone (carbonyl) [1], alkene [1], testosterone would be
expected to decolourize bromine water (or will give a
crystalline solid with 2,4-DNP) [1].   (b) alkyne [1], alcohol
[1], amine [1] phenyl [1]. Norethynodrel 'fools' the female
reproductive system by mimicking the action of progesterone in
true pregnancy [1], prevent the release of an egg [1], inhibits
sperm from reaching the egg/makes the uterus unreceptive to
any fertilized egg [1].

2. (a) Fats (or lipids) **[1]**. **(b)** Vegetable **[1]**, all three fatty acid residues are unsaturated **[1]**. **(c)** Oil **[1]**, the structure around the double bonds prevents close packing **[1]**. **(d)** Any two from: energy source/insulation/cell membrane **[2]**.

3. (a)(i) alkene **[1]**, alcohol **[1]**. **(ii)** Retinol is fat soluble **[1]**, as it has a long non-polar hydrocarbon chain **[1]**, vitamin C is water soluble as it contains many –OH groups which can hydrogen bond with water molecules **[1]**. **(iii)** Vitamin A: night blindness/xerophthalmia **[1]**, vitamin C: bleeding of gums etc. **[1]**, scurvy **[1]**. **(b)** 0.0559 mol **[1]**, each mole of oil contains four C=C double bonds **[2]**.

4. (a) Form coloured compounds **[1]**, form complex ions **[1]**.
(b)(i) 55% - 25% = 30% **[1]**. **(ii)** Carbon dioxide and lactic acid are acidic so lower the pH **[1]**, haemoglobin is less able to carry oxygen at lower pH **[1]**.

Option D – Environmental chemistry
1. (a) 1000:1 **[1]**. **(b)** Sulfuric acid (or sulfurous acid)**[1]**, $S + O_2 \rightarrow SO_2$ (or $2SO_2 + O_2 \rightarrow SO_3$) **[1]**, $SO_2 + H_2O \rightarrow H_2SO_3$ (or $SO_3 + H_2O \rightarrow H_2SO_4$) **[1]**. **(c)** Nitric acid **[1]**, any two from catalytic converter/richer petrol:air mixture/thermal exhaust system/switching to a different fuel e.g. solar power **[2]**. **(d)** $CaCO_3 + 2H^+ \rightarrow Ca^{2+} + CO_2 + H_2O$ **[1]**.

2. (a)(i) As the temperature increases the solubility of oxygen decreases **[1]**, **(ii)** As organic pollutants decompose they use up available oxygen so the amount of dissolved oxygen decreases **[1]**, **(iii)** Nitrates and phosphates act as nutrients and increase the growth of algae. As the algae die they use up dissolved oxygen (eutrophication) **[1]**. **(b)** The quantity of oxygen (in ppm) utilised when the organic matter in a fixed volume of water is decomposed biologically over a set time period (usually five days) **[1]**. **(c)** 8.00×10^{-3} g dm^{-3} (or 8 mg dm^{-3})**[3]**.

3. (a) Carbon dioxide **[1]**, combustion of fossil fuels **[1]**, large scale combustion is causing an increase in atmospheric CO_2 levels **[1]**. **(b)** The molecules absorb infrared radiation emitted from the Earth **[1]**, this absorption prevents the heat from escaping into outer space and causes global warming **[1]**. **(c)** Particulates **[1]**, counteracts the greenhouse effect by reflecting sunlight **[1]**.

4. (a) 2.41×10^{-7} m (241 nm) **[3]**. **(b)** From the Lewis structures **[1]** the bond is weaker in ozone as it is equivalent to 1.5 bonds rather than the double bond found in oxygen so requires less energy/longer wavelength to break **[1]**.
(c)(i) $CCl_2F_2 \rightarrow CClF_2^\bullet + Cl^\bullet$ (UV light causes radical formation **[1]**, $Cl^\bullet + O_3 \rightarrow ClO^\bullet + O_2$ (chlorine radicals react with ozone)**[1]**, $ClO^\bullet + O^\bullet \rightarrow Cl^\bullet + O_2$ (chlorine radicals regenerated)**[1]**. **(ii)** During the arctic winter small amounts of water vapour freeze into ice crystals **[1]**, the surface of the ice crystals acts as a catalyst to produce species such as Cl_2, when the winter is over the Sun converts this into chlorine radicals **[1]**.

5. (a)(i) Lethal dose (amount needed to kill) **[1]** 50% of the population **[1]**. **(ii)** Advantage: gives good indication of toxicity **[1]**, disadvantage: does not help to give safe level for humans (or ethical considerations) **[1]**. **(b)** Pb: source; paints/gasoline (petrol)/lead pipes in plumbing **[1]**, effect: brain damage **[1]**, reduction: unleaded gasoline (petrol)/ lead-free paints/copper or plastic pipes **[1]**. Nitrates: source: leaching of fertilizers into rivers/intensive farming **[1]**, effect: cancer/'blue baby' syndrome/affects haemoglobin in young **[1]**, reduction: use less fertilizer, avoid use before rain is due/ tertiary treatment of sewage e.g. algal ponds **[1]**.

Option E – Chemical industries
1. (a)(i) Cryolite lowers the melting point of the alumina **[1]**.

(ii) $2O^{2-} \rightarrow O_2 + 4e$ **[1]**. **(iii)** The oxygen produced reacts with the carbon electrode $C + O_2 \rightarrow CO_2$ **[1]**. **(b)** Alumina acts as an acidic oxide and reacts with NaOH **[1]**, the basic impurities do not react with the NaOH **[1]**. **(c)** Low density/aviation industry; resistance to corrosion/window frames, electrical conductor/transmission lines etc. **[2]**.
(d) Save energy, less damage to the environment **[2]**.

2. (a) Crude oil was formed from marine organisms which contained S (in their amino acids) **[1]**. **(b)** S can poison catalysts used in refining (or when oil is burned the SO_2 produced leads to acid rain) **[1]**. **(c)** $C_{10}H_{22} \rightarrow C_8H_{18} + C_2H_4$ **[1]**, larger molecules are broken down into smaller, more useful molecules **[1]**.
(d) Isomerization involves the rearrangement to form another isomer **[1]**, to raise the octane rating of a fuel **[1]**.
(e)(i) $C_6H_{14} \rightarrow C_6H_6 + 4H_2$ **[1]**. **(ii)** Haber process **[1]**.

3. (a) Exothermic **[1]**, K_c decreases with temperature so heat must be evolved as a product **[1]**. **(b)** Yield too low at higher temperature **[1]**, rate too slow at lower temperature **[1]**, catalyst increases the rate at which equilibrium is reached by providing an alternative pathway with a lower activation energy **[1]**.

4. (a)

$[3]$

(b) Moderate pressure, 60 °C, Ziegler catalyst (e.g. TiCl$_4$) **[2]**, high density polythene proceeds by an ionic mechanism whereas low density polythene involves a radical mechanism **[1]**.

5. (a) Mercury(II) oxide can be decomposed to mercury and oxygen by heating above approximately 750 K **[1]**, the value of ΔG for the reaction $2Hg + O_2 \rightarrow 2HgO$ becomes positive above this temperature **[1]**, above this temperature the reaction is spontaneous in the opposite direction **[1]**. **(b)** The decomposition of lead(II) oxide does not occur spontaneously because even at very high temperatures the value of ΔG for the reaction $2Pb + O_2 \rightarrow 2PbO$ is positive **[1]**, the reaction between lead(II) oxide and carbon occurs spontaneously above about 625K **[1]** at this temperature the lines for the reactions $2Pb + O_2 \rightarrow 2PbO$ and $2C + O_2 \rightarrow 2CO$ intersect **[1]**, above this temperature the reaction $PbO + C \rightarrow CO + Pb$ has $\Delta G < 0$ **[1]**. **(c)** Al cannot be obtained by heating Al_2O_3 with C even at very high temperatures **[1]** because the curves do not cross **[1]**, the reaction $Al_2O_3 + 3C \rightarrow 2Al + 3CO$ never has $\Delta G < 0$ **[1]**.

Option F – Fuels and energy
1. (a) –128 kJ **[2]**. **(b)** $CH_3OH + 1\frac{1}{2}O_2 \rightarrow CO_2 + 2H_2O$ **[1]**.
(c) 1.57×10^5 kJ **[1]**. **(d)** 7.03 kg **[2]**.

2. (a) Methane **[1]**, **(b)** carbon monoxide or hydrogen **[1]**,
(c) any two from: particulates (soot)/ C_xH_y (hydrocarbons)/carbon monoxide **[2]**. **(d)** Fossil fuels are running out or it is produced from quick growing crops so there is no need to use foreign exchange to import oil or other fuels **[1]**.

3. (a) Splitting of an unstable nucleus **[1]**. **(b)** There is a small mass loss which is converted into a large amount of energy **[1]**.
(c) X = n, Y = Pu, Z = e, a = 92, b = 1, c = 0, d = 0, e = –1 **[4]**.
(d) To prevent radioactivity from escaping **[1]**, one from: water/heavy water/sodium/carbon dioxide **[1]**.
(e) 1.7×10^4 years (17190 years) **[2]**.

4. (a) In diamond all the electrons are localized in strong bonds between the carbon atoms **[1]**, in germanium and silicon the bonds are weaker and a few electrons can move from the localised bonds to become delocalized **[1]**. **(b)(i)** When gallium is added 'holes' are formed as gallium only has three

outer electrons [1], so it becomes a p-type conductor [1].
(ii) When arsenic is added an extra electron is available as it contains five outer electrons [1], so it becomes an n-type conductor [1].

5. The ratio increases from 1:1 for the first twenty elements [1], to around 1.5:1 for elements with atomic numbers of about 80 [1], more neutrons are required to help counteract the increasing forces of repulsion between increasing numbers of protons [1].   **(b)** The repulsive forces between protons are too large [1], $_1^1p \rightarrow {}_0^1n + {}_{+1}^0e$ or $_1^1p + {}_{-1}^0e \rightarrow {}_0^1n$ [1].

Option G – Modern analytical chemistry
1. (a) Bonds vibrate (bend/stretch) [1], change in dipole moment required [1], different bonds (functional groups) absorb in different regions [1], precise absorption affected by neighbouring atoms [1].   **(b)(i)** Ethanoic acid: C=O 1680–1750 cm^{-1}, O–H 2500–3300/3580–3650 cm^{-1}, C–H 2840–3095 cm^{-1} [2], methyl methanoate: C=O 1680–1750 cm^{-1}, C–H 2840–3095 cm^{-1}, C–O 1000–1300 cm^{-1} [2].   **(c)** O–H in ethanoic acid could be used or C–O absorption in ester [1], other absorptions cannot be used as they occur in both spectra [1].   **(d)** O–H [1], 3.03×10^{-4} cm [1].

2. (a) R_f = distance travelled by solute divided by distance travelled by solvent [1].   **(b)(i)** Measure distance travelled by centre of blue spot and solvent from the origin [2],   **(ii)** Each dye has different attractions for the paper and the solvent [2].   **(iii)** Negligible attraction between the dye and paper compared with that of dye and solvent [2].

3. (a) Many conjugated double bonds [1] leading to extended delocalized π bond [1].   **(b)** 1.54×10^{-5} mol dm^{-3} [2].
(c) Gas phase chromatography allows separation of the components of the mixture [1], mass spectrometry allows accurate identification of each component separated [1].

4. (a) As a reference [1].   **(b)** quartet [1].   **(c)** 3:3:2 [1].
(d)(i) $RCOOCH_2R$ and $C_6H_5OCOCH_3$ [1].   **(ii)** $RCOOCH_2R$ is only one possible [2], no protons due to phenyl group present in spectrum or other structure has hydrogen atoms in only two different chemical environments [1].
(e) A: C–H [1], B: C=O [1], C: C–O [1].   **(f)** 88: $C_4H_8O_2^+$ [1], 73: $C_3H_5O_2^+$ or $(M - CH_3)^+$ [1], 59: $C_2H_3O_2^+$ or $(M - C_2H_5)^+$ [1].
(g) Ethyl ethanoate [1], $CH_3COOC_2H_5$ [1].

Option H – Further organic chemistry
1. (a)(i) Electrophilic substitution [1].   **(ii)** $C_2H_5 - Br$ $AlBr_3$ → $C_2H_5^+ + AlBr_4^-$ [2]

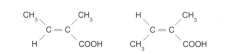

[2]
(b)(i) Ultraviolet light [1].   **(ii)** Free radical substitution [1], Br–Br → 2Br• [1].
Br• + $C_6H_5C_2H_5$ → $C_6H_5C_2H_4^•$ + HBr [1], $C_6H_5C_2H_4^•$ + Br_2 → $C_6H_5C_2H_4Br + Br^•$ [1].

(c)(i) Elimination [1], The OH⁻ ion acts as a strong base [1], removing H⁺ from the bromoalkane [1].
(ii) $C_6H_5CH_2CH_2OH$ [1]. Nucleophilic substitution [1].

2. (a) $CH_3C(CH_3)BrCH_3$ 2-bromo-2-methylpropane [1], $CH_3CH(CH_3)CH_2Br$ 1-bromo-2-methylpropane [1].
(b) When a hydrogen halide adds to the C=C double bond the hydrogen atom bonds to the carbon atom which is already bonded to the greatest number of hydrogen atoms [1].
(c) 2-bromo-2-methylpropane [1].
(d)

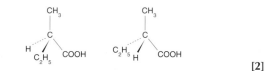

[2]
(e) The 2-bromo isomer involves a tertiary carbocation intermediate [1]. This is much more stable than the primary carbocation intermediate which would be formed if the 1-bromo isomer were the product [1]. The inductive effect of the CH_3- groups 'pushes' electrons towards the carbocation. This disperses the positive charge and makes it more stable [1].
(f) $CH_3C(CH_3)ClCH_2I$ [1].

3. (a)

[2]

(b) $CH_3CH=C(CH_3)COOH + H_2 \rightarrow CH_3CH_2CH(CH_3)COOH$ [1].
(c) 2-methylbutanoic acid [1]   **(d)** A 50:50 mixture of the two different enantiomers of an optically active compound [1].

(e)

[2]

(f) Physical properties are identical [1], except for their ability to rotate the plane of plane-polarized light in opposite directions [1].

Origin of individual questions

The questions detailed below are all taken from past IB examination papers and are © IBO. The questions are from the May (M) or November (N), 1998 (98), 1999 (99) and 2000 (00) paper 1 (P1), paper 2 (P2), or paper 3 (P3) with the question number in brackets. All other questions are IB style questions made by the author for this book.

STOICHIOMETRY
13. N98SLP2(1).

ATOMIC THEORY
3. M99HLP1(5);   4. M98HLP1(5); 5. M98HLP1(5);
6. N99HLP1(5); 7. N99HLP1(7); 8. M98HLP1(6);
9. N99SLP1(8);   10. M99SLP1(7);   11. M99HLP1(6),
12. M00HLP1(6);   13. N98HLP1(6); 14. N98HLP1(7);
15. M98HLP1(8).

PERIODICITY
3. N98SLP1(9);   4. M00SLP1(10);   6. M99SLP1(8);
7. M99SLP1(10);   8. N98SLP1(8); 9. N98SLP1(10),
10. N99HLP1(9);   11. HLM99P1(10); 12. M00HLP1(8);
13. N99HLP1(10);   14. M00HLP1(10);   15. M99HLP1(9);
16. M98HLP1(10).

BONDING
1. M99HLP1(11);   4. N99SLP1(12);   5. M99SLP1(12);
6. N98SLP1(13); 7. M99SLP1(14); 8. M00SLP1(14);
10. N98SLP1(14);   11. M99HLP1(13); 12. N99HLP1(12);
13. N99HLP1(13);   14. N98HLP1(11);   16. N98HLP1(12).

STATES OF MATTER
1. N99SLP1(15); 2. M99HLP1(15); 3. N99HLP1(16);
4. M98HLP1(16); 5. N98HLP1(17); 6. M00SLP1(15);
8. M98SLP1(15); 9. M98HLP1(15); 10. M99SLP1(15);
11. N98HLP1(16); 12. M00HLP1(16);   13. M99HLP1(16).

ENERGETICS
1. N99SLP1(16);   2. M98SLP1(16); 3. M99SLP1(16);
4. M98SLP1(17); 5. M99SLP1(17); 6. N99SLP1(17);
7. M98SLP1(18); 8. M99HLP1(20); 9. N99HLP1(21);
10. N98HLP1(19); 11. M98HLP1(18); 12. M98HLP1(20);
13. N99HLP1(20);   14. M00HLP1(20);

KINETICS
1. M00SLP1(19);   2. M98SLP1(19);   3. M99SLP1(19);
4. N99SLP1(19);   5. M98SLP1(22);   6. N98SLP1(19);
7. N98SLP1(18);   8. M00SLP1(20);   9. M98SLP1(20);
10. M99SLP1(20);   11. M00HLP1(20);   12. M98HLP1(23);
13. N98HLP1(23);   14. N99SLP1(20);   15. N99HLP1(23).

EQUILIBRIUM
1. M00SLP1(21);   2. N99SLP1(21);   3. M00HLP1(26);
4. M99SLP1(21);   5. M00SLP1(22);   6. M98SLP1(21);
7. M99SLP1(22);   8. N99HLP1(27);   9. N98SLP1(21);
10. M99HLP1(26);   11. M99HLP1(27);   12. N98HLP1(27);
13. M98HLP1(27).

OPTION A – HIGHER PHYSICAL ORGANIC CHEMISTRY
1. M00SLP3(A1);   2. M00SLP3(B2)/M00HLP2(1b);
3. N99HLP2(2).

OPTION B – MEDICINES AND DRUGS
1. M98SL(S2);   2. M98SL(S1);   3. M99SL(S1).

OPTION C – HUMAN BIOCHEMISTRY
1. N98HLP3(C3);   2. M98HLP3(C1);   3. N99HLP3(C1,C2b);
4. N98HLP3(C2).

OPTION D – ENVIRONMENTAL CHEMISTRY
1. N00HLP3(D1);   2. N00SLP3(D2);   3. M98HLP3(D2);
4. N00HLP3(D3);   5. M00HLP3(D3).

OPTION E – CHEMICAL INDUSTRIES
1. M99HLP3(E1);   2. N00HLP3(E2);   3. N99HLP3(E2);
4. N98HLP3(E2);   5. N98HLP3(E3).

OPTION F – FUELS AND ENERGY
1. N99HLP3(F1);   2. N00HLP3(F1);   3. N00HLP3(F2)
5. N99HLP3(F5).

OPTION G – MODERN ANALYTICAL CHEMISTRY
1. M99HLP3(G2);   2. M00HLP3(G2);   3. N98HLP3(G2).

OPTION H – FURTHER ORGANIC CHEMISTRY
1. N98HLP3(H1);   2. N98HLP3(H2);   3. N99HLP3(H1).

Index

Periodic Table for use with the IB

GROUP

Key:
- Atomic Number
- Element
- Atomic Mass

1	2											3	4	5	6	7	0
1 **H** 1.01																	2 **He** 4.00
3 **Li** 6.94	4 **Be** 9.01											5 **B** 10.81	6 **C** 12.01	7 **N** 14.01	8 **O** 16.00	9 **F** 19.00	10 **Ne** 20.18
11 **Na** 22.99	12 **Mg** 24.31											13 **Al** 26.98	14 **Si** 28.09	15 **P** 30.97	16 **S** 32.06	17 **Cl** 35.45	18 **Ar** 39.95
19 **K** 39.10	20 **Ca** 40.08	21 **Sc** 44.96	22 **Ti** 47.90	23 **V** 50.94	24 **Cr** 52.00	25 **Mn** 54.94	26 **Fe** 55.85	27 **Co** 58.93	28 **Ni** 58.71	29 **Cu** 63.55	30 **Zn** 65.37	31 **Ga** 69.72	32 **Ge** 72.59	33 **As** 74.92	34 **Se** 78.96	35 **Br** 79.90	36 **Kr** 83.80
37 **Rb** 85.47	38 **Sr** 87.62	39 **Y** 88.91	40 **Zr** 91.22	41 **Nb** 92.91	42 **Mo** 95.94	43 **Tc** 98.91	44 **Ru** 101.07	45 **Rh** 102.91	46 **Pd** 106.42	47 **Ag** 107.87	48 **Cd** 112.40	49 **In** 114.82	50 **Sn** 118.69	51 **Sb** 121.75	52 **Te** 127.60	53 **I** 126.90	54 **Xe** 131.30
55 **Cs** 132.91	56 **Ba** 137.34	57† **La** 138.91	72 **Hf** 178.49	73 **Ta** 180.95	74 **W** 183.85	75 **Re** 186.21	76 **Os** 190.21	77 **Ir** 192.22	78 **Pt** 195.09	79 **Au** 196.97	80 **Hg** 200.59	81 **Tl** 204.37	82 **Pb** 207.19	83 **Bi** 208.98	84 **Po** (210)	85 **At** (210)	86 **Rn** (222)
87 **Fr** (223)	88 **Ra** (226)	89‡ **Ac** (227)															

† Lanthanides

58 **Ce** 140.12	59 **Pr** 140.91	60 **Nd** 144.24	61 **Pm** 146.92	62 **Sm** 150.35	63 **Eu** 151.96	64 **Gd** 157.25	65 **Tb** 158.92	66 **Dy** 162.50	67 **Ho** 164.93	68 **Er** 167.26	69 **Tm** 168.93	70 **Yb** 173.04	71 **Lu** 174.97

‡ Actinides

90 **Th** 232.04	91 **Pa** 231.04	92 **U** 238.03	93 **Np** (237)	94 **Pu** (242)	95 **Am** (243)	96 **Cm** (247)	97 **Bk** (247)	98 **Cf** (251)	99 **Es** (254)	100 **Fm** (257)	101 **Md** (258)	102 **No** (259)	103 **Lr** (260)